Lecture Notes in Computer Science 14778

The series Lecture Notes in Computer Science (LNCS), including its subseries Lecture Notes in Artificial Intelligence (LNAI) and Lecture Notes in Bioinformatics (LNBI), has established itself as a medium for the publication of new developments in computer science and information technology research, teaching, and education.

LNCS enjoys close cooperation with the computer science R & D community, the series counts many renowned academics among its volume editors and paper authors, and collaborates with prestigious societies. Its mission is to serve this international community by providing an invaluable service, mainly focused on the publication of conference and workshop proceedings and postproceedings. LNCS commenced publication in 1973.

Angel A. Juan · Javier Faulin ·
David Lopez-Lopez
Editors

Decision Sciences

Second Decision Science Alliance
International Summer Conference, DSA ISC 2024
Valencia, Spain, June 6–7, 2024
Proceedings, Part I

Springer

Editors
Angel A. Juan
Universidad Politécnica de Valencia
Valencia, Spain

Javier Faulin
Public University of Navarre
Pamplona, Spain

David Lopez-Lopez
ESADE Business School
Sant Cugat, Spain

ISSN 0302-9743 ISSN 1611-3349 (electronic)
Lecture Notes in Computer Science
ISBN 978-3-031-78237-4 ISBN 978-3-031-78238-1 (eBook)
https://doi.org/10.1007/978-3-031-78238-1

This Springer imprint is published by the registered company Springer Nature Switzerland AG
The registered company address is: Gewerbestrasse 11, 6330 Cham, Switzerland

Preface

It is with great pleasure that we present the proceedings of the Decision Science Alliance International Summer Conference 2024 (DSA ISC 2024), held at Universitat Politècnica de València (UPV), Valencia, Spain, on June 6–7, 2024. This conference, co-chaired by Angel A. Juan (UPV) and Raffaele Maccioni (DSA), stands as a premier European event for decision-makers, industry experts, academic professionals, and anyone interested in the future of decision science and technological innovation. The event builds on the previous edition of the DSA International Summer Conference, which was held at the University of Milan, Italy, and chaired by Giovanni Righini.

At the core of DSA ISC 2024 were in-depth discussions and analyses across a spectrum of technological domains. Notably, experts shared their knowledge on areas such as Artificial Intelligence & Machine Learning, Mathematical Optimization, Operational Research & Management Science, Statistics, Simulation, and Decision Processes Analysis. Each of these areas represents a key aspect of decision science, contributing to the interdisciplinary nature of the conference. Attendees explored how these advancements impact decision-making in Business & Marketing, Transportation & Logistics, Production & Manufacturing, Finance & Insurance, Healthcare, Agrifood Industry, and other sectors.

DSA ISC 2024 allowed participants to position themselves at the forefront of innovation and gain a solid understanding of the dynamics that will shape the future of strategic decisions and the technological world. Over the course of two days, attendees benefited from advanced strategies and insights for solving complex problems and enhancing decision-making processes, presented by world leaders in decision science. Networking opportunities arose, enabling participants to form professional alliances and collaborations with other experts and thought leaders, thereby expanding their professional network. Additionally, access to special workshops provided practical tools and valuable insights applicable to their fields of work or research.

The Program Committee Chairs, together with other members of the DSA such as Emanuele Carpanzano, Alice Noris, Raffaele Cerulli, and Giovanni Righini, played a key role in shaping the scientific program of DSA ISC 2024, ensuring its quality and relevance. The peer review process, conducted in a single-blind manner with at least three reviews per paper, upheld rigorous standards of evaluation. All papers, including invited contributions, underwent thorough peer review by anonymous and independent reviewers, ensuring the integrity of the conference proceedings. A total of 71 papers were accepted from the 101 submitted documents.

DSA ISC 2024 also featured a collection of insightful contributions from distinguished keynote speakers, moderated by Vanesa Juan. Guillermo Franco (Guy Carpenter) provided a speech titled "Uninsurable America?". Canan Gunes Corlu (Boston University) presented "The Nation's Fight Against Illegal Supply Chains: Decision Science Can Help". Marco Maccalli (Maccalli & Pezzoli SRL) discussed "Algorithms, AI & Patents". Andrea De Martin (OVS) shared strategies in his presentation "Leveraging

AI and Automation to Overcome Distribution Challenges: Insights from the Fashion Retail". Finally, Elena Meneghetti examined the "Insights on Implications of Machine Advancements for Human Language, Cognition, and Behaviour". Additionally, the conference served to introduce the DSA official magazine, IDA (inform, decide, and act), with Luca Lisci as the Editor-in-Chief.

The DSA ISC 2024 program encompassed a series of regular sessions, special workshops, and a round table titled "Boosting Decision Making with Artificial Intelligence", which was chaired by Patricia Rodriguez. Regular sessions included theory and applications of Decision Science in Marketing and Business, Industry 4.0, Artificial Intelligence, Manufacturing, Computing, Business Analytics, Energy Systems, Shape Analysis, Internet of Things, Statistical Methods, Finance and Insurance, Sustainable Transportation, Optimization Models, Mathematical Methods, Corporate Strategy, Transport Infrastructure, Healthcare and Hospitals, Sustainable Development Goals, Bioengineering, Climate Change, and Agrifood.

We would like to express our special gratitude to the sponsors of DSA ISC 2024: Guy Carpenter, Divina Seguros, Baobab Soluciones, NTT Data, Optit Optimal Solutions, and Excentriq. Their generous support made this event possible. Special thanks are also due to the team at UPV and the members of the DSA who have worked tirelessly to organize and coordinate every aspect of the conference. In particular, we extend our appreciation to Maddalena Mazzali, Xabier A. Martin, Veronika Tsertvedze, and Celia Osorio for their exceptional efforts. Without any doubt, they have been the heart and soul of this conference. Finally, special thanks to Alberto Conejero and Israel Griol from the IDEASUPV department for their support during the conference.

All in all, DSA ISC 2024 welcomed over 175 participants from 11 countries including Spain, Italy, Ireland, Germany, Austria, Turkey, the UK, Switzerland, Argentina, the USA, and Canada. They represented a total of 29 universities and 28 companies. We hope that the proceedings of DSA ISC 2024 serve as a valuable resource for researchers, practitioners, and students, inspiring further advancements in the field of decision science.

June 2024

Angel A. Juan
Javier Faulin
David Lopez-Lopez

Organization

General Chairs

Angel A. Juan	Universitat Politècnica de València, Spain
Raffaele Maccioni	Decision Science Alliance, Italy

Program Committee Chairs

Angel A. Juan	Universitat Politècnica de València, Spain
Javier Faulin	Public University of Navarra, Spain
David Lopez-Lopez	ESADE Business School, Spain

Steering Committee

Raffaele Maccioni	Decision Science Alliance, Italy
Giovanni Righini	University of Milan, Italy
Emanuele Carpanzano	SUPSI, Switzerland
Maddalena Mazzali	Decision Science Alliance, Italy
Raffaele Cerulli	University of Salerno, Italy
Alice Noris	SUPSI, Switzerland
Cristina Caracciolo	Decision Science Alliance, Italy
Luca Lisci	BAX Group AG, Italy

Program Committee

Eva Barrena	Universidad Pablo de Olavide, Spain
Luis Cadarso	Universidad Rey Juan Carlos, Spain
David Canca	Universidad de Sevilla, Spain
Patricia Carracedo	Universitat Politècnica de València, Spain
Paula Carroll	University College Dublin, Ireland
Emanuele Carpanzano	SUPSI, Switzerland
Raffaele Cerulli	University of Salerno, Italy
Alberto Conejero	Universitat Politècnica de València, Spain
Adrian Colomer	Universitat Politècnica de València, Spain
Canan G. Corlu	Boston University, USA

Cristofer Exposito	Universidad de La Laguna, Spain
Javier Faulin	Public University of Navarra, Spain
Miguel Figliozzi	Portland State University, USA
Pau Fonseca	Universitat Politècnica de Catalunya, Spain
Guillermo Franco	Guy Carpenter, USA
Alvaro Garcia	Baobab Soluciones, Spain
Ximo Gual	Universitat Jaume I, Spain
Roberto Guidotti	Guy Carpenter, USA
Jose L. Guisado	Universidad de Sevilla, Spain
David Hervas	Universitat Politècnica de València, Spain
Patrick Hirsch	BOKU, Austria
Angel A. Juan	Universitat Politècnica de València, Spain
Vanesa Juan	Solbeatz, Spain
Peter Keenan	University College Dublin, Ireland
Yvonne Kummer	BOKU, Austria
David Lopez-Lopez	ESADE Business School, Spain
Fermin Mallor	Public University of Navarra, Spain
Xabier A. Martin	Universitat Politècnica de València, Spain
Raffaele Maccioni	Decision Science Alliance, Italy
Sean McGarraghy	University College Dublin, Ireland
Maddalena Mazzali	Decision Science Alliance, Italy
Sara Moll	Universitat Politècnica de València, Spain
Jose A. Moreno	Universidad de La Laguna, Spain
Josefa Mula	Universitat Politècnica de València, Spain
Valery Naranjo	Universitat Politècnica de València, Spain
Armando Nieto	Divina Seguros, Spain
Alice Noris	SUPSI, Switzerland
Stephan Onggo	University of Southampton, UK
Javier Panadero	Universitat Autònoma de Barcelona, Spain
Elena Perez-Bernabeu	Universitat Politècnica de València, Spain
Raul Poler	Universitat Politècnica de València, Spain
Givanni Righini	University of Milan, Italy
Patricia Rodriguez	Venture Capital, Spain
Loris Roveda	SUPSI, Switzerland
Raquel Sanchis	Universitat Politècnica de València, Spain
Wolfgang Seiringer	University of Applied Sciences Upper Austria, Austria
Fabio Silva	University College Cork, Ireland
Francisco Salas	Universitat Politècnica de València, Spain
Veronika Tsertsvadze	Universitat Politècnica de València, Spain
Fatos Xhafa	Universitat Politècnica de Catalunya, Spain

Additional Reviewers

Lucia Agud (UPV)
Francisco Ballestin (UV)
Yun Bao (UdL)
Marc Bara (EAE)
Laura Calvet (UAB)
Giovanni Campuzano (UTWENTE)
Carmine Cerrone (UNIGE)
Victor Gallego (EUNCET)
Alisson Garcia (UPNA)
Neus Garrido (UPV)
Elnaz Ghorbani (UPV)
Juan F. Gomez (decide4AI)
Cristina Gonzalez (ULL)
Guillermo Gonzalez (Baobab)
Antoni Guerrero (Baobab)
Irene Izco (UPNA)
Mauro Laguna (UPV)
Pol Llagostera (UdL)
Almudena Llorens (UPV)
Jonas F. Leon (UOC)
Griselda Lopez (UPV)
Wenwen Chen (UPV)
Marta Cildoz (UPNA)
Marc Escoto (UPV)
Carmen Fonseca (UPC)
Pau Font (UdL)
Alfons Freixes (EUNCET)
Noelia Fuster (UPV)
Paolo Marone (Spindox)
Yangchongyi Men (UPV)
Juan Moreno (UPV)
Isa Moll (ESADE)
Miguel Ortega (UPM)
Mohammad Peyman (UPV)
Antonio Rodriguez (decide4AI)
Miguel Saiz (UOC)
Sandra Oltra (UPV)
Celia Osorio (UPV)
David Peidro (UPV)
Raquel Soriano (UPV)
Carles Torrecilla (ESADE)
Hongtao Wang (UPV)

Sponsors

GuyCarpenter

baobab soluciones

divinaseguros

OPTIT optimal solutions

NTT DATA

EXENTRIQ

Contents – Part I

Science

Engineering

Contents – Part II

Management

Engineering

Management

Combining LLMs and Simheuristics: An Application to the Project Portfolio Selection Problem

Miguel Saiz[1(✉)] and Laura Calvet[2]

[1] Computer Science Department, Universitat Oberta de Catalunya, 08860 Barcelona, Spain
msaiz@weoptimize.net

[2] Telecommunication and System Engineering Department, Universitat Autònoma de Barcelona, 08202 Sabadell, Spain

Abstract. Large Language Models (LLMs) are very large deep learning models, which are pre-trained on vast amounts of data, able to achieve general-purpose language understanding and generation. GPT-4 is an increasingly popular example from OpenAI. We introduce a novel framework called Generative Heuristics, which hybridize (meta)heuristics for its quantitative capabilities with the qualitative proficiency of LLMs. To illustrate this framework, we address a rich version of the Project Portfolio Selection Problem with a Generative Heuristic relying on a Simheuristic Algorithm. Our solving approach explores candidate portfolios, optimizing quantitative outcomes, such as the Net Present Value, while maintaining a focus on non-quantifiable strategic contributions. To maximize realism and applicability, the strategic plan from the LIFE program (EU's funding instrument for environment and climate action) and project summaries from actual LIFE-funded projects are utilized. A computational experiment is carried out to illustrate the feasibility and potential of our approach, which enhance quantitative search procedures with cognitive capabilities, offering decision-makers both a more comprehensive and versatile set of options.

Keywords: Project Portfolio Selection Problem · Large Language Models · Simheuristics · Hybrid Optimization · Generative Heuristics

1 Introduction

Project Portfolio Selection and Optimization has emerged as a critical discipline within Project Portfolio Management (PPM), particularly in response to the escalating complexities of modern organizational landscapes [8]. The Project Portfolio Selection Problem (PPSP) involves identifying the set of projects that maximize benefits while adhering to predefined constraints, ultimately aiming to enhance competitive advantage and achieve strategic business objectives [17].

A. A. Juan et al. (Eds.): DSA ISC 2024, LNCS 14778, pp. 3–13, 2025.
https://doi.org/10.1007/978-3-031-78238-1_1

Despite being a subject of extensive study over several decades, the PPSP continues to evolve due to dynamic project dynamics, organizational structures, and market conditions [8]. This complexity underscores the need for innovative methodologies and decision-making techniques to address stochastic uncertainties inherent in project portfolios [5].

The literature on PPSP spans diverse methodologies and algorithms, with a notable focus on quantitative techniques for optimization [22]. While traditional approaches predominantly emphasize quantitative objectives, recent studies have advocated for multi-objective optimization frameworks that incorporate qualitative attributes such as culture, sustainability, and reputation [3,6]. Quantitative methods employed in PPSP encompass a wide spectrum, including fuzzy multi-objective programming, discrete linear programming, nonlinear integer programming, and mixed-integer linear programming [12,13,19]. Additionally, simulation techniques like Monte Carlo Simulation (henceforth referred to as MCS) have gained popularity for their ability to model uncertainties effectively [21].

However, the conventional assumption of project independence and the lack of detailed risk consideration pose challenges in existing models [4]. Integrating risk management into portfolio decision-making processes requires a comprehensive understanding of operational activities and their impact on overall portfolio performance [15].

One approach gaining traction in addressing stochastic problems is the integration of simulation techniques with heuristic methods, known as simheuristics [14]. Simheuristics offer a pragmatic solution to NP-hard problems like the PPSP by blending the flexibility of simulation with the efficiency of heuristic algorithms [9]. This amalgamation allows for a more realistic representation of uncertainties and interdependencies among projects within a portfolio [18].

By simulating real-world scenarios and incorporating heuristic algorithms, simheuristics provide a realistic framework for optimizing PPM. This, coupled with the findings of [16] and [7] that Large Language Models (subsequently abbreviated as LLMs) can generate logical responses in uncertain environments, raises the following research question:

Could simheuristics, along with the use of LLMs, provide a more robust framework for optimizing project portfolios in dynamic and uncertain environments using tangible and intangible variables?

This study constitutes what we consider to be the first approach that combines the advantages of: (i) simheuristic methods to achieve near-optimal results when the parameters of the problem are quantitative and include probability distributions of uncertain variables; and (ii) the use of LLMs to incorporate less tangible aspects, which also implies the consideration of extensive documents written in natural language, like for example strategic plans and project proposal documents.

2 The Stochastic Project Portfolio Selection Problem

The stochastic PPSP aims to maximize the expected Net Present Value (hereinafter, NPV) of a portfolio:

$$\max E[NPV] = \max \sum_{i=1}^{n} x_i \cdot \sum_{t=1}^{T} \frac{R_{i,t}}{(1+W)^t} \tag{1}$$

being R_i the reward associated with the execution of each project i, subject to uncertainty and therefore expressed as a random variable, and W a random variable expressing an uncertain weighted average cost of capital (WACC). $x_i \in \{0, 1\}$ is the decision variable, where $x_i = 1$ if project i is selected and 0 otherwise. The constraint related to the overall cost is:

$$p\left(\sum_{i=1}^{n} x_i \cdot C_i \leq C_{max}\right) \geq F_{min} \tag{2}$$

This equation represents the probability that the total cost of a set of projects, each having an uncertain cost denoted by C_i, does not exceed the maximum budget C_{max}. The probability that the total cost remains within budget must be at least F_{min}, which is normally set by the decision makers, depending on the risk appetite of their organization or industry. F_{min} is described as an estimator of the reliability level of the portfolio.

The random variable C_i is made up of several random variables as follows:

$$C_i = b_i + L_i \cdot f \tag{3}$$

where the duration of each project is L_i, calculated through the critical path method [10], subject to the uncertainty related to the activity duration D_{ij} (modeled as a triangular distribution).

Building upon this classical modelization of the stochastic PPSP model, we propose the inclusion of an additional objective function: maximizing the strategic alignment of projects within a portfolio to the funding organization's strategic plan. This alignment will need to be quantified through a strategic score (S) ranging from 1 to 100.

A solution to this problem is the composition of the best portfolio found, accompanied of metrics such as expected NPV and reliability, as well as the justification of the selection of each project and an executive summary.

3 Solving Methodology

We introduce a generative heuristic leveraging a genetic algorithm (GA) based simheuristic for solution generation, evaluated through the expected NPV, coupled with an LLM to gauge alignment with project proposals and the strategic plan. The outlined methodology is visually depicted in Fig. 1 and elaborated upon in subsequent sections.

3.1 Overview

Our methodology initially prioritizes maximizing the expected NPV while ensuring the portfolio budget is consistently met. This results in a Hall of Fame (HoF) of portfolios and a shortlisted set of projects. The HoF comprises portfolios with the highest expected NPV potential, while the shortlist excludes the projects not featured in any of the HoF portfolios. The HoF is then quantitatively and qualitatively assessed. Additionally, the LLM reads the project proposals (one plain txt file per project, copying from the EU LIFE website), evaluating each portfolio's alignment with the strategic plan (also in plain txt format, taken from the EU Regulation from their official publication website), resulting in a strategic score (S), which is a quantitative assessment. Each HoF portfolio proposal is then reported with its expected NPV, strategic score, and score justification. Lastly, the Pareto front is extracted to present only the non-dominated portfolios to the Decision Makers (DM) and the LLM for the executive summary creation.

3.2 Simheuristic Component

The HoF is obtained through a simheuristic, which starts using MCS to simulate the uncertainty in the schedule (which is assigned a cost per day the project is under execution) and in the cash flows related to the project benefits. Subsequently, all the possible outcomes are aggregated on top of a baseline budget, generating two probability density functions that represent the probabilities of the different cost and benefit outcomes. This enables a point estimate under a specific level of confidence of budget fulfilment at project level, which is then fed into the GA for optimization purposes.

The GA then generates a random population of portfolios, evaluates their fitness (in terms of expected NPV and fulfilment of a level of reliability of the whole portfolio in terms of budget), filters the best individuals, and then creates a new generation of individuals generated by crossover and mutation of the offspring. This is done through several iterations, extracting several near-optimal results from expected NPV perspective for further evaluation.

This stage is executed prior to the LLM stage, so that only promising portfolio candidates are evaluated during the LLM search stage. As underscored by [2], this strategy is beneficial not only to mitigate the usage cost of the LLM-API but also because costs associated with LLMs tend to escalate rapidly when domain-specific data is used in conjunction with queries. In the context of our study, the domain-specific data comprises the strategic plan and project proposals.

3.3 LLM Component

Figure 2 shows the detail on how the LLM evaluates the near-optimal portfolios generated coming from the simheuristic. First, there is an initial prompt that provides the Strategic Plan respect to which the portfolios must be evaluated, and also the proposal documents related to each of the shortlisted projects. Next, the LLM is fed with the information on which projects are included in each of

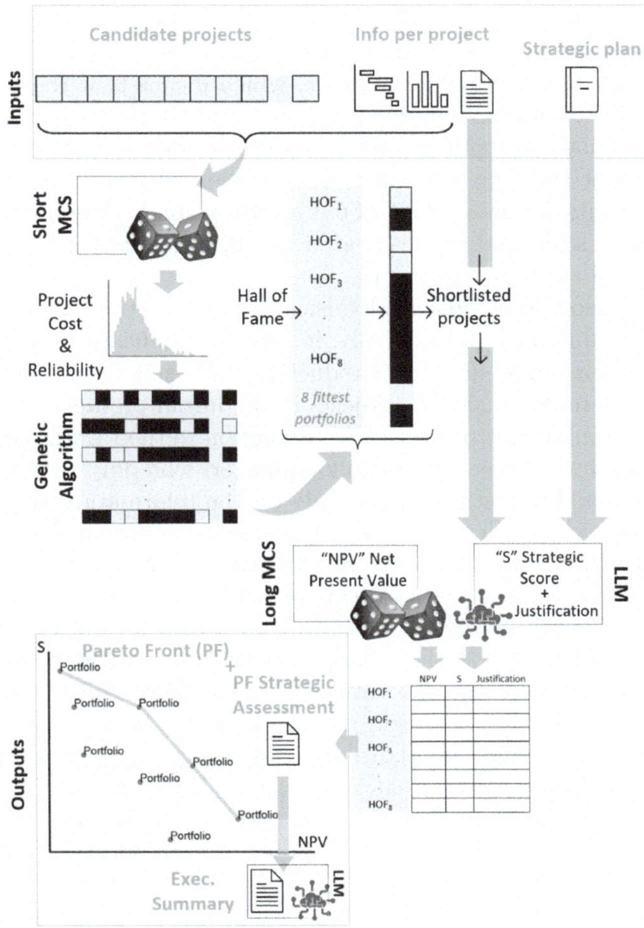

Fig. 1. Overview of the generative heuristic proposed.

the HoF portfolios. This is done through an iterative process, with one prompt for each portfolio that is in the HoF, and getting in return the quantitative appraisal in form of a strategic score (S) and a justification of the assessment. S is a number ranging from 1 to 100 and is directly generated by the LLM following the prompt we include at the end of this section. Both, S and the justification, are stored together with the expected NPV, so that the Pareto Front of non-dominated portfolios is extracted. A portfolio is dominated if there is another portfolio that has a better expected NPV and a better S. Only the portfolios that belong to the Pareto front are included in a DM-report. Such report is then fed again to the LLM, so that it can generate a simplified executive summary that takes the portfolio with highest strategic score, summarizes its advantages, and then reports the other non-dominated portfolios in a differential way, explaining the aspects from strategic alignment that would be compromised if deciding to go for other non-dominated portfolios with higher expected NPV.

LLMs are inherently non-deterministic, which means that the outputs they generate may differ between requests [1]. This variability in output is modulated by a parameter known as 'temperature'. Lower temperature values typically yield more deterministic and focused results, while higher values lead to more random and diverse outputs. In the context of our study, we have chosen a temperature value of 0.1, on a scale that ranges from 0 to 2.0, to prioritize the repeatability of our results and ensure consistency.

The project shortlist is beneficial to save computing time at both the MCS and the LLM evaluations. Additionally, it saves a significant percentage (comparable to the percentage of project reduction) of tokens, as it does not input the proposals related to excluded projects in the initial prompt.

The initial prompt that provides the instruction to the LLM is the following: "You are a reliable and ojective portfolio manager, who only writes statements that can be reasonably justified based only on the information you have available. You are provided with a proposal of a portfolio consisting of several project for which you will be provided with an executive summary. You will compare the portfolio choice respect to a strategic plan I am providing at this message. Your evaluation should be on a scale from 1 to 100, with 100 being the maximum alignment. Provide first the rating (only the value without saying '/100') and then the justification for the rating, detailing how well the combination of projects aligns with the strategic plan and the key factors that influenced your assessment. The strategic plan is the following:" + strategic_plan.read() + "And the promising projects that you must evaluate are the following (take into account that each project summary is preceded by its project number so that you remember which projects are proposed at each potential portfolio):" + Summary_Promising_Projects.read().

4 Computational Experiment

The proposed algorithm has been implemented in Python [20], and all experiments have been performed on a standard personal computer equipped with an Intel Core i7 CPU at 2.9 GHz, 16 GB of RAM memory, and Windows 10 as the operating system. The LLM side is a Microsoft Azure Deployment of Opean AI's GPT4_turbo 1106-Preview, released in November 6, 2023 only for developers and featuring a 128k context window (which is 4x longer than the best-performing LLM, GPT-4-32k, in terms of understanding long contexts, according to [11]). The GPT-4 parameters are: 0.1 *temperature* and 0.95 *top_p*. The code, processed datasets and results, are available at the following Github repository: https://github.com/weOptimize/MS4_genHeuristics. As reference for the strategic plan and the project proposals, a realistic situation has been considered, where the strategic plan is the EU Regulation 1293/2013, which establishes the Programme for the Environment and Climate Action (LIFE), and the project proposals are the ones related to real projects executed under such

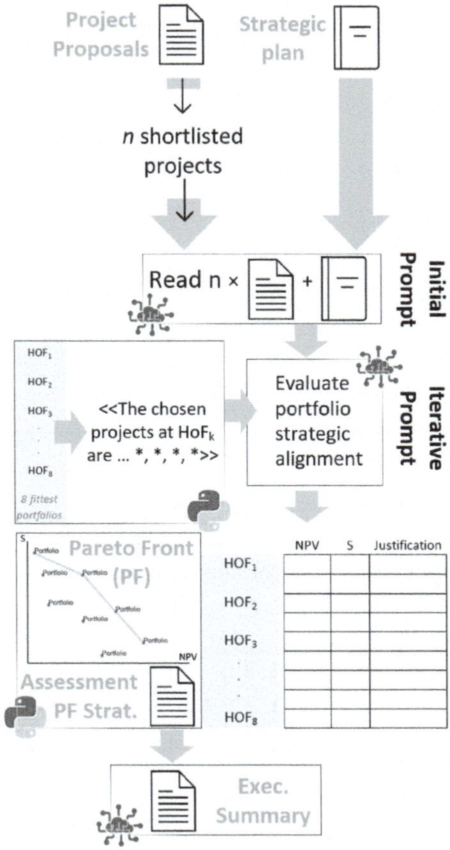

Fig. 2. Outline of the interactions with the LLM.

programme. The projects were taken from the European Comission Webpage, https://webgate.ec.europa.eu/life/publicWebsite/search (consulted on January the 25th, 2024, under the option of "Latest Updated Projects") and are listed at Table 2 (placed at the end of this document). Due to the lack of detailed information on each project's budgets, economic benefits and original schedules (including triangular estimates for each activity duration), we have generated them ourselves in an illustrative way.

5 Results and Analysis

Figure 3 represents the assessment of expected NPV and Strategic Score for the portfolios evaluated. Only six out of the eight portfolios from HoF are depicted, as the remaining two perform markedly worse and are distinct from the rest. As observed in the Figure, the Pareto Set comprises four portfolios: 0, 1, 2, and 4. These non-dominated portfolio are further detailed in Table 1.

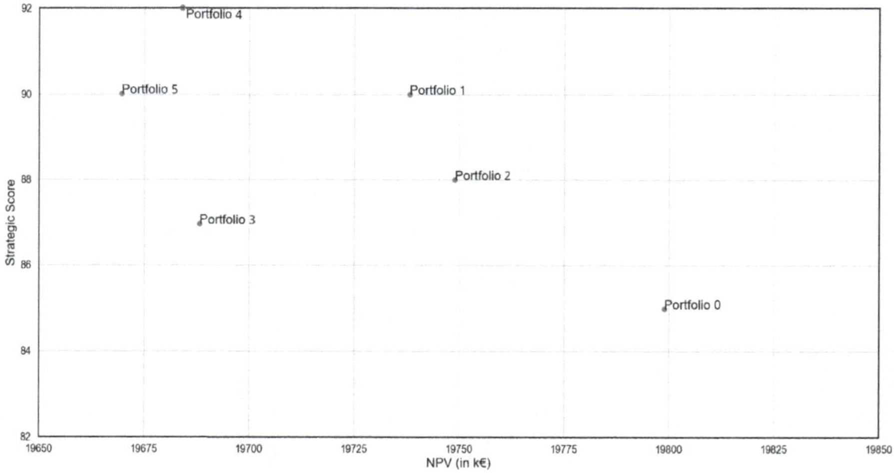

Fig. 3. Evaluation of portfolios from the Hall of Fame.

Table 1. Pareto set of non-dominated portfolios.

Id.	Project Selection	S	NPV(k€)
4	[1, 2, 3, 4, -, 6, 7, 8, -, -, 11, 12, -, 14, -, 16, -, -, -, -]	92	19,683.02
1	[-, 2, 3, 4, -, 6, 7, 8, 9, -, 11, 12, 13, 14, -, -, -, -, -, -]	90	19,738.64
2	[1, 2, 3, 4, -, 6, 7, 8, -, -, 11, 12, -, 14, -, -, -, -, -, 20]	88	19,748.88
0	[1, 2, 3, 4, -, 6, 7, 8, -, -, 11, 12, 13, 14, -, -, -, -, -, -]	85	19,796.66

The most strategically aligned option is portfolio 4, which according to the executive summary outlined by the LLM, "*excels in biodiversity and ecosystem conservation, with projects that manage natural habitats and protect endangered species. It promotes sustainable resource management in the winemaking and dairy industries, directly contributing to climate change mitigation through enhanced forest management. This portfolio supports EU directives and policies, showcases innovative solutions like grape residue utilization and whey protein extraction, and encourages transnational cooperation. It emphasizes stakeholder engagement and governance improvements, socio-economic benefits, and addresses specific environmental issues like waste reduction in the wine industry*". The following option in the Pareto set is portfolio 1, which removes the wine-related project (proj.1) and also the construction of the blue ecological corridor (proj.13). This is also identified at the executive summary generated by the LLM, removing references to wine in the advantages and mentioning the following disadvantage related to project 13: "*it lacks the direct climate change mitigation focus provided by forest management in the top-rated portfolio*". The rest of the LLM-generated analysis at the executive summary is consistent with the portfolio choices, as well as the more extensive document

Table 2. Projects taken for proposal analysis by the LLM.

Nr.	Ref.	Project Name
1	LIFE21-ENV-IT-[...]/101074362	Circular-based alternative protein extraction from grape seeds for the wine-making process
2	LIFE22-NAT-FR-[...]/101113629	LIFE Mobile Marine Species
3	LIFE23-PRE-CZ-[...]/101148569	Conservation of Natural Heritage for Life in Ukraine
4	LIFE22-NAT-PT-[...]/101113906	Creating socioecological conditions for the expansion of Iberian wolf and Iberian lynx
5	LIFE23-PRE-IT-[...]/101148077	Buildings and Education in Wood Ecosystem for the New European Bauhaus
6	LIFE22-NAT-LT-[...]/101113698	Baltic wide web for Osmoderma eremita: the creation of an ecological network in the Baltic region
7	LIFE22-NAT-BE-[...]/101113519	Restoration of Wet and Transition Biotops in the Atlantic Valleys of Wallonia
8	LIFE20 ENV/NL/000200	Sustainable recycling of (Carbon) Fibre Composites and biomass waste to valued added CarbonGreen products
9	LIFE18 ENV/FR/000029	SOLI FOOD WASTE: Inclusive, green and circular economy for unsold food collection and high value valorization
10	LIFE20 CCA/ES/001783	HYBRID SYSTEM FOR RENEWABLE ENERGY PRODUCTION FROM DESALINATION BRINE
11	LIFE19 ENV/PT/000689	Digital circular economy market place for food waste prevention
12	LIFE18 NAT/DK/000732	New habitats for Bombina bombina other amphibians and birds to counteract problems from climate change in coastal areas
13	LIFE11 NAT/PL/000424	The construction of the blue ecological corridor along the valley of Riga river and its tributaries
14	LIFE16 ENV/SI/000335	Reuse of waste acid whey for extraction of high added value bioactive proteins
15	LIFE19 CCM/IT/001201	Promotion and Enhancement of Forest System through different Ecosystem Services aimed to increased Carbon Stock
16	LIFE17 NAT/PT/000554	Decreasing socio-ecological barriers to connectivity for wolves south of the Douro river
17	LIFE18 NAT/UK/000995	SWAforLIFE: Scottish Wildcat Action Phase 2 Wildcat recovery through threat mitigation and translocation
18	LIFE17 PRE/DE/000003	LIFE e-Natura2000.edu
19	LIFE16 ENV/ES/000375	Enhanced Nitrogen and phosphorus Recovery from wastewater and Integration in the value Chain
20	LIFE19 PRE/DE/000005	Transformation of business accounting standards and practices to reinforce enterprise decisions and the environment

with the individual justification for the valuations of Strategic Score included for the four non-dominated portfolios. Both files can be found under the corresponding Github repository under the names "GPT4_128k_ParetoSET_exec_summary" and "GPT4_128k_Evaluation" respectively.

6 Conclusions

This study constitutes the first approach that combines a 128k LLM with a simheuristic algorithm, showing the potential of the interaction between both.

In particular, we have proposed a novel approach to address the Stochastic Project Portfolio Selection Problem. Our model aims to maximize the expected net present value, which constitutes a powerful objective function for portfolio optimization due to its focus on present value, risk-return balance, decision-making clarity, flexibility, and widespread use. The approach could be easily adapted in order to consider other objective functions as the Sharpe Ratio and the Sortino Ratio, or even a multi-objective function. Focusing on generative heuristics, this work opens several lines of further research such as: (i) integration of the LLM assessment within the optimization iterations of the GA, incorporating the Strategic Score inside the fitness evaluation; (ii) evaluation of different strategies, such as preliminary calibrations of the LLM by providing portfolios that exceed the budgetary restrictions or offer poor portfolio alternatives to assess their impact on LLM assessments; (iii) repeatability and reproducibility tests at various *temperature* levels and *top_p* values; (iv) validation of the proposed framework in a real-world decision-making environment, involving actual decision-makers; and (v) evaluation of these generative heuristic approaches to contexts beyond the Project Portfolio Selection Problem.

Acknowledgements. This project has been partially supported by the Spanish Ministry of Science and Innovation (RED2022-134703-T).

References

1. OpenAI Platform Documentation. https://platform.openai.com/docs/guides/text-generation/faq. Accessed 25 Mar 2024
2. Arefeen, M.A., Debnath, B., Chakradhar, S.: LeanContext: cost-efficient domain-specific question answering using LLMs. Nat. Lang. Process. J. **7**, 100065 (2024)
3. Bhattacharyya, R., Kumar, P., Kar, S.: Fuzzy R&D portfolio selection of interdependent projects. Comput. Math. Appl. **62**(10), 3857–3870 (2011)
4. Branger, N., Muck, M., Seifried, F.T., Weisheit, S.: Optimal portfolios when variances and covariances can jump. J. Econ. Dyn. Control **85**, 59–89 (2017)
5. Chica, M., Juan Pérez, A.A., Cordon, O., Kelton, D.: Why simheuristics? Benefits, limitations, and best practices when combining metaheuristics with simulation. SORT **44**(2) (2017)
6. Fouladgar, M.M., Yazdani-Chamzini, A., Zavadskas, E.K., Yakhchali, S.H., Ghasempourabadi, M.H.: Project portfolio selection using fuzzy AHP and VIKOR techniques. Transf. Bus. Econ. **11**(1) (2012)
7. Grohs, M., Abb, L., Elsayed, N., Rehse, J.R.: Large language models can accomplish business process management tasks. In: International Conference on Business Process Management, pp. 453–465. Springer (2023)
8. Hansen, L.K., Svejvig, P.: Seven decades of project portfolio management research (1950–2019) and perspectives for the future. Proj. Manag. J. **53**(3), 277–294 (2022)
9. Hatami, S., Calvet, L., Fernández-Viagas, V., Framinan, J.M., Juan, A.A.: A simheuristic algorithm to set up starting times in the stochastic parallel flowshop problem. Simul. Model. Pract. Theory **86**, 55–71 (2018)
10. Kelley Jr, J.E., Walker, M.R.: Critical-path planning and scheduling. In: Papers presented at the December 1-3, 1959, eastern joint IRE-AIEE-ACM Computer Conference, pp. 160–173 (1959)

11. Li, J., Wang, M., Zheng, Z., Zhang, M.: Loogle: can long-context language models understand long contexts? arXiv preprint arXiv:2311.04939 (2023)
12. Li, X., Wang, Y., Yan, Q., Zhao, X.: Uncertain mean-variance model for dynamic project portfolio selection problem with divisibility. Fuzzy Optim. Decis. Mak. **18**, 37–56 (2019)
13. Nesticò, A., Sica, F.: The sustainability of urban renewal projects: a model for economic multi-criteria analysis. J. Property Invest. Financ. **35**(4), 397–409 (2017)
14. Panadero, J., Doering, J., Kizys, R., Juan, A.A., Fito, A.: A variable neighborhood search simheuristic for project portfolio selection under uncertainty. J. Heuristics **26**, 353–375 (2020)
15. Project Management Institute: A guide to the Project Management Body of Knowledge (PMBOK guide) - Seventh Edition (2021)
16. Rane, N.: Role and challenges of ChatGPT and similar generative artificial intelligence in business management (2023). Available at SSRN 4603227
17. Şahin Zorluoğlu, Ö., Kabak, Ö.: A literature survey on project portfolio selection problem. Multiple Criteria Decis. Mak. 387–411 (2021)
18. Saiz, M., Lostumbo, M.A., Juan, A.A., Lopez-Lopez, D.: A clustering-based review on project portfolio optimization methods. Int. Trans. Oper. Res. **29**(1), 172–199 (2022)
19. Senjuntichai, A., Luong, H.T.: The research, development and acquisition selection policies. Adv. Sci. Lett. **19**(1), 252–259 (2013)
20. Van Rossum, G., Drake, F.L.: Python 3 Reference Manual. CreateSpace, Scotts Valley, CA (2009)
21. Xingmei, L., Zailing, L., Qiuhong, Z.: The local sensitivity analysis of project portfolio selection problem with divisibility. Syst. Eng. Theory Pract. **36**, 1816–1825 (2016)
22. Zarjou, M., Khalilzadeh, M.: Optimal project portfolio selection with reinvestment strategy considering sustainability in an uncertain environment: a multi-objective optimization approach. Kybernetes **51**(8), 2437–2460 (2021)

Applying Machine Learning in Marketing: An Analysis Using the NMF and K-Means Algorithms

Victor Gallego[1(✉)], Alfons Freixes[1], and Jessica Lingan[2]

[1] Euncet Business School, Barcelona, Spain
vgallego2598@gmail.com, alfons.freixes@gmail.com
[2] Data Science Research Group, Ramon Llull University, La Salle International School of Commerce and Digital Economy, Barcelona, Spain
bettyjessica.lingan@salle.url.edu

Abstract. The integration of machine learning (ML) techniques into marketing strategies has become increasingly crucial in modern business. Utilizing scientific manuscripts indexed in the Scopus database, this article explores how this integration is being carried out. Initially, a focused search is undertaken for academic articles containing both the terms "machine learning" and "marketing" in their titles, which yields a pool of 58 papers. These papers have been processed using the Supabase platform. The process has included steps like text refinement and feature extraction. In addition, our study uses two key ML methodologies: topic modeling through NMF and a comparative analysis utilizing the k-means clustering algorithm. Through this analysis, three distinct clusters emerged, thus clarifying how ML techniques are influencing marketing strategies, from enhancing customer segmentation practices to optimizing the effectiveness of advertising campaigns.

Keywords: Machine learning · digital marketing · algorithms · artificial intelligence

1 Introduction

The impact of marketing on businesses is crucial to ensure their growth and development. Marketing enables brand awareness and recognition, customer acquisition and loyalty, differentiation from competitors, and increased sales, thus contributing to the sustainability of business profitability [1]. Systematic data management and data processing is essential to obtain timely and accurate information about the market, products and customers, as well as the business environment, as it facilitates decision-making in the different marketing processes. With the advancement of information technologies, companies need to apply new methods for data processing and analysis [2]. Machine learning (ML) can be used to gain new insights into consumer behavior, improve the performance of marketing actions [3], formulate strategic marketing decisions [4],

A. A. Juan et al. (Eds.): DSA ISC 2024, LNCS 14778, pp. 14–26, 2025.
https://doi.org/10.1007/978-3-031-78238-1_2

identifying competitors, effectively manage products and prices, etc. [5]. The use of machine learning offers great possibilities for companies, marketing professionals and academics, but at the same time it has become a challenge to identify the methods, techniques and tools that are being applied to different marketing processes, as well as to determine the level of application and technological adaptation by companies to marketing activities, commercial management, improving customer experience, optimizing operations or creating new business models.

This study analyzes the academic literature and case studies to investigate the impact of ML techniques on marketing practices. To this end, the following objectives have been established: (i) to identify the ML methods, techniques, and tools that are more frequently employed in marketing practices; (ii) to find out marketing processes in which ML techniques are being applied in an intensive manner; and (iii) to establish successful cases of the use of ML techniques in marketing. To achieve the aforementioned research objectives, the study makes use of the Scopus data base using two precise keywords, "machine learning" and "marketing", in order to obtain relevant data. After identifying some of the most representative scientific manuscripts on the subject, they are systematically processed. Two analysis methodologies are employed: (i) topic modeling using the non-negative matrix factorization (NMF) algorithm [6]; and (ii) a comparative analysis using the k-means clustering algorithm. The NMF has allowed us to discover and understand the predominant themes in the dataset of selected articles. The purpose of using the k-means clustering methodology is to analyze and validate the results of the NMF.

The rest of the document is structured as follows: Sect. 2 identifies the research methodology that explains the process of obtaining and systematizing the information based on the main keywords of the study. Section 3 introduces the first cluster of papers obtained, which describes the applications of ML algorithms in marketing. Section 4 refers to a second cluster, which refers to the application of big data and ML techniques in marketing processes and their impact on consumer behavior. Section 5 describes the results obtained in a third cluster, which refers to the applications of ML in digital marketing. Finally, Sect. 6 summarizes the main findings of the article.

2 Methodology

This section describes the methodology employed in our study, from the data collection process to data management and preprocessing, feature extraction, topic modeling, comparative analysis, and evaluation.

2.1 Data Collection Process

The first step of this study is the identification of scholarly articles in which ML has been employed to develop marketing applications. We utilized the Scopus database, known for its wide range of peer-reviewed literature, to conduct the search. In order to ensure that only the most relevant studies are included, the search criteria were precise. Hence, we searched for articles including both terms, "machine learning" and "marketing" in their title, also refining our search with several filters: (i) document type was set as 'Article', thus excluding non-peer-reviewed materials; (ii) source type was restricted to 'Journal', thus ensuring sources were reputable academic journals; (iii) language was set to 'English', hence ensuring the findings are widely accessible; and (iv) the period considered is from 2018 to 2023. This approach resulted in a collection of 63 papers, marking a significant corpus of work at the intersection of ML and marketing.

2.2 Data Management and Storage

After identifying relevant literature, we organized and stored this data systematically. We chose Supabase (https://supabase.com/), an open-source platform for database storage, to manage our collection. We created a SQL database table to catalog each article, including fields for: (i) ID (a unique identifier for each paper); (ii) created-at (the date and time of the record's database entry); (iii) title (the article's title as it appears in the source); (iv) abstract (a summary of the research and its outcomes); (v) year (the publication year of the article); (vi) cited (the citation counts of the article, indicating its influence); (vii) title-clean (a formatted version of the title for analysis); and (viii) abstract-cleaned (a standardized version of the abstract for uniform analysis). Figures 1 and 2 provide insight into the scholarly impact and trends, including the distribution of citations received and the count of published articles per year. The distribution of cites is consistent with what might be expected in a broad range of academic disciplines, where most articles contribute incrementally to literature, while a few have an outstanding impact. The number of publications per year suggests a significant peak in 2022, with a general trend of increasing volume over the years followed by a sharp decline in 2023.

2.3 Data Preprocessing: Text Data Cleaning

A critical step in our methodology was the thorough cleaning of text data from article titles and abstracts. This involved several steps: tokenization, which breaks down text into basic elements; normalization, to ensure all text is in the same case; removal of numbers and punctuation; and stemming, to simplify words to their base form. We also identified and removed custom stop words, which included terms specific to our field that didn't add value to our thematic analysis. This detailed cleaning process greatly improved the text data's quality and prepared it for analysis.

2.4 Feature Extraction: Generation and Storage of Vector Embeddings

In the fourth step of our methodology, we converted the cleaned text data into vector embeddings using the "text-embedding-ada-002" model and the "cl100k-base" encoding scheme. The "text-embedding-ada-002" model, a part of OpenAI's suite of NLP tools, turns text into high-dimensional vectors that reflect

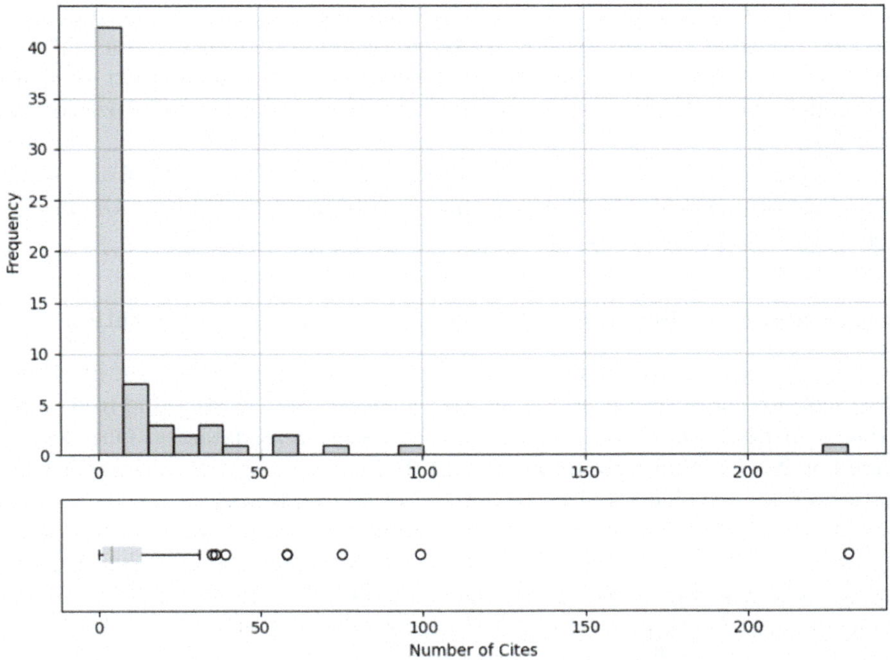

Fig. 1. Distribution of cited articles.

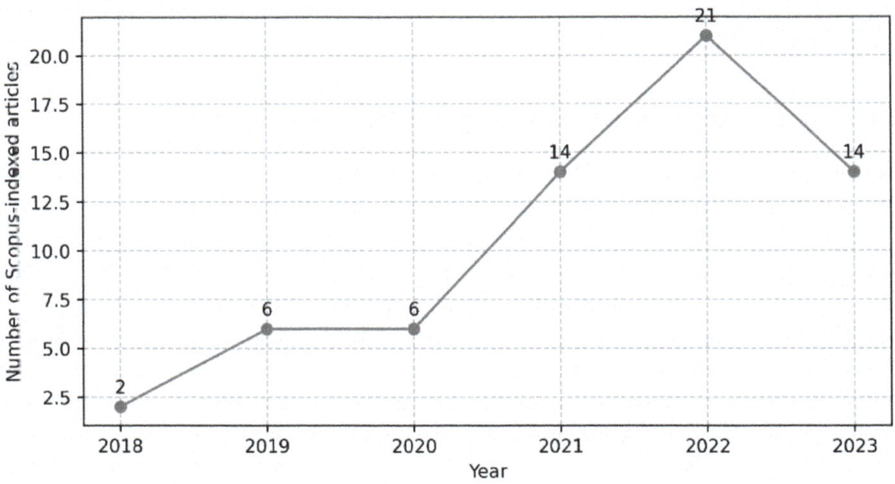

Fig. 2. Number of articles published per year in Scopus-indexed journals.

its underlying meaning. This model, which is based on deep learning, effectively captures linguistic aspects like context and tone, making it easier for machine learning algorithms to work with textual data. The "cl100k-base" encoding works hand in hand with the model to ensure that the text data is represented in the best possible way, preserving its semantic fullness even in large datasets. We imposed a maximum token limit of 8000 for each piece of text to match the model's processing power, employing tiktoken for efficient tokenization and encoding.

We then stored these vector embeddings in our Supabase database, taking advantage of its capabilities for real-time management to facilitate the integration of NLP tasks smoothly.

2.5 Topic Modeling Using Non-Negative Matrix Factorization

In our effort to uncover dominant themes within a large collection of cleaned abstracts, we chose Non-Negative Matrix Factorization (NMF) due to its effectiveness in revealing hidden topics from extensive text datasets. Our decision leaned on NMF's ability to decompose TF-IDF vectorized text into distinct thematic topics, each characterized by a group of related words. This feature makes NMF particularly useful for delving into complex themes found in machine learning and marketing research studies. Our approach started with detailed data preparation, which included identifying common words and vectorizing the text, before moving on to apply the NMF model. The use of NMF enabled us to gain structured insights into the primary research themes by summarizing the topics identified and associating documents with their nearest thematic categories. Moreover, this process ensured that the interpretation of these topics was straightforward, enhancing our overall grasp of the scholarly conversation in our area of study.

2.6 Comparative Analysis Using K-Means Clustering

To complement our use of NMF, we also applied K-Means clustering to the vector embeddings to both validate and deepen our thematic analysis. We chose to use K-Means alongside NMF because of its ability to efficiently sort features into clear groups based on their vector similarity. This made it an effective comparison tool for spotting natural document clusters. This method allowed us to carefully track the clustering process, from preparing the data to the final step of assigning documents to their clusters. It also helped us spot patterns and themes relevant to machine learning's role in marketing, such as customer segmentation and the optimization of advertising efforts. By exporting the dataset organized by clusters for additional study, K-Means clustering provided a quantitative perspective for our data review. This reinforced the thematic frameworks identified by NMF and laid a solid groundwork for our comparative studies. Our dual-methodology strategy ensured a thorough review of how machine learning intersects with marketing, offering insightful revelations into how machine learning techniques are transforming marketing strategies.

2.7 Evaluation of Topic Modeling and Clustering Results

The final step in our approach was to compare the topics from NMF with the clusters found through K-Means clustering. By running several trials with different numbers of topics and clusters, we pinpointed an ideal thematic categorization that consistently brought out three clear and understandable categories using both methods. This comparative review didn't just confirm our thematic discoveries; it also emphasized the benefit of using multiple unsupervised learning methods to explore the thematic foundations of our dataset. We then combined the results of this detailed analysis, spotlighting major scholarly interests and proposing directions for future investigations in the intersection of machine learning and marketing.

2.8 Software Tools and Scripting Workflow

Throughout this research we employed Python, which is recognized for its robust ecosystem that includes powerful libraries for data manipulation, text processing, and machine learning. The series of Python scripts described next constitute a workflow that was adjusted to meet the specific requirements of the study: (i) script 'v2-01-save-data-supabase.py' initiates our data management process – it automates the collection and secure storage of literature within Supabase, utilizing the platform's database management efficiency to systematically organize the collected literature with comprehensive metadata; (ii) script 'v2-02-save-embeddings-to-supabase.py' handles the transformation of cleaned textual data into vector embeddings using the 'text-embedding-ada-002' model and 'cl100k-base' encoding scheme; (iii) scripts 'v2-03-modelling-nmf.py' and 'v2-04-clustering-kmeans.py' apply the NMF algorithm and the k-means clustering, respectively; and (iv) script 'v2-05-evaluate-results.ipynb' enables the comprehensive evaluation of the thematic analysis and clustering outcomes, supporting a comparative and iterative approach to refine our topic categorizations and cluster identifications. The deployment of these Python scripts, together with the use of Supabase for efficient data management and the robust Python ecosystem for data analysis, contributes to the development of a solid methodology.

3 Cluster 0: Machine Learning Algorithms in Marketing

Cluster zero consists of a total of four items. The four articles analysed focus on the use of machine learning algorithms in the field of marketing, specifically in customer segmentation, predicting consumer behaviour and optimising marketing campaigns in the banking sector.

The objective of the firts article is to provide a lead prediction algorithm (PCPA) using big data analysis and machine learning techniques. [7] The authors of the second article investigate the performance of machine learning techniques on a bank marketing dataset. [8]. The authors of the third article investigate the effectiveness of machine learning algorithms in predicting consumers' application for personal loans. [7]. Finally, the fourth article proposes a supervised and unsupervised machine learning modelling approach for bank marketing in the field of E-CRM [9].

Within the cluster, it has been concluded that these articles address two themes: Applications of machine learning techniques in marketing and evaluation and comparison of algorithms:

Applications of Machine Learning Techniques in Marketing
Each article deals with the application of machine learning algorithms to strengthen marketing strategies. The authors investigate different techniques, e.g. Gradient Boosting, Random Forest, k-Nearest Neighbour, BP neural networks, SVM, as well as clustering algorithms, such as k-means, to improve marketing effectiveness, customer acquisition and consumer behaviour prediction.

Evaluation and Comparison of Algorithms
Each article evaluates and compares different machine learning algorithms based on their performance metrics such as accuracy, precision, recall, F-measure and cross-validation scores. The authors highlight the effectiveness of certain algorithms in specific contexts, such as Random Forest, which consistently shows high accuracy in predicting various marketing-related outcomes.

In summary, the four papers demonstrate that the use of machine learning algorithms is effective in improving marketing strategies in the banking sector, enabling better customer segmentation, more accurate predictions of customer behaviour and optimisation of marketing campaigns to attract premium customers.

Table 1 shows a summary of the articles in cluster 0. In this case, we have added all the articles in the cluster.

Table 1. Summary of cluster 0 table

Authors	Title	Date	Source	Citations
El Koufi, N., Belangour, A., Sdiq, M.	Research on Precision Marketing based on Big Data Analytics and Machine Learning: A Case Study in Morocco	2022	International Journal of Advanced Computer Science and Applications, 13(10), pp. 58–63	3
Saengthongrattanachot, W., Na-Udom, A., Rungrattanaubol, J.	Una Comparison of Machine Learning Techniques for Classification in Bank Marketing Data	2022	Thai Journal of Mathematics, 2022(Issue Especial), pp. 157–168	0
Zhang, M.	Research on Precision Marketing based on Consumer Portrayal from a Machine Learning Perspective	2022	Wireless Communications and Mobile Computing, 2022, 9408690	4
Eltahir, A., Ahmed, T., Abdelmageed Mohammed, A., Hilal, A.	A Supervised and Unsupervised Machine Learning Model Approach to E-CRM in Bank Marketing	2022	International Journal of Computer Science and Network Security, 22(4), pp. 625–636	2

4 Cluster 1: Application of Big Data and Machine Learning in Marketing Processes and Their Impact on Consumer Behaviour

The second cluster consists of fourteen articles that analyze the use of big data and ML in marketing and their impact on consumer behavior. The main objective of this study is to determine the potential of these technologies in marketing processes, as well as the challenges and opportunities that marketers face in connecting with consumers. Marketing is a dynamic and constantly changing business function. To achieve its goal of delivering the right product to the right customers at the right time, it must manage information systematically, timely and accurately about the market, products and customers, as well as the business environment. [10]. The adoption of these technologies in marketing shows their full potential in different aspects:

- Social networks: Big data helps gather real-time consumer info. ML is used to analyze reviews and spot fake ones or unusual patterns.
- Understanding consumer behavior: Big data and fuzzy support vector machines show what consumers prefer when searching for info or comparing products. This is handy for launching new products where people need to pick what suits them best. Precision marketing, with big data, targets customers accurately with tailored offers [10].
- Advertising: Big data suggests TV/Internet content and predicts what people will watch and buy, making ads more accurate. It also helps in creative decision-making and predicts sales from text, audio, and video data.
- Targeted digital promotions: Companies boost email campaigns by setting clear goals, personalizing content, and tracking engagement [11].
- Direct marketing: Predicting customer behavior helps tailor messages and improve campaign effectiveness [12]. ML helps companies understand customers better and run more targeted campaigns.
- Customer experience: Technology enhances marketing value and creates better customer experiences. Meeting customer needs and expectations requires using technology to understand behavior and assess marketing impact. Big data aids in decision-making and understanding consumer behavior in real time.

All in all, the combination of big data and ML allows for enhancing customer relationship management, and for developing effective marketing strategies. In that way, businesses can understand customer needs, segment audiences, predict behavior, and personalize marketing efforts. Big data also plays a key role in identifying target audiences for advertising campaigns, ensuring maximum impact and engagement. Figure 3 displays three cases of companies that are applying big data, ML and artificial intelligence in their marketing processes.

The Table 2 shows the first two articles of the cluster and the last article.

	PortAventura World Parks & Resort	Starbucks	Cruzcampo
Objective	Increase the volume of visitors and new customers by exploring new audiences.	Improve the personalised experience and increase the average value of a customer.	Repositioning the brand by focusing on innovation, while emphasising the importance of tradition, the value of pride in what is local, the power of identity and differences in the face of new generations.
Strategy	AI and statistical calculations for the creation of an audience architecture to work with, in order to propose a media mix.	Use of big data. IA and predictive analytics to find out what, where and how your customers shop. This information is used to personalise promotions and other customer-centric actions..	AI for the creation of advertising messages: Cruzcampo "resurrects" Lola Flores.
Conclusion	More than 30% increase in new users and 68% increase in total transactions over the previous year.	Significantly improve customer experiences. Increasing the effectiveness of your rewards programme, encouraging consumers to spend more and visit your shops more often.	Throughout 2021 the ad has generated 614 million impacts, which translates into the number of times it has reached a viewer's eyes, either on TV or digital media.

Fig. 3. Use of big data, machine learning and artificial intelligence by brands. Source: Own elaboration

Table 2. Summary of cluster 1 table

Authors	Title	Date	Source	Citations
Ellickson, P. B., Kar, W., & Reeder III, J. C.	Estimating Marketing Component Effects: Double Machine Learning from Targeted Digital Promotions	2023	Marketing Science, 42(4), 704–728	18
Ebrahimi, P., Basirat, M., Yousefi, A., Nekmahmud, M., Gholampour, A., and Fekete-Farkas, M	Social Networks Marketing and Consumer Purchase Behavior: The Combination of SEM and Unsupervised Machine Learning Approaches	2022	Big Data and Cognitive Computing, 6(2), 35	49
Miklosik, A., and Evans, N.	Impact of Big Data and Machine Learning on Digital Transformation in Marketing: A Literature Review	2020	Ieee Access, 8, 101284–101292	107

5 Cluster 2: Machine Learning Applied to Digital Marketing

The third cluster encompasses a collection of thirty-seven articles. These articles are focused on two primary themes: general applications of ML in digital marketing and innovation and technological advancements [13]. One significant application of ML in digital marketing is the identification and recommendation of key indicators within brand communities and public behavioral data. Techniques such as random forest, AdaBoost, and XGboost are utilized to distinguish between active and passive behaviors on Facebook pages of digital product brands and platforms. Additionally, text mining and ML methods such as content and sentiment analysis are employed to identify patterns within public and community behavior data of brands. The integration of AI further strengthens the processing of unstructured information, enhancing data quality through sampling methods like oversampling and undersampling. Furthermore, ML aids in analyzing consumer demand for information and uncovering hidden unstructured elements within data available on social networks, thereby improving marketing effectiveness. AI-based data analysis also helps in understanding the impact of community content on public engagement and how participatory posts enhance interactive experiences.

In digital marketing, ML finds many applications in analyzing the vast amount of data generated by social media and online platforms. NLP techniques enable AI systems to analyze human language across various mediums such as blogs, product reviews, and social media posts. This analytical capability is key in understanding and responding to customer perceptions online. Moreover, ML techniques aid in specific applications such as fraud detection, stock market prediction, customer relationship management, and automated generation of content summaries. Integration of ML tools with marketing strategies enhances performance by leveraging statistics and artificial intelligence to make informed decisions. [14] propose a framework comprising three key steps: (i) text mining involves the examination of unstructured information to identify meaningful patterns; (ii) use of ML tools to better understand and interpret data; and (iii) perform a data analysis that provides efficient marketing insights. Beyond social media, ML applications extend to various domains within digital marketing, including chatbots, advertisement refinement, email marketing, customer behavioral targeting, lead generation, and automated content creation. These applications enable enhanced customization, deeper understanding of user behavior, and optimization of marketing strategies for improved efficiency [15].

Innovation and technological development in digital marketing are underscored by intelligent marketing strategies based on simulation with ML and mobile computing. Studies reveal significant improvements in company performance metrics, including efficiency gains of approximately 20% and revenue increases exceeding 30% compared to traditional strategies [16]. Additionally, advancements in ML facilitate automatic content classification in marketing, with neural networks demonstrating superior performance in multi-label classification of news articles compared to other algorithms like random forest and k-nearest neighbors [17]. Their ability to handle high-dimensional data without dimensionality reduction techniques highlights their efficacy in enhancing marketing strategies.

The Table 3 shows the first two articles of the cluster and the last article.

Table 3. Summary of cluster 2 table

Authors	Title	Date	Source	Citations
Brei, V.A.	Machine Learning in Marketing	2020	Foundations and Trends in Marketing, 14(3), pp. 173–236	23
Akter, S., Dwivedi, Y.K., Sajib, S., ...Bandara, R.J., Michael, K.	Algorithmic bias in machine learning-based marketing models	2022	Journal of Business Research, 144, pp. 201–216	30
Mackey, T., Kalyanam, J., Klugman, J., Kuzmenko, E., Gupta, R.	Solution to Detect, Classify, and Report Illicit Online Marketing and Sales of Controlled Substances via Twitter: Using Machine Learning and Web Forensics to Combat Digital Opioid Access	2018	Journal of Medical Internet Research, 20(4), e10029	56

6 Conclusions

The use of Machine Learning (ML) in marketing is a vast and evolving area that covers various methods, techniques, and tools. These tools help in 1) gathering data from various marketing activities; 2) processing, recording, and identifying patterns in the data; 3) transforming data into valuable insights and opportunities; and 4) predicting customer behavior and aiding in marketing decisions. The pursuit of more precise and adaptable marketing strategies has led to significant changes through the integration of Artificial Intelligence (AI), Machine Learning (ML), and Business Analytics (BA). This shift is driven by several factors. Firstly, today's consumers are more demanding and seek personalized products and services. Secondly, in our data-rich world, businesses face both numerous opportunities and challenges. When used together, AI, ML, and BA provide the capabilities to handle large data sets, uncover patterns, and predict customer behavior with remarkable precision.

ML techniques in marketing give marketers a chance to 1) uncover new insights into consumer behavior and 2) enhance the efficiency of marketing tasks. On the consumer front, machine learning algorithms allow for better customer segmentation and more accurate predictions of their actions, offer recommendations based on behavior data, analyze sentiments, evaluate brand perception impacts, analyze demand, generate leads, etc. This method improves understanding of consumer behaviors, enables more effective personalization, and optimizes marketing strategies. Regarding marketing tasks, ML is used in product development, pricing, distribution, and promotion. Common applications include personalizing products and services, enhancing customer service, dynamic pricing, optimizing e-commerce, predicting conversion success, forecasting advance purchases, managing social media engagement, automating image analysis, optimizing search engines, and improving customer experiences, among others.

It's worth noting that many companies implement ML in their marketing strategies by developing solid data management practices. These practices ensure that marketing information aligns with business goals and that decisions are data-driven. ML techniques are crucial for understanding customer feedback, enhancing products, and customer experiences. The importance of product reviews in e-commerce is also highlighted, underscoring their role in assisting customers with their buying decisions and guiding companies in their marketing strategies. Reviews are vital for driving sales, maintaining a positive reputation, and enhancing brand recognition. Thus, companies like PortAventura World, Starbucks, Cruzcampo, and others leverage ML to add value to their marketing efforts and create more engaging customer experiences.

References

1. Ye, G., Hudders, L., De Jans, S., De Veirman, M.: The value of influencer marketing for business: a bibliometric analysis and managerial implications. J. Advert. **50**(2), 160–178 (2021)
2. Ngai, E.W., Wu, Y.: Machine learning in marketing: a literature review, conceptual framework, and research agenda. J. Bus. Res. **145**, 35–48 (2022)
3. Herhausen, D., Bernritter, S.F., Ngai, E.W., Kumar, A., Delen, D.: Machine learning in marketing: recent progress and future research directions. J. Bus. Res. **170**, 114254 (2024)
4. Akter, S., Dwivedi, Y.K., Sajib, S., Biswas, K., Bandara, R.J., Michael, K.: Algorithmic bias in machine learning-based marketing models. J. Bus. Res. **144**, 201–216 (2022)
5. Volkmar, G., Fischer, P.M., Reinecke, S.: Artificial intelligence and machine learning: exploring drivers, barriers, and future developments in marketing management. J. Bus. Res. **149**, 599–614 (2022)
6. Guo, Y.T., Li, Q.Q., Liang, C.S.: The rise of nonnegative matrix factorization: algorithms and applications. Inf. Syst. 102379 (2024)
7. Zhang, M.: Research on precision marketing based on consumer portrait from the perspective of machine learning. Wirel. Commun. Mob. Comput. **2022** (2022)
8. Saengthongrattanachot, W., Na-udom, A., Rungrattanaubol, J.: A comparison of machine learning techniques for classification in bank marketing data. Thai J. Math. 157–168 (2022)
9. Sarker, I.H.: Machine learning: algorithms, real-world applications and research directions. SN Comput. Sci. **2**(3), 160 (2021)
10. Miklosik, A., Evans, N.: Impact of big data and machine learning on digital transformation in marketing: a literature review. IEEE Access **8**, 101284–101292 (2020)
11. Ellickson, P.B., Kar, W., Reeder, J.C., III.: Estimating marketing component effects: double machine learning from targeted digital promotions. Mark. Sci. **42**(4), 704–728 (2023)
12. Choi, Y., Choi, J.W.: Assessing the predictive performance of machine learning in direct marketing response. Int. J. E-Bus. Res. (IJEBR) **19**(1), 1–12 (2023)
13. Chen, Y.: Comparing content marketing strategies of digital brands using machine learning. Humanit. Soc. Sci. Commun. **10**(1), 1–18 (2023)
14. Arasu, B.S., Seelan, B.J.B., Thamaraiselvan, N.: A machine learning-based approach to enhancing social media marketing. Comput. Electr. Eng. **86**, 106723 (2020)

15. Boddu, R.S.K., Santoki, A.A., Khurana, S., Koli, P.V., Rai, R., Agrawal, A.: An analysis to understand the role of machine learning, robotics and artificial intelligence in digital marketing. Mater. Today: Proc. **56**, 2288–2292 (2022)
16. Mao, S., Huang, R.: Complexity construction of intelligent marketing strategy based on mobile computing and machine learning simulation environment. Complexity **2021**, 1–11 (2021)
17. Salminen, J., Yoganathan, V., Corporan, J., Jansen, B.J., Jung, S.G.: Machine learning approach to auto-tagging online content for content marketing efficiency: a comparative analysis between methods and content type. J. Bus. Res. **101**, 203–217 (2019)

A Systematic Literature Review on Conversational AI's Influence in Consumer Behavior

Marc Bara Iniesta[1](✉) and David López[2]

[1] EAE Business School, Barcelona, Spain
marc.bara@campus.eae.es
[2] ESADE Business School, Barcelona, Spain
david.lopez1@esade.edu

Abstract. In the dynamic realm of digital marketing, the integration of conversational artificial intelligence (AI) has emerged as a pivotal force reshaping consumer behaviors and market strategies. This paper presents a comprehensive overview of the research landscape surrounding conversational AI's influence on consumer decision-making processes. Through a systematic literature review (SLR), we meticulously analyze a diverse array of scholarly works, providing insights into the temporal, disciplinary, and geographical dimensions of this burgeoning field. Furthermore, we discuss the implications of highly cited research and identify emergent trends, shedding light on critical gaps and future directions for scholarly inquiry. Our findings underscore the transformative potential of conversational AI in driving strategic marketing initiatives and enhancing consumer experiences in the digital age.

Keywords: Conversational AI · Digital Marketing · Consumer Decision-Making · Systematic Literature Review · Strategic Marketing · Consumer Behavior · Consumer influence

1 Introduction

The proliferation of conversational AI technologies, including chatbots, virtual assistants, and natural language processing (NLP) algorithms, has revolutionized the landscape of digital marketing [20]. These advancements offer unprecedented opportunities for marketers to engage with consumers in personalized, interactive, and real-time interactions, thereby influencing their decision-making processes [22]. Understanding the intricate dynamics between conversational AI and consumer behaviors is essential for devising effective marketing strategies and staying ahead in today's competitive market environment.

In this paper, we examine the role of conversational AI in shaping consumer decision-making within digital marketing contexts. Leveraging the gold standard methodology of systematic literature review (SLR), we meticulously survey a corpus of scholarly articles to provide a current and comprehensive synthesis of existing research. Our review

A. A. Juan et al. (Eds.): DSA ISC 2024, LNCS 14778, pp. 27–39, 2025.
https://doi.org/10.1007/978-3-031-78238-1_3

covers the temporal evolution, disciplinary perspectives, and geographical distribution of studies, offering valuable insights into the progression of academic exploration in this domain.

Furthermore, we examine highly cited research articles to identify foundational insights and emerging trends that delineate the forefront of academic inquiry. By critically analyzing both seminal and recent works, our goal is to map out the current research landscape and suggest potential areas for future study.

Through this paper, we aim to contribute to the academic discussion on the impact of conversational AI on consumer decision-making processes and strategic marketing practices. By highlighting key findings, pinpointing research gaps, and proposing future research directions, this paper endeavors to enrich our understanding of the transformative potential of conversational AI in driving consumer-centric marketing strategies.

In the subsequent sections, we provide a detailed account of our methodology (2), present descriptive analyses of our findings (3), discuss insights from highly cited research (4), illustrate significant examples of AI application in Marketing and consumer behavior (5) and delineate future directions for scholarly inquiry (6). Through this comprehensive exploration, we seek to offer actionable insights and stimulate further research endeavors in this dynamic and rapidly evolving field.

2 Methodology

To develop a current and comprehensive summary of the research landscape regarding conversational AI's role in consumer decision-making within digital marketing, we conducted a systematic literature review (SLR). SLRs are regarded as the gold standard for methodically reviewing and synthesizing a broad array of literature, offering insights and evaluations that are critical for academic research [1]. They provide a detailed, organized, and critical approach to review literature, making them an indispensable tool for uncovering research voids [2].

Our research data was meticulously compiled from Scopus, renowned for its extensive archive of peer-reviewed academic content. Currently indexing over 2.4+ billion references [3], Scopus presented itself as a superior choice to Web of Science (WoS). This preference was based on two main considerations: firstly, Scopus offers significantly broader coverage, approximately 38% more extensive than that of WoS in terms of publication records [4]. Secondly, to avoid the dilution of data quality that often accompanies the use of multiple databases, our SLR relied solely on Scopus [5].

To navigate this database, we pinpointed a series of keywords pertinent to AI within our designated research domains. The keywords were strategically chosen to cover a wide range of terms directly related to the technology and its application in consumer behavior and digital marketing, arranged in Boolean queries as follows:

1. Technology-focused terms:

("conversational AI" OR "chatbots" OR "ChatGPT" OR "GPT-4" OR "virtual assistants" OR "natural language processing" OR "NLP")

These keywords focus on various aspects and advancements in AI technologies that engage with users via natural language. Specifically:

- "Conversational AI" covers the broad application of AI technologies capable of simulating conversational human interactions.
- "Chatbots" are a practical implementation of conversational AI, commonly used in customer service and engagement.
- "ChatGPT" and "GPT-4" represent the latest in generative conversational models, offering advanced capabilities that are critical to understanding state-of-the-art developments.
- "Virtual assistants" are included to capture studies on AI applications that assist users through interactive interfaces.
- "Natural language processing (NLP)" is crucial for dissecting the underlying technology that enables machines to understand and process human language.

2. Consumer behavior-focused terms:

("consumer decision-making" OR "consumer behavior" OR "purchase intention" OR "customer engagement" OR "user experience")

These terms aim to capture how conversational AI influences consumer actions and perceptions, crucial for understanding its impact on marketing strategies.

- "Consumer decision-making" explores how AI influences choices and preferences in the buying process.
- "Consumer behavior" assesses broader patterns and tendencies in how consumers interact with AI-driven marketing efforts.
- "Purchase intention" is targeted to understand the direct impact of AI interactions on consumers' readiness to buy.
- "Customer engagement" focuses on the depth and quality of interactions between consumers and AI applications.
- "User experience" evaluates the effectiveness and satisfaction of consumers when interacting with AI-enabled interfaces.

3. Marketing context-focused terms:

("digital marketing" OR "online marketing" OR "e-commerce" OR "social media marketing" OR "internet marketing")

These keywords help to focus the search on studies that explore the integration of conversational AI within various online marketing platforms and environments.

- "Digital marketing" and "Internet marketing" encompass the broad application of digital technologies which include AI tools.
- "Online marketing" specifically refers to marketing activities conducted through the Internet which use AI capabilities.
- "E-commerce" is crucial as many conversational AI tools are implemented to enhance the online shopping experience.
- "Social media marketing" focuses on the use of AI in social platforms where customer interaction is key.

By combining these keywords through AND operators, we formulated a comprehensive search query that allowed us to capture relevant research intersections across

AI technology, consumer behavior, and digital marketing contexts. These search criteria were applied to the "title," "abstract," and "keywords" sections of Scopus, focusing on studies published up to December 31st, 2023. By excluding conference proceedings, book chapters, and non-English publications, we identified 78 articles. We then extracted their metadata, including authorship, titles, the country of the corresponding author, publication count, citation numbers, journal names, keywords, and geographical distribution.

3 Results – Descriptive Analysis

The systematic search conducted within the Scopus database yielded a multifaceted dataset, which provides an informative snapshot of the research landscape concerning conversational AI's influence on consumer behavior. First, we conduct a descriptive analysis divided into temporal, disciplinary, and geographical dimensions.

3.1 Analysis by Year

The analysis of publication dates shows an increasing trend in research on conversational AI's impact on consumer behavior. Initially, the topic saw sparse attention, but from 2014, interest began to significantly grow, peaking in 2023 with 27 documents. This surge reflects the field's expanding academic and practical focus, driven by technological advancements and broader adoption in the market (Fig. 1).

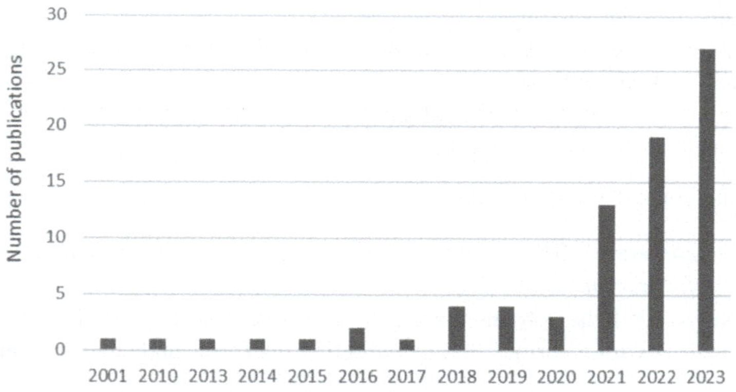

Fig. 1. Annual Publication Trend

3.2 Analysis by Subject Area

The document distribution highlights a strong focus on Computer Science (56 documents), showcasing AI's technical foundation. Decision Sciences, Business, Management, and Accounting are also well-represented with 24 and 22 documents, underlining AI's role in strategic and organizational decision-making. Engineering and Mathematics contributions (with significant numbers) illustrate the interdisciplinary approach of AI research, combining technical and theoretical aspects (Fig. 2).

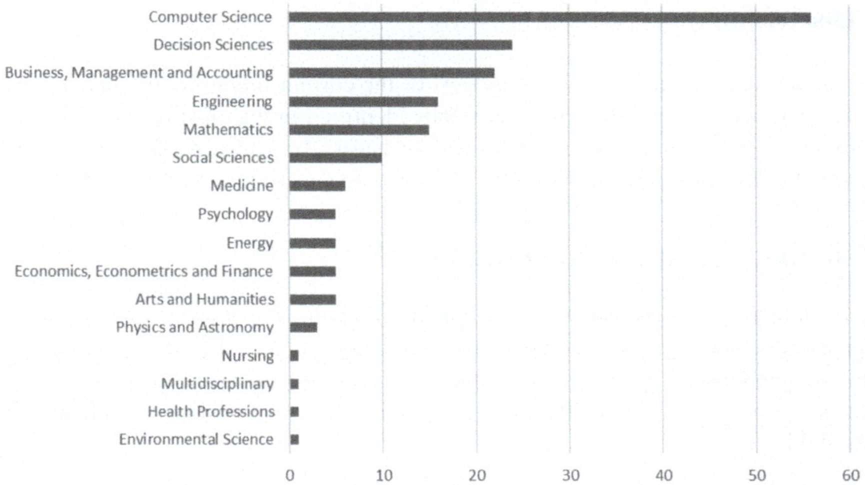

Fig. 2. Document Distribution by Subject Area

3.3 Analysis by Country

Geographical analysis shows a wide international involvement in conversational AI research, with India, the United States, and China leading in publications. This suggests a concentrated research effort in these nations, crucial to the global AI dialogue. Europe, notably Portugal, and other regions also contribute, highlighting conversational AI's worldwide impact across diverse markets and cultures (Fig. 3).

Fig. 3. Geographical Distribution of Publications

4 Discussion

This section delves into two critical aspects of the current literature: the insights from highly cited research and the emergent trends identified in the most recent articles. By examining both the foundational studies and the latest advancements, we aim to provide a comprehensive overview of the field's current state and its future directions.

4.1 Insights from Highly Cited Research

The examination of seminal works within the domain of conversational AI and its influence on consumer behavior uncovers significant findings. The most cited articles, although foundational, span a broader range of publication years, from 2010 to 2021. This spectrum of research dates indicates a continuous interest and evolving understanding in the field:

- Lee D., Hosanagar K., Nair H.S., "**Advertising content and consumer engagement on social media: Evidence from Facebook**" (2018, Cited by 485 times) [6]: This foundational study investigates how advertising content on platforms like Facebook influences consumer engagement, establishing a critical understanding of digital marketing's effectiveness.
- Decker R., Trusov M. "**Estimating aggregate consumer preferences from online product reviews**" (2010, Cited by 265 times) [7]: Presenting a novel methodology for analyzing consumer preferences through online reviews, this research showcases the significant role of AI in deciphering complex consumer sentiment data.
- Liu X., Shin H., Burns A.C., "**Examining the impact of luxury brand's social media marketing on customer engagement: Using big data analytics and natural language processing**" (2021, Cited by 200 times) [8]: Assessing the effects of luxury brands' social media strategies on customer engagement, this study emphasizes the strategic incorporation of big data and NLP in marketing initiatives.
- Yen C., Chiang M.-C., "**Trust me, if you can: a study on the factors that influence trustworthiness of conversational agents**" (2021, Cited by 96 times) [9]: This article delves into the determinants of trust in conversational agents, elucidating key factors that shape consumer perceptions of AI's reliability.
- Jabbar J., Urooj I., Junsheng W.; Azeem N., "**Real-time sentiment analysis on E-Commerce applications through big data**" (2019, Cited by 41 times) [10]: Focusing on real-time sentiment analysis in e-commerce, this research highlights the operational benefits of AI technologies in agilely adapting to and managing consumer feedback.

The publication years of these highly cited articles indicate a gap in addressing the latest conversational AI advancements, such as ChatGPT and GPT-4, introduced in late 2022 and 2023. The novelty of these technologies means they're yet to be widely cited, marking a promising direction for new research. This observation underscores the importance of examining the latest 2023 studies, especially those incorporating these advanced GPT models, to capture the field's current innovations.

4.2 Thematic Analysis from Most Recent Articles

The following thematic analysis synthesizes insights from the five most recent articles in our dataset, underscoring the forefront of academic inquiry into conversational AI technologies.

- **Combatting Misinformation through AI:** A notable trend in recent research is the focus on employing conversational AI to identify and mitigate the effects of fake reviews and manipulated content within e-commerce platforms. For instance, a study by [11] delves into the latent characteristics of fake reviews, demonstrating AI's capacity to discern authenticity in online consumer feedback, a critical factor in maintaining trust and integrity in digital marketplaces. Similarly, [12] present an advanced AI model that detects manipulated reviews, highlighting the escalating challenge of ensuring review authenticity in the age of ubiquitous online retail.
- **Enhancing Consumer Experience with Emotionally Intelligent AI:** The incorporation of emotional intelligence in AI design represents another emerging theme. Research by [13] explores the impact of emotions in avatar design, suggesting that emotionally responsive AI can significantly increase consumer engagement and satisfaction. This shift towards emotionally intelligent AI avatars indicates a broader trend of humanizing digital interactions to foster deeper consumer connections.
- **Optimizing E-commerce Operations with AIOps:** The application of artificial intelligence in optimizing operations (AIOps) within e-commerce signifies a growing emphasis on backend efficiency. [14] investigate how AIOps can enhance system performance, suggesting that AI-driven operational improvements can lead to better consumer experiences through faster response times and more reliable service delivery.
- **Advanced AI Models for In-depth Consumer Insight Analysis:** Finally, the development and application of sophisticated AI models for text analysis are highlighted in recent studies. [15] introduce a deep learning-based model that utilizes hybrid features for sentiment analysis, showcasing the potential of AI in generating nuanced insights from consumer feedback. This research direction points to the increasing complexity and capability of AI technologies in extracting strategic insights from vast quantities of textual data.

Our analysis reveals emerging themes and advancements that are shaping the capabilities of conversational AI within the digital marketing sphere. We have seen how AI technologies, particularly in combating misinformation, enhancing emotional intelligence, optimizing e-commerce operations, and providing in-depth consumer insight analysis, play pivotal roles in refining consumer interaction and business strategies.

5 Example of Significant Applications of AI in Marketing and Consumer Behavior

Building on the detailed exploration of conversational AI's influence on consumer behavior from Sect. 4, this section aims to expand our focus to the extensive applications of AI in marketing more broadly. As we transition from discussing specific influences of conversational AI, we now explore how these technologies are revolutionizing market research, marketing strategy, and marketing actions more generally.

We will showcase examples that illustrate AI's transformative influence across various facets of marketing. This broader examination helps us appreciate the depth and breadth of AI's role in shaping modern marketing practices and consumer experiences. As illustrated in Fig. 4, the conversational AI agent, while a significant focus of our earlier discussion, is intricately connected within the entirety of the marketing AI strategy, underscoring its value as a pivotal component of a larger framework.

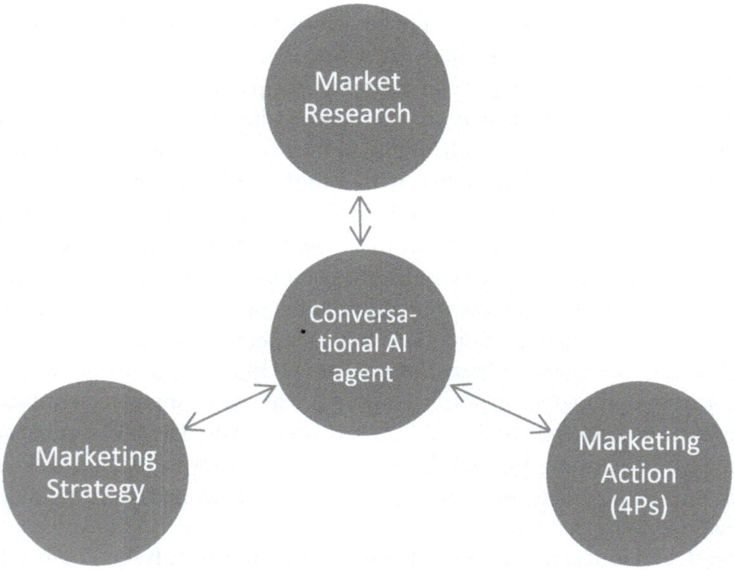

Fig. 4. Relationship between Conversational AI agents and marketing strategy, market research and marketing actions.

Thanks to the unprecedented evolution of AI, some authors [16–18] argue that it excels above human capability in various areas of marketing, which we can group into the 3 stages of the traditional strategic marketing planning framework [19] shown in Fig. 5: market research – marketing strategy – marketing action cycle.

5.1 Market Research

In the market research stage, AI is used for market intelligence, data collection, market analysis, and understanding of the customer by reducing human errors, as it enables:

- **Analysis of large amounts of market data:** AI analyzes large volumes of market data and predicts the action a user is likely to take next. It understands billions of search queries and helps determine how close a user is to making a purchase.
- **Streamlines real-time data processing:** AI accelerates data processing compared to human interaction, ensuring accuracy and security, and can collect and track tactical data in real-time, allowing marketers to make decisions in the present rather than waiting for campaigns to conclude.

Fig. 5. The 3 stages of the traditional strategic planning framework. Adapted from [19]

- **Provides valuable and relevant insights:** AI technologies make things easier by examining each new piece of data and offering more relevant information about customers based on their preferences.
- **Makes customer-centric decisions:** The insights AI gathers for organizations are valuable resources for better understanding their consumers and making customer-centric decisions and predicting their future behaviors.

5.2 Marketing Strategy

In this strategic stage, marketers leverage AI to enhance both strategic and operational aspects of marketing through the key decisions of segmentation, targeting, and positioning, thereby eliminating biases:

- **Segmentation:** AI provides exceptional flexibility in segmentation, capable of disaggregating the market into increasingly narrow segments, even to the extent of individual customer levels—each becoming a segment. Conversely, it can aggregate dispersed long tails into a single segment. This agility in both aggregation and disaggregation enables marketers to accurately determine the optimal segment size and develop meaningful, potentially unique customer profiles. While segmentation helps in strategically placing the brand in the market landscape, the operational application of AI here aids in the real-time adjustment and dynamic adaptation of these segments.
- **Targeting:** Unlike the more systematic task of segmentation, selecting the appropriate segment to target entails not just data processing but also strategic insight, judgment, and intuition. AI supports this process through advanced modeling algorithms that predict the most effective segment to target, ensuring that the marketing efforts meet the intended audience precisely. This is where the operational capabilities of AI intersect with strategic decision-making.
- **Positioning:** Once the target audience is identified, AI ensures that this group receives the most compelling message, delivered at the right time and through the appropriate channels. This operational action supports the strategic goal of establishing a strong and effective brand position in the consumer's mind.

Such integrations of AI in marketing not only align with what is traditionally considered strategic marketing but also deeply involve aspects of operational marketing, consistent with the definitions provided by J. J. Lambin [21].

5.3 Marketing Action (4Ps)

In this stage of executing a marketing program, marketers can use AI to automate and standardize processes, personalize actions, or even build relationships with consumers. Depending on which benefit is desirable, AI enables companies to offer their customers an optimal and seamless experience at every point of their customer journey.

- **Product/Brand:** AI provides emotional support to customers by training bots and virtual assistants to have a brand personality. By asking meaningful questions, chatbots identify consumers' moods and offer social support and emotional comfort with real-time conversations mimicking their communication styles. Once matching personalities (bot-consumer) are adjusted, recommendation programs (cross-selling and up-selling) are offered according to the viewer's mood and the behavior of other similar consumers (lookalike audiences).
- **Price:** Even if information about prices is incomplete, AI can dynamically adjust online prices in real-time, optimize customized prices by product, channel, or even initiate price negotiations with customers.
- **Promotion/Communication:** AI optimizes and reduces marketing budgets using automation in scheduling and media buying (programmatic advertising), consumer website navigation, real-time posting, bidding, and promotional content updating, and notification to consumer devices. Additionally, based on analytical profiling of customers, AI tracks audience sentiments and creates and customizes campaigns by individualizing the content of advertising messages. It can engage with customers by creating commercial scripts and tracking and measuring campaigns.
- **Place/Omnichannel:** AI enables maximizing omnichannel efficiency, helping decide which products, at what prices, and what promotions should be placed in each channel and for how long. Likewise, it allows adjusting email marketing campaigns (Marketing Automation) to redirect the audience to the most convenient channel at any given time.

5.4 Contribution of the Conversational AI Agent to Marketing Components

As we conclude this exploration into the multifaceted role of AI in marketing, it's critical to synthesize how conversational AI agents specifically contribute to the enhancement of marketing components. The diagram provided in Fig. 4 illustrates the symbiotic relationships between the conversational AI agent and the core areas of marketing.

- **Market Research**: The conversational AI agent transcends traditional data analysis methods by actively analyzing consumer queries to identify market trends. This proactive approach to market research allows for a deeper understanding of shifting consumer patterns and preferences. Furthermore, it is instrumental in collecting real-time feedback, which is invaluable for continuous product development and improvement.

- **Marketing Strategy:** At the strategic core, the conversational AI agent serves as an ally in dynamic segmentation. It refines the segmentation process based on rich consumer interaction data, ensuring that targeting is not only accurate but also relevant and personalized. These insights lead to more effective positioning as the AI agent helps to tailor the brand's messaging and offerings to align with the consumer's evolving preferences.
- **Marketing Action (4Ps):** When it comes to tactical marketing action, the conversational AI agent proves to be a versatile tool that adapts to the 4Ps of marketing:

  - Product/Brand: The agent's capacity to understand and respond to consumer emotions and history allows for tailored recommendations, enhancing the consumer's experience with the brand.
  - Price: The agent facilitates dynamic pricing strategies that reflect current demand and consumer behaviors, thus enabling more agile market responsiveness.
  - Promotion/Communication: In promotional efforts, the agent automates content delivery, personalizes marketing messages, and ensures timely communication, leading to increased efficiency and engagement.
  - Place/Omnichannel: Lastly, the agent contributes to omnichannel efficiency by optimizing the distribution of products across various channels based on consumer behavior and preference data.

In light of these contributions, it becomes evident that the conversational AI agent is not merely a tool but a dynamic participant in the marketing process, consistently enhancing the decision-making framework and execution of marketing activities. As we leverage these intelligent agents, we are not only streamlining processes but also creating more intimate and personalized consumer experiences. This integrated approach heralds a new era in marketing, where the alignment of strategy and action with consumer insights becomes seamless and automatic, thanks to the advancements in conversational AI technology.

6 Future Directions and Conclusion

The exploration of conversational AI's impact on consumer behavior and digital marketing has not only highlighted significant advancements but also exposed notable gaps in the current literature. These gaps underscore a series of avenues ripe for further exploration, emphasizing topics not sufficiently covered or entirely absent from existing studies.

- **Ethical Considerations and Consumer Privacy:** The ethical implications of conversational AI, particularly concerning consumer privacy, stand out as a critical area needing more depth in the literature. Future research should aim to develop frameworks that balance the benefits of personalization with robust privacy protections, fostering a deeper level of trust among consumers in AI-driven interactions.

- **Cross-Cultural Consumer Responses:** The current body of work reveals a scarcity in research on the impact of cultural differences on consumer responses to conversational AI. Investigating how cultural nuances influence engagement and trust in AI will fill a significant gap, providing invaluable insights for crafting globally resonant marketing strategies.
- **Longitudinal Studies on Consumer Behavior:** A notable omission in existing studies is the longitudinal examination of consumer behavior in response to conversational AI. Such research is essential for a more nuanced understanding of the sustained effects of AI on consumer decision-making processes and the dynamic nature of consumer-AI relationships.
- **Integration of Advanced AI Models:** While recent advancements such as GPT-4 have been acknowledged, there is a clear gap in understanding how these sophisticated AI models can be leveraged to enhance consumer engagement and personalization in digital marketing. Future research should focus on bridging this gap, exploring the potential of these technologies to revolutionize the consumer experience.
- **Combating Misinformation and Enhancing Operational Efficiency:** The recent themes also highlight the absence of substantial research on using AI to combat misinformation and enhance e-commerce operations. Addressing these gaps by exploring the ethical implications of misinformation, integrating emotional intelligence in AI, and the operational benefits of AI in e-commerce are pivotal for future exploration. Additionally, the opportunity to develop advanced AI models for in-depth consumer insight analysis suggests a fertile area for research, promising to deepen our understanding of consumer behavior in the digital era.
- **Quantitative Impact Measurement:** A notable gap exists in quantitatively measuring the impact of conversational AI on key marketing metrics. Future studies should aim to develop and apply rigorous models for quantifying the effects of AI technologies on conversion rates, customer lifetime value, and brand loyalty, among other metrics.

Our SLR highlights key areas for future research in conversational AI's role within consumer behavior and digital marketing. Addressing ethical concerns, cultural impacts, and leveraging the latest technological progress in conversational AI are crucial. While our review comprehensively covers key areas of current interest, the scope of our study is naturally limited by the selection of sources and the specific focus areas we chose to investigate. Therefore, further studies could provide additional insights and a broader exploration of the field. These studies could not only address the highlighted gaps but also expand into areas that were outside the purview of this review, thus capturing the evolving dynamics of conversational AI applications in marketing and their broader implications across different consumer segments and cultural contexts.

Disclosure of Interests. The authors have no competing interests to declare that are relevant to the content of this article.

References

1. Ginsberg, A., Venkatraman, N.: Contingency perspectives of organizational strategy: a critical review of the empirical research. Acad. Manag. Rev. **10**(3), 421–434 (1985)

2. Paul, J., Criado, A.R.: The art of writing literature review: what do we know and what do we need to know? Int. Bus. Rev. **29**(4), 1–7 (2020)
3. Scopus Homepage. http://www.elsevier.com/products/scopus. Accessed 12 Feb 2024
4. Singh, V.K., Singh, P., Karmakar, M.: The journal coverage of Web of Science, Scopus and Dimensions: a comparative analysis. Scientometrics **126**(2) (2022)
5. Galati, F., Bigliardi, B.: Industry 4.0: emerging themes and future research avenues using a text mining approach. Comput. Ind. **109**, 100–113 (2019)
6. Lee, D., Hosanagar, K., Nair, H.S.: Advertising content and consumer engagement on social media: evidence from Facebook. Manag. Sci. **64**(11), 5105–5131 (2018)
7. Decker, R., Trusov, M.: Estimating aggregate consumer preferences from online product reviews. Int. J. Res. Mark. **27**(4), 293–307 (2010)
8. Liu, X., Shin, H., Burns, A.C.: Examining the impact of luxury brand's social media marketing on customer engagement: using big data analytics and natural language processing. J. Bus. Res. **125**, 815–826 (2021)
9. Yen, C., Chiang, M.-C.: Trust me, if you can: a study on the factors that influence trustworthiness of conversational agents. Behav. Inf. Technol. **4**(11) (2021)
10. Jabbar, J., Urooj, I., Junsheng, W., Azeem, N.: Real-time sentiment analysis on E-Commerce applications through big data. In: 2019 IEEE 16th International Conference on Networking, Sensing and Control, ICNSC 2019 (2019)
11. Kumar, R., Mukherjee, S., Rana, N.P.: Exploring latent characteristics of fake reviews and their intermediary role in persuading buying decisions. Inf. Syst. Front. (2023)
12. Al Wahshat, H., Abu-Ulbeh, W., Yusoff, M.H., Zakaria, M.D., Hamzah, W.M.A.F.W., Stenin, N.P.: The detection of e-commerce manipulated reviews using GPT-4. In: 2023 International Conference on Computer Science and Emerging Technologies (2023)
13. Cortes, B., Teles, J., Duarte, E.: Exploring emotions in avatar design to increase adherence to Chatbot technology. In: Marcus, A., Rosenzweig, E., Soares, M.M. (eds.) HCII 2023. LNCS, vol. 14033, pp. 273–282. Springer, Cham (2023). https://doi.org/10.1007/978-3-031-35708-4_21
14. Ramu, V.B., Yeruva, A.R.: Optimising AIOps system performance for e-commerce and online retail businesses with the ACF model. Int. J. Intellect. Prop. Manag. **13**, 412–429 (2023)
15. Kaur, G., Sharma, A.: A deep learning-based model using hybrid feature extraction approach for consumer sentiment analysis. J. Big Data **10**(1) (2023)
16. Tariq, S., Iftikhar, A., Chaudhary, P., Khurshid, K.: Is the 'technological singularity scenario' possible: can AI parallel and surpass All human mental capabilities? World Futures **79**(2), 200–266 (2023)
17. Manis, K.T., Madhavaram, S.: AI-enabled marketing capabilities and the hierarchy of capabilities: conceptualization, proposition development, and research avenues. J. Bus. Res. **157**, 113485 (2023)
18. Wu, C.W., Monfort, A.: Role of artificial intelligence in marketing strategies and performance. Psychol. Mark. **40**(3), 484–496 (2023)
19. Tafesse, W., Wood, B.: Hey ChatGPT: an examination of ChatGPT prompts in marketing. J. Mark. Anal. (2024)
20. McTear, M.: Conversational AI: Dialogue Systems, Conversational Agents, and Chatbots. Springer, Cham (2022). https://doi.org/10.1007/978-3-031-02176-3
21. Martínez-López, F.J., Anaya-Sánchez, R., Fernández Giordano, M., Lopez-Lopez, D.: Behind influencer marketing: key marketing decisions and their effects on followers' responses. J. Mark. Manag. **36**(7–8), 579–607 (2020)
22. Lambin, J.-J., Schuiling, I.: Market-Driven Management: Strategic and Operational Marketing, 3rd edn. Palgrave Editorial (2012)

Creativity, Innovation and Decision-Making in Marketing

Isa Moll(✉) and Jordi Montaña

ESADE Business School, Avenida de Pedralbes 60-62, 08034 Barcelona, Spain
{Isa.moll,Jordi.montana}@esade.edu

Abstract. In the realm of business strategy, the intertwined concepts of marketing and innovation have long been acknowledged as pivotal drivers of organizational success. However, while considerable attention has been devoted to innovation within marketing, there remains a notable gap in understanding the sources and dynamics of marketing-led innovation, particularly in areas such as product design and packaging.

This paper synthesizes existing literature and empirical evidence to elucidate the multifaceted nature of marketing innovation, emphasizing its iterative nature and encompassing processes ranging from ideation to commercialization. By incorporating insights from scholars like Grimpe, Sofka, Bhargava, and Chatterjee, we address the prevalent confusion surrounding marketing innovation, providing clarity on its antecedents and consequences.

Moreover, we underscore the pivotal role of creativity and decision-making in driving successful marketing innovations. Drawing on case studies, including Kodak's missed opportunity with digital cameras and Sony's triumph with the Walkman, we highlight the transformative potential of creative thinking in shaping market outcomes.

Furthermore, we explore various techniques for fostering creativity and innovation, such as Design Thinking and proactive strategic approaches. Through illustrative examples like Absolut Vodka's entry into the U.S. market, we elucidate how companies can leverage external pressures and internal capabilities to drive innovation and gain a competitive edge.

Ultimately, this paper advances a comprehensive understanding of marketing innovation as a strategic imperative for organizations navigating today's dynamic business landscape. By recognizing the symbiotic relationship between creativity and innovation, businesses can harness novel opportunities and cultivate sustainable growth in an increasingly competitive marketplace.

Keywords: Innovation · Marketing · Decision-Making

1 Introduction

There is abundant literature on innovation in marketing that dates back many years. As far back as 1954, Peter Drucker wrote: "Because the purpose of business is to create a customer, the business enterprise has two – and only two – basic functions: marketing and innovation" [1].

A. A. Juan et al. (Eds.): DSA ISC 2024, LNCS 14778, pp. 40–49, 2025.
https://doi.org/10.1007/978-3-031-78238-1_4

In reality, as Purchase and Volery point out, innovation is "an iterative process initiated by the perception of a problem or unmet customer needs, which leads to the development, production, and marketing of an offering addressing this opportunity. Innovation, therefore, includes not only market research to identify those problems or customers needs but also invention, production, development, commercialization, and subsequent product adaptation and upgrading" [2]. However, there is little attention dedicated to new sources of marketing strategies that can be real drivers of business innovation such as product design and packaging.

Precisely, Christoph et al. noted that "little attention has been paid to new products that do not result from R&D investments but rather from novel marketing strategies" according to Grimpe, Sofka, Bhargava, Chatterjee [3]. These authors conduct a systematic review of the subject to address the existing confusion between academic literature and empirical work, which often leads marketing managers to experience difficulties in understanding the nature of marketing innovation, its antecedents, and consequences.

Our paper clarifies the process of innovation in marketing based on experiences and literature and adds a relevant factor, which is creativity and decision-making in the innovation process of marketing.

2 Innovation, Creativity and Marketing

Innovation is the introduction into the market of something new, whether it's new for the company or new for the market. At what moment or moments is the decision to innovate made in the company?

According to this idea of innovation, which is directly related to the market, it seems evident that innovation has a lot to do with marketing. Precisely in the context of the course titled "Innovation in Marketing: Creativity, New Products, and Design"[1] at ESADE Business School this relationship between innovation with marketing is analyzed in depth keeping in mind that innovation is a vitally important element in any organization and that the popular business saying "innovate or die" is radically true. It's a risky decision.

Innovation is not invention. Innovation requires the market. Without it, innovation does not exist. Kodak was a major company focused primarily on the development of film for cameras. A leading company, a giant with strong research and development departments. They invented the digital camera but didn't bring it to market: they thought it could never surpass the quality of their products, so the invention stayed in a drawer and never reached the market. They didn't innovate. In 2012, they filed for bankruptcy. They didn't make the decision to launch the digital camera, perhaps thinking it would cannibalize their traditional products, or perhaps thinking digital photography wouldn't match the quality of traditional photography. In any case, they missed the opportunity. A bad decision.

Years ago, when digital cameras already existed in the market, HP developed a technologically extraordinary product called Capshare. A product slightly larger than a

[1] Moll, I., Montaña, J.: Innovation in Marketing: creativity, new products and design. ESADE Business School. Barcelona (2023) [https://prodesade.esade.edu/gea/generate/9A_EN/report.html?dgr=102634422&year=2023&wfl=n&sby=109282639].

handheld calculator that allowed scanning book pages even when not completely flat and introducing them into a computer through an infrared system. Hence the name "capture and share." It was a mechanically and electronically complex product and therefore expensive. It was fully developed, promotional and advertising materials were prepared. The utilities it could offer to different segments were studied, and the decision was made not to launch the product. The potential benefits at its cost didn't outweigh the value offered by digital cameras that could do something similar. It was developed too late, but they made a good decision.

Akio Morita, the founder and president of Sony, propelled a small electronics company at the end of World War II in a devastated Japan. One of the company's successes was a tape recorder that represented an innovation in the market. It was the beginning of the company's introduction to the United States and of "Made in Japan" as a mark of quality. Mr. Morita liked music and noticed that people liked to carry music with them. It was normal in the United States to see someone with a huge music device on their shoulder booming through the streets. It was a good target for Sony's recorders, but the recorders were too large. Mr. Morita proposed eliminating the recording part and the speakers, replacing them with headphones to reduce the weight and size of the conventional recorder. Sony's marketing team didn't see the product clearly, but Mr. Morita was the president, and the product was put on the market. It was a great success. The only thing Mr. Morita didn't decide, and initially didn't like at all but accepted when he saw its acceptance, was the product name: Walkman. A successful invention in the market: an innovation. A timely launch. In reality, Sony initially made successful decisions, but the opportunities were immense: a devastated country, Japan, where everything was needed. They succeeded by combining technical capacity and financing capacity [4].

Some years ago, the strategists of the US Armed Forces in the Pentagon defined the post-Cold War times as VUCA (volatility, uncertainty, complexity, and ambiguity): volatile, uncertain, complex, and ambiguous. Today they continue to be so, and together with the acceleration of changes, they are driving innovation in all companies in all sectors to ensure their survival. This dizzying pace of change together with the uncertainty of the environment makes innovation a necessary but enormously risky activity. It requires quick and courageous decisions.

Traditionally, it was calculated that 80% of innovations failed. Today it can be said that in many sectors, the risk rate reaches 95% [5]. Add to this the cost of innovation, no matter how small it may be. Cost of ideation, development, production, and introduction into the market. From billions of euros for a new car to hundreds of thousands for a consumer product like a new cookie. Add the time to generate innovation: years for the car but a few months for a new candy that will have to fight to find a free space in distribution and convince thousands of consumers to choose the new cookie. If so, many innovations fail, let's generate many more innovations, but this seems very burdensome. Perhaps it's better to have and study many ideas and select the ones that seem most convenient: to increase creativity in the company. The first decision to make.

The process of developing new products, the basis of innovation, goes through various steps: idea exploration, idea selection, product concept development, concept testing, technical design and development, pilot production runs, product testing, marketing plans, market testing, production launch, introduction, and release, etc. Each phase, more or less abbreviated, requires a series of unilateral or shared "stop and go" decisions. But at the outset, a great source of inspiration is needed, a great capacity for unrestricted ideation, without prior decision.

Creativity is a new paradigm. Indeed, creativity (which is not the same as innovation) is a necessary step for innovation. An old study by Booz, Hamilton, and Allen [6], based on the analysis of a hundred successful products, found that the idea leading to the successful product had required sixty ideas that were left out in the development process. In other words: the probability of success if the first idea that comes up in the development process is implemented, something that often happens, is 1/60, or 1.6%. Not much.

3 Creativity and Decision-Making in Innovation

Creativity is defined in many ways. Ferran Adrià, possibly the world's most creative chef, often quotes a phrase from a famous French chef, Jacques Maximin, a fiery, creative, generous, and strong-willed character (hence he is known as the Napoleon of the stoves). The phrase, which is often attributed to Ferran Adrià himself, who absolutely believes it and makes it his own, is: "to create is not to copy."

There are many interpretations of what creativity is. The simplest is the ability to create new ideas. Others define it as the ability to find new things from existing ones. Einstein is attributed the sentence: "creativity is seeing what everyone else sees and thinking what no one else has ever thought." In any case, it is a consequence of the evolution of the human brain and is one of the activities governed by it.

Creativity is developed based on three principles: knowledge, motivation, and an appropriate environment. Take, for example, the creative process of an artist, a musician, a composer. Without mastery of musical instruments, basic composition rules, harmony, rhythm, etc., it will hardly be possible to compose something decent, and in addition, one must know the rules of solfeggio to write it. They will also need to know what other composers have done to achieve something original. Let's not talk about the necessary creativity in a scientist who needs years of study and knowledge of what has already been researched and the foundations of what they are researching.

Motivation is necessary to endure the hardship of mastering previous knowledge and wanting to be a successful composer or scientist. And while it is true that one can create in many environments, it is also true that it is difficult to create something decent in the schoolyard during recess or in the subway car. What applies to the musician and scientist applies to a painter, a writer, or a company executive. Knowledge of techniques and materials for the painter, willingness to spend hours with brushes, and a space to store canvases, paints, and to mix pigments. A visit to Joan Mirò's studio at his foundation in Palma de Mallorca reveals an ideal space for creation, with a warehouse of hundreds of objects, from stones to shells, through branches of curious shapes and bird feathers, which served as inspiration. Writers may need less space, but each has their preferences for the order and disorder of their workplace. Some, like Hemingway, wrote standing up, others in bed, like Pla; most, sitting down. Creativity in business also requires motivation and an appropriate environment.

Creativity in business cannot be confined to a specific group and space. It must be a general atmosphere, an open structure in both a real and figurative sense. Vertical, hierarchical structures do not encourage free expression. Preferring that people don't ask for permission and, if necessary, it's better that they ask for forgiveness, is typical of open organizations and, consequently, more creative ones. A study of two external research laboratories working for NASA showed the importance of environment and atmosphere for creativity and innovation. The less efficient laboratory was a closed space with individual cubicles, while the more efficient one had a common space with a coffee maker that researchers often shared. In the latter, ideas flowed from one to another, and the figure of some researchers who were more open to the outside world was discovered; they read more articles and shared them with others. They were called "technological gatekeepers." [6].

4 Creating New Ideas

elBulli, Ferran Adrià's restaurant, which was considered the best in the world during the years it was open, only operated for six months. The other six months of the year, it had a space, elBulli Taller, which was a real laboratory. In it, two teams worked constantly in an extremely bright and pleasant environment of an old palace near La Rambla in Barcelona. More than five thousand recipes were experimented with to achieve a menu of forty dishes. Thousands of them were decided to be discarded [7].

What techniques were used to generate ideas? In elBulli Taller, food was crossed with cooking techniques and experimented with one by one. Confronting two opposing concepts and analyzing their intersections, one of the simplest yet most effective techniques. This technique is actually one of the most proper to the management science, the matrices like Igor Ansoff's growth-share matrix [8], the General Electric- McKinsey matrix[2] first introduced in 1970, Porter's competitive strategies matrix [9], etc.

Besides the matrices, other powerful techniques have been devised, such as the well-known brainstorming, Edward De Bono's lateral thinking, the Six Thinking Hats technique by the same author, or the Toyota's Why? technique for continuous improvement, among many other methods. As said, there are many techniques that can be used depending on the specific objectives pursued.

Another methodology for idea generation is the popular Design Thinking. Initiated at Stanford by David Kelley, it was implemented as a working method at IDEO, initially a design services company for the development of new products and later a true innovation factory with global reach. Design thinking consists of three phases: inspiration, ideation, and implementation [10]. The inspiration phase is the phase of knowledge: defining the problem to be solved and gathering as much information as possible about it. Secondary information, primary information, expert consultation, user analysis, consumer behavior, etc. Fundamentally, it involves observing how the product is used.

Defining the problem is fundamental. The origin of Nespresso is in Nestlé's concern, the world leader in soluble coffee, about not being a leader in whole bean coffee [7]. This led to an analysis of coffee consumption in bars, cafes, hotels, and also in homes. The certainty that the best coffee is barista-made coffee led to the development of Nespresso capsules and machines. The obsession with quality and the lack of control over capsule

[2] https://www.mckinsey.com/capabilities/strategy-and-corporate-finance/our-insights/enduring-ideas-the-ge-and-mckinsey-nine-box-matrix.

stock in traditional distribution led to the creation of their own channel and, ultimately, to create a different business and company, Nespresso, one hundred percent owned by Nestlé. Years of preliminary research were needed to make an appropriate decision.

The second phase, ideation, is fundamentally based on the brainstorming technique, with the particularity that participating members form an interdisciplinary team with very different specialties. The goal is to put as many ideas as possible on a board, without prior selection, wild ideas that may in turn generate new ideas. The selection and decision of which idea to develop will come at the end. Absence of hierarchy in this phase is important; the session leader is neither a boss nor the most knowledgeable, but the one who best knows how to manage the session and direct the group. In the end, the group selects the best ideas.

The third phase, implementation, is a phase consisting of designing quick prototypes that physically illustrate what the selected idea consists of, in order to contrast it with the initial problem: it's better to fail early to succeed later.

The popularity of the method has led to hundreds of varieties and hundreds of publications. But we insist, the basis of innovation is creativity but it is not everything.

Perhaps the first question we should ask ourselves is what forces drive companies to innovate. Probably the first answer is competition. Being the first in the market, although riskier, also yields more long-term benefits. Other forces include product, service, or process decline in their life cycle, legal regulations, lack of raw materials or sudden price increases, changes in consumer habits, changing demographics, energy crises, climate change, and many other factors beyond the company's decision-making scope. Yes, that VUCA environment we talked about is the driving force behind innovation. Innovate or die.

In the face of this, what can companies do? Anticipate the environment, have a long-term strategy, and a structured process for innovation. Creativity, strategy, and method.

5 Creativity, Strategy and Method

The business case of Absolut Vodka is an insightful example to illustrate the relevance of creativity, strategy and method in achieving the necessary innovation to address the forces and demands of the environment. Absolut Vodka is an original product from Sweden where alcohol regulations are very strict and its distribution is in the hands of the state in a monopoly with regulated stores, the Systembolaget. It was obvious that the company's growth was limited in Sweden and they decided to export. The United States was an opportunity where vodka consumption was high, although dominated by national companies. The import market was very limited. They had a problem: the image of vodka is clearly of Russian origin, even national brands were identified with a Russian name (Smirnoff, Romanoff…), and Sweden, in the American consumer's mentality, did not have the tradition of producing vodka, although it had an image of quality products.

The first decision was to think about the product concept: who the consumer will be, the benefit it will provide, when and where the product will be consumed, what price level it will have, with what product category it will be associated. The answer was consumers in large cities, in sophisticated environments, with an aspirational benefit, premium price, and not associated with Russian vodka: a neutral concept. The first decision point was the packaging. The idea was found in the evocation of a pharmaceutical product bottle seen by a designer. The brand name unrelated to the product's origin but with connotations of quality: Absolut Vodka. Intelligent communication seeking complicity with the consumer. The result is that by adapting to different markets, the basic strategy is the same and it is a market leader. A proactive strategy, ahead of its competitors.

The first strategic decision is to decide whether to have a proactive strategy or a reactive strategy [6]. Being the first or being a follower. Neither is better nor worse. One has more risk and if successful, more profit, and the other the opposite. It will depend on the capabilities and strengths of the company. A proactive strategy can be based on research and development capabilities, Bayer, or on a great marketing capacity, Nestlé. Both need research and marketing capabilities although the strategic emphasis is different. These capabilities can be acquired by acquiring a company that already has them, Pixar, for Disney, or by developing internally the capacity to undertake: Nespresso, for Nestlé.

Reactive strategy can be very effective for a "me too" that has skills to differentiate through design or technological capabilities to be the "second but better". Many companies, especially B2B ones, innovate by directly responding to their customers' demands.

Opportunities for innovation can be explored methodically. Larry Keeley explains ten types of innovation to be systematically analyzed in each sector: business model, networks, structure, processes, product, systems, service, channels, brand, and customer involvement [11]. For each of these types, he has a list of possibilities that help to think about possible opportunities. It is nothing more than an aid to creativity prior to innovation.

In the context of the already mentioned subject "Innovation in Marketing: Creativity, New Products and Design" Keely's framework was adapted to deepen the understanding of the sources of innovation driven by marketing. As it is shown in the figure below, the adapted model, the Marketing Innovation Drive, focuses on marketing innovation, differentiating between the company's offering and the consumer experience. In the company's offering, we understand the product and its functions, systems, aesthetics, and sustainability. Within the experience, we have the service, channels, brand, and communication. Innovation is also found by crossing these concepts (Fig. 1).

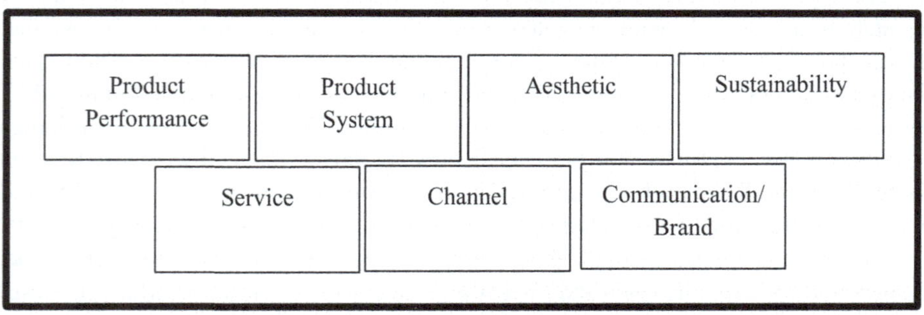

Fig. 1. The Marketing Innovation Drive, an adapted framework

The students of the 23–24 course have worked on examples of innovation in each of the sources highlighted in the Marketing Innovation Drive. There are some of them next:

In terms of Product Performance, the following best practices were highlighted: Spotify that stands out for its good features and customer retention; Apple, particularly Vision Pro for its spatial capabilities and high resolution; Beats by Dre for differentiation from other headphones in quality and appearance.

In the category of Product System, Apple's ecosystem in products and services was pointed out emphasizing how Apple places the consumer at the center of product development and the effectiveness of its marketing resources by integrating them across online and offline channels; Sony Playstation, for its ability to interconnect all Sony products, console, controllers, headphones, and monitor along with online services, as well as for the excellent gaming experience; Microsoft, leveraging the Windows operating system to create laptops and tablets, Xbox gaming consoles, and software to enhance Office 365 productivity, always with a focus on user experience.

The innovations based on Aesthetics, not covered in Keeley's model, aim to create new visual codes, new messages. It's a type of innovation based on design. The Tropicana products with their packaging and marketing campaigns highlighting sustainability are a great example; as well as IKEA for its simple and functional design with an attractive aesthetic that has democratized furniture and accessories design. Other best practices cited were the fashion brand Maison Margiela, which focuses on the deconstruction of unconventional materials for brand differentiation with an undeniable aesthetic, Louis Vuitton, CocaCola's packaging and advertising campaigns, and Barbie dolls as innovations that have endured over time with constant renewal to maintain their leadership.

The category of Sustainability is considered a major source of innovation in marketing although it is not included in Keeley's original model. The examples chosen are Bayer Crop Science and the ECOSIA search engine to make agriculture more sustainable; White:Space, a clothing brand with sustainability as its core values throughout the process; Nike, Adidas, and Puma, three competing brands with different sustainability-based solutions; Beyond Meat, a plant-based meat alternative promoted through partnerships with NBA stars and major fast-food chains; and Tesla, the reference electric car.

Service is a great source of innovation utilized by companies in such diverse sectors as Starbucks and Netflix, as well as in traditional sectors like Nordstrom, personalizing the shopping experience; Grap Holding Inc., a leading app in Southeast Asia operating in delivery, mobility, and digital financial services with spectacular growth; Spotify and its Spotify Wrapped campaign, or Tinder and Bumble services, seemingly now in difficulties.

As for Channels, another important source of innovation as demonstrated by Amazon's continuous improvement of the customer experience and diversification of its offerings, or its Chinese counterparts like Alibaba or Pinduoduo.

Lastly, Branding and Communication as sources of innovations were added to the model due to their importance in marketing innovation and their undeniable creative source. Great examples within this category are brand extensions like Disney, that creates content for different segments of different audiences; Dove, a Unilever brand with a unique and innovative brand strategy promoting authentic beauty, self-confidence, and inclusivity; Ryanair, which takes a friendly approach to users by humorously addressing their weaknesses and maintaining a very active presence on social media to increase brand recognition.

6 Conclusion

The application of the Marketing Innovation Drive shows that it's a powerful model to explore sources of innovation driven by marketing. More potential sources of innovation opportunities can be found in search of more creativity: the United Nations' 17 Sustainable Development Goals for 2030 offer a broad spectrum for new ideas, as does the concept of the doughnut economy, which proposes a new way of viewing economic science [9]. Furthermore, artificial intelligence, ChatGPT and many other emerging tools open up a truly exciting field of possibility. While it is true that humans are still beings who govern themselves and produce, and design, new things for their benefit by observing themselves and others, all these new tools emerging can be seen as superpowers at humanity's service, helping us surpass our current capabilities. This topic opens up a very interesting line of research about how we can harness these new tools as powerful sources of creativity.

In any case, it has been verified the relevance of innovation in the competitive landscape as well as the close relationship between creativity and innovation. Creativity is a human process that requires knowledge, motivation, and an appropriate environment. It is a structured process that follows certain steps and that can be facilitated through the use of tools. They can be classic creativity techniques, like brainstorming, or newer ones like AI. These tools serve as aids in the creative process. However, ultimately, it is the human factor—the marketing executive—who decides which insights are most suitable for their purposes and who is responsible for the decision-making process. Innovation is necessary to compete, but there will be no innovation without creativity and decision-making.

Disclosure of Interests. The authors have no competing interests to declare that are relevant to the content of this article.

References

1. Drucker, P.: The Practice of Management. Haper & Row, New York (1954)
2. Purchase, H., Volery, T.: Marketing innovation: a systematic review and research agenda. J. Bus. Res. **100**, 409–423 (2019)
3. Grimpe, C., Sofka, W., Bhargava, M., Chatterjee, R.: R&D, marketing innovation, and new product performance: a mixed methods study. J. Prod. Innov. Manag. **34**(3), 360–383 (2017)
4. Morita, A., Reingold, E.M., Shimomura, M.: Made in Japan: Akio Morita and Sony. Dutton, New York (1986)
5. Christensen, C., Hall, T., Dillon, K., Duncan, D.S.: Market understanding. In: Mastering Disruption and Innovation in Product Management, pp. 1–18. Springer (2016)
6. Montaña, J.: Marketing de nuevos productos. Hispano Europea, Barcelona (1992)

7. Moll, I.: Orientación al Diseño y Orientación al Mercado. La relación entre la optimización del diseño como recurso empresarial y la orientación al mercado de las empresas. (Doctoral dissertation). Universitat de Barcelona, Barcelona (2006)
8. Ansoff, H.I.: Strategies for diversification. Harv. Bus. Rev. **35**(5), 113–124 (1957)
9. Porter, M.E.: Competitive Strategy: Techniques for Analyzing Industries and Competitors. Free Press (1980)
10. Kelley, T.: The Art of Innovation: Lessons in Creativity from IDEO America's Leading Design Firm. Harper Collins, New York (2001)
11. Keely, L., Walters, H., Pikkel, R., Quinn, B.: Ten Types of Innovation: The Discipline of Building Breakthroughs. Wiley, Hoboken (2013)
12. Raworth, K.: Doughnut Economics: Seven Ways to Think Like a 21st-Century Economist. Penguin Books, London (2022)

AI-Boosted Decision Techniques for Strategy Formulation and Implementation

Lander Gamboa(✉)

Globant, Barcelona, Spain
lander.gamboa@globant.com

Abstract. Decision science, by combining techniques like Generative Intelligence and Retrieval-Augmented Generation, System Dynamics, and Strategy Mapping with the Balanced Scorecard, can become a powerful tool for businesses to not only make informed decisions but also gain a deeper understanding of their resources as identified by the Resource-Based View of the Firm. The understanding can become quantitative by creating either a higher-level System Dynamics model or a detailed Digital Twin. Either way, this integrated approach fosters a more comprehensive picture of a company's assets and capabilities and enables a more sophisticated and effective formulation of the company strategy. Finally, the principles and techniques of the Balanced Scorecard can be adapted based on the capabilities of modern AI systems to maintain a tighter control on strategy implementation.

Keywords: Generative AI · Retrieval Augmented Generation · Resource Based View · System Dynamics · Balanced Scorecard

1 Introduction: Boosting Strategy Decision Techniques

Strategic Management is hard work. It requires a disciplined and systematic understanding of current reality, the ability to envision an attractive future, and the capacity to manage the disconnect between them and at the same time formulate a practical course of action capable of moving the organization in the desired direction. In this regard, decision science is an enormously practical body of knowledge that can help managers make their efforts worthwhile. Decision science stands at the crossroads of multiple disciplines, harnessing the power of analytical techniques, business understanding coming from experience, and domain-specific knowledge to transform information landscapes into actionable insights and strategic blueprints for organizations. Its core objective is to steer entities toward informed decisions, paving the way for impactful initiatives and sustainable growth. Within this context, the integration of Generative Intelligence techniques with Retrieval-Augmented Generation (RAG) is helping improve the way businesses extract strategic insights from available data, both internal and external, to underpin strategic decisions.

A. A. Juan et al. (Eds.): DSA ISC 2024, LNCS 14778, pp. 50–59, 2025.
https://doi.org/10.1007/978-3-031-78238-1_5

Generative Intelligence, particularly using Large Language Models (LLM), has recently established itself as the most exciting frontier in AI innovation, offering remarkable capabilities in language processing and support for key business processes. According to Kampik [1], these advanced models, trained on extensive unstructured datasets, possess the ability to understand, summarize and interpret both existing documentation and real time communications, mirroring human comprehension in text, speech, and visual language.

RAG techniques, in turn, help feed the AI system with relevant content so that the model can answer questions about the business. According to Lewis [2], these techniques enhance the capabilities of large language models by integrating context-specific knowledge bases, such as financial documents, product catalogs, asset inventories, operating procedure manuals, customer databases, and human resources master files, into the workings of the AI system. This approach allows for more nuanced and better-informed responses to user queries, leveraging relevant information from internal company documents that extend beyond the original model training. This synergy between generative intelligence and pertinent information retrieval allows buildup of a sophisticated and timely understanding to the resources of the company, and so facilitates dynamic adaptation to industry shifts and specific organizational needs, essential in a rapidly evolving business environment.

The combination of Generative AI techniques with RAG is enabling businesses not only to extract necessary strategic insights for critical decision-making more effectively but also to adapt and respond in a more agile way to market changes and internal demands. As the companies in the services sector navigate the complexities of digital transformation, the role of innovative technologies like RAG will continue to evolve. Its effective utilization hinges on flexibility, resilience, and seamless integration with other AI advancements, ensuring that businesses' AI systems are not only cutting-edge but also deeply aligned with their unique informational ecosystems, paving the way for more informed, efficient, and customer-centric services.

2 Possible Applications

For example, a company might use decision science to identify its most profitable customers by analyzing data on customer purchases, demographics, and behavior. This information could then be used to guide the strategic focus of the company by developing both marketing campaigns and product and service offerings geared to acquire, serve and retain these desirable customer segments.

Another example of successful implementation of RAG technologies is shown by financial institutions like Allianz Bank Italy [3] and NatWest [4]. Allianz optimized knowledge-based processes for its financial consultants by leveraging RAG to provide relevant information from internal databases, enabling more informed client interactions. Similarly, NatWest enhanced customer service through real-time data updates from RAG systems, improving response times and the overall customer experience. These examples underscore RAG's ability to tailor AI model responses to the specific needs of individual company processes and departments by integrating highly specialized and frequently updated internal information.

While the advantages of RAG are substantial, its adoption is not without challenges. The inherent complexity of RAG implementations, the need for ongoing maintenance to adapt to changing data and usage patterns, and the continuously evolving landscape of AI technology all pose obstacles [5]. This situation underscores the importance of appropriately designed dynamic data source management systems and secure data retrieval processes, ensuring that sensitive information is protected while being made readily available to authorized users.

However, RAG's ability to efficiently handle large volumes of company data and its seamless integration with advanced search capabilities make it an invaluable solution for institutions with complex documentation and database needs.

The dynamic landscape of the modern service businesses, characterized by intensifying tech-enabled competition, evolving regulatory demands, and shifting customer expectations, presents a fertile ground for the application of Generative AI and RAG. These technologies offer a strategic advantage by enabling companies to harness their vast repositories of unstructured data for enhanced business processes such as risk underwriting, claims management, and regulatory compliance. By leveraging advanced natural language processing capabilities, RAG systems can extract relevant information from complex legal documents, policy manuals, and industry reports, facilitating more accurate risk assessments and streamlining compliance efforts.

The implementation of Generative AI in business processes extends beyond operational efficiencies to redefine how companies design their business model, according to Brynjolfsson [6]. We have seen how through advanced service interfaces like virtual assistants these technologies offer real-time, contextually relevant interactions, improving customer service channels and personalizing the customer journey. But maybe the most important impact on the business is that they allow the application of intelligence to the design of the business processes. By analyzing customer data, transaction histories, and communication logs, modern Artificial Intelligence (AI) systems can create new ways to segment customers, anticipate the evolution of customer needs, provide recommendations for possible new products and services, and identify what are the bottlenecks that need to be removed or managed to continue the growth of the business.

Furthermore, the evolution of Generative AI and RAG to include integration with advanced machine learning algorithms and semantic search techniques enables ever more sophisticated data analysis, per Shakeri [7]. As RAG systems incorporate advanced techniques like natural language understanding, knowledge graph construction, and federated learning, they will become better equipped to handle the complexities of financial data and provide more nuanced and accurate insights.

2.1 The Resource-Based View of the Firm

The Resource-Based View of the Firm emphasizes how companies use their internal resources to compete in the market and how they invest efforts and cash flows to build and improve those resources, according to Barney [8]. In this framework, strategic management involves two main processes:

Investment to Build and Improve Resources: It involves recognizing the unique resources a company possesses—which could be tangible assets like technology, machinery, or financial capital, or intangible assets like brand reputation, company culture, or proprietary knowledge. Once identified, companies must invest in these resources, both financially and through efforts such as training, research and development, and organizational development. The goal is to enhance these resources' value, rarity and inimitability that determine a resource's potential to provide a competitive advantage.

Leverage of Resources to Compete in the Market: After developing these resources, companies must then leverage them effectively to compete in the market. This involves deploying these resources in a way that maximizes their impact, such as using advanced employee skills to offer services that are significantly better than those of competitors or leveraging a strong brand reputation to expand into new markets. The key is to use these resources strategically to create barriers to competition, differentiate from competitors, and secure a market position that is defensible over time.

Formulating a resource-based strategy requires a deep understanding of how resources can be built and improved over time to reinforce a company's competitive position. This involves not only the initial identification and development of resources but also ongoing investment in and management of these resources to ensure they continue to provide a competitive edge. This strategic approach emphasizes the importance of internal capabilities and resources in achieving and maintaining success in the competitive market.

By incorporating Generative AI's ability to analyze vast amounts of data and RAG's targeted information retrieval, businesses can discover hidden strengths and underutilized resources within the organization.

2.2 System Dynamics

To further enhance this understanding and guide strategic decision-making, system dynamics modeling offers a powerful solution.

A system dynamics model represents the key structural elements, cummulative effects and feedback loops that drive the behavior and performance of the company, and it allows strategists to simulate how the build-up of resources will enhance the company's competitive position, attract customers from competitors, and grow the customer base and cash flow over time, as shown by Warren [9].

By simulating the dynamic behavior of the business system, these tools allow businesses to visualize resource flows, understand interdependencies, experiment with scenarios and optimize resource utilization:

System Dynamics models can map out how resources (e.g., personnel, financial capital, intellectual property) interact and flow within the organization, highlighting potential bottlenecks and inefficiencies.

These models expose the complex relationships between resources and business outcomes, anticipating how changes in one resource might ripple through the system, impacting other areas.

Managers can test different resource allocation strategies and strategic decisions in the simulated environment.

Based on simulation findings, managers can make data-driven decisions to maximize the utilization of their resources.

System dynamics helps analyze this information to understand their potential impact. With a clearer picture of their resources, companies can make strategic decisions based on the RBV framework. Decision science, informed by Generative AI insights, can guide investments in building upon existing strengths or developing new capabilities. This empowers companies to formulate an effective resource strategy based on a continuous invest-and-leverage cycle, so as to achieve and maintain a sustainable competitive advantage.

To build such a model of the business and its environment in a way that is both representative and useful, the modelers need to analyze and synthesize information from a large number of sources, in order to extract valuable insights about the interplay of the company's resources and capabilities with the competition for customer acquisition and loyalty.

This model, by integrating resource growth, competitive strategies, and market dynamics, will allow the strategists to identify those lead indicators that can detect emerging changes in the dynamic of competition, and whose frequent measurement can thus serve to trigger early strategic action.

For example, a few years ago this author helped a major beverage company to apply system dynamics to analyze the complex dynamics of consumer behavior. Their model captured factors such as:

Changing consumption habits of existing customers.

The impact of advertising, in-store marketing, and product availability.

Social validation, where a brand's perceived popularity influences adoption.

Competition from new entrants, distribution channels, and packaging formats.

Price sensitivity and brand switching.

Possible lead indicators of competitor growth.

This model provided the company with a deeper understanding of the long-term consequences of its decisions and the market forces at play, aiding in the development of sophisticated pricing and packaging strategies to maintain market shares. The model required the investment of several months of effort to understand the dynamics, gather the data, build the equations, and calibrate the parameters. Using current AI techniques this model could be built today in a much shorter time.

2.3 Digital Twins

An even more powerful tool for applying decision science to strategy formulation is a digital twin of a business and its market environment:

A digital twin allows businesses to create continuously updated virtual representations of their operations, products, services, and market dynamics. This enables them to simulate various scenarios, test different strategies, and analyze their potential outcomes without disrupting real-world operations. By running simulations, companies can explore the impact of decisions before implementing them, reducing risks, and increasing confidence in their strategic choices, as per Pavlovich [10].

Digital twins can integrate real-time data from multiple sources, including IoT sensors, customer feedback, market research, and historical data. This wealth of data can be analyzed using advanced analytics, machine learning, and artificial intelligence techniques to uncover detailed insights, patterns, and trends that would be difficult to discern through other means.

By modeling the business and market environment in painstaking detail, digital twins can help optimize resource allocation, supplier management, and product launch planning. Companies can identify bottlenecks, inefficiencies, and areas for improvement, and then test different approaches to address them. This can lead to cost savings, increased productivity, and better resource utilization, contributing to the success of the chosen strategy.

Digital twins can facilitate collaboration across different departments and stakeholders within an organization. By providing a shared virtual environment, teams can visualize the impact of their decisions on the overall system, achieving better communication and alignment.

2.4 Strategic Performance Management with the Strategy Maps and the Balanced Scorecard

The Resource Based View (RBV) emphasizes the significance of a firm's resources and capabilities in achieving a competitive advantage. It highlights that these resources should be valuable, rare, inimitable, and non-substitutable to be a defensible source of competitive advantage. To conduct a resource based strategic performance management in a systematic way, two pragmatic tools that have proven useful for many years are the Balanced Scorecard (BSC) and Strategy Maps.

According to Kaplan [11], the Balanced Scorecard translates strategy into action by providing a framework that measures performance across various perspectives. These perspectives include financial, customer, internal business processes, and learning and growth.

The BSC can complement the Resource Based View in several ways:

By outlining the strategic objectives across different perspectives, it helps identify the valuable, rare, and inimitable resources and capabilities required to achieve those objectives. This fosters strategic thinking and ensures that the strategy considers the resource implications.

By translating strategy into specific metrics and targets, it facilitates better communication and alignment across the organization. This ensures everyone understands how their individual efforts contribute to the overall strategic goals.

It helps break down the strategy into actionable initiatives, making it easier to implement. By assigning responsibilities and setting timelines, it fosters a more focused and coordinated approach to strategy execution.

It provides a framework for monitoring progress and measuring performance against strategic objectives. This allows for timely identification of deviations and course correction, ensuring the strategy remains dynamic and adaptable to changing circumstances.

In essence, the Balanced Scorecard bridges the gap between the Resource Based View's strategic resource focus and the practicalities of strategy execution. By providing a holistic view of performance across various dimensions, it helps organizations leverage their valuable resources and capabilities to achieve a sustainable competitive advantage.

Strategy Maps provide a clear representation of the organization's strategy, showing how intangible assets are transformed into tangible outcomes. This clarity is essential when using a resource-based strategy, as it helps to identify and leverage the unique resources and capabilities that provide a competitive advantage, also per Kaplan [12].

Strategy Maps depict the cause-and-effect linkages between strategic objectives across the four BSC perspectives. This visualization helps to understand how the effective use of resources (learning and growth perspective) can improve internal processes, which in turn can enhance customer satisfaction and lead to better financial performance.

They serve as a communication tool that can be used to explain the strategy to all levels of the organization. This ensures that everyone understands how their role and the resources they manage contribute to the overall strategy, fostering alignment and engagement.

By highlighting the strategic objectives, they can guide resource allocation regarding where to invest to achieve the greatest impact on the strategy. This is particularly important in a resource-based strategy, where the focus is on investing in and developing unique resources that drive competitive advantage.

When used in conjunction with the BSC, they provide a framework for measuring performance in pursuing strategic objectives. This allows organizations to track the effectiveness of their resource-based strategy and make necessary adjustments to stay on course.

They enable organizations to understand the impact of changes in resource allocation on strategic objectives. This facilitates a feedback loop for continuous improvement and strategic learning.

Overall, the Balanced Scorecard and Strategy Maps complement the Resource Based View so that, when used together, they can provide a powerful framework for formulating, communicating, implementing, and following up on a dynamic resource-based strategy.

2.5 Recent Advances in AI-Driven Strategy Decision-Making

Strategic decision making has typically been seen as a high-level intellectual ability and so intrinsically human. But in a world where computers not only compute better and faster, but can also beat as at chess, go, or other games, and to boot understand language and compose prose, poetry, music, or create enjoyable music, graphic art, and even tell stories using video, it is a question of time for AI to gain a seat at the strategy table.

Indeed, it has been shown by Jarrahi [13] that AI can "extend humans' cognition when addressing complexity", which makes it a tool for augmenting the mind of the strategist. At the same time, humans can still offer "a more holistic, intuitive approach". The result will be the dawn of a new approach of hybrid decision making, using human soft skills and transforming the job of the strategist.

Advances in explainable AI can also help in leveraging artificial intelligence for the strategy process, as it becomes easier for the strategist to reflect on the ideas generated by the LLM if he or she can understand how they were generated.

Explainable AI enables trust and confidence, uncovers potential biases that can be corrected, provides deeper insights, and allows for iterative refinement. Well understood AI thought processes can help advance human knowledge by revealing patterns, relationships and reasoning that may not be obvious or intuitive to humans.

An explainable AI system for strategic decision making will not only provide a recommendation, but also provide the rationale, advantages, risks, and key unknowns to look out for. According to Liao [14], providing explanations that allow users to understand the AI has become a key concern in human-AI interaction.

A new and different approach for the use of AI in strategic decision making in shown by Doshi [15], that tried generative AI to produce evaluations of business models. They found that while each single evaluation tends to be inconsistent, by calculating the average of several evaluations with different LLMs, the resulting rankings were less biased and closer to those of human experts.

These new techniques offer the possibility of leveraging AI technology to make better and faster strategic decisions. Marr [16], while skeptic of AI skills as a strategist, considers that routine and analytic tasks are likely to be automated, leaving more time for creative strategic thinking.

An alternative approach is offered in his latest book by Wharton Professor Ethan Mollick [17] recommends managers to use an LLM to critique their strategy formulation ideas following the style of a relevant entrepreneur, leader, artist, or philosopher, to gain a broader perspective than just asking his or her fellow managers or a paid consultant. So we can see that artificial intelligence gives us many ways to skin the strategy cat.

3 Conclusion

The dynamic insights obtained either through a higher-level system dynamics model or through a state-of-the-art digital twin simulator can become the basis for strategic management. They can enable the construction of a management information system that reflects the strategic position of the company through Key Performance Indicators (KPI). These KPIs would serve as constructive elements of a Balanced Scorecard that is forward-looking, emphasizing not only current performance but also factors that will drive future success. It facilitates strategic feedback and learning by providing a framework for monitoring performance so that managers can follow the evolution of the business and its environment and implementing strategic adjustments as needed.

Leveraging RAG to build a Balanced Scorecard involves utilizing its capabilities to extract, summarize, and synthesize relevant information from a wide array of company documents. This can streamline the process of gathering insights for the BSC's four perspectives: Financial, Customer, Internal Business Processes, and Learning and Growth. In particular, the ability of GenAI systems to handle unstructured information enable the scorecard builders to include richer information about the non-financial perspectives.

This capability is invaluable in sectors such as banking, where companies compete over time to capture and retain large customer portfolios by managing a very complex system based on multiple products, processes and decision-making nodes.

By using decision science, system dynamics, generative intelligence, and RAG techniques, businesses can gain a better understanding of their resources and how to use them to gain a competitive advantage.

As companies grapple with the challenges of digital transformation, the strategic application of Generative AI and RAG emerges as a pivotal factor in navigating this complex landscape. By offering tailored solutions that leverage the unique informational ecosystems of these institutions, these technologies not only enhance operational efficiencies and customer engagement but also equip businesses with the agility and resilience needed to thrive in an ever-changing environment. As AI capabilities continue to advance, the integration of RAG with other cutting-edge technologies will play a crucial role in shaping the future of the financial services industry. And as the technology develops further and strategy practitioners get more comfortable with artificial intelligence, it will gain an increasing role in strategy decision making.

In this paper I propose a novel approach to strategic decision-making by integrating diverse AI techniques like Generative AI, RAG, System Dynamics, and Digital Twins. This allows for a comprehensive understanding of a company's resources and market dynamics. The paper proposes using AI to evaluate resources based on the Resource-Based View, providing an objective assessment of competitive advantage. The emphasis on Explainable AI is crucial for building trust and collaboration between humans and AI systems in strategic decision-making. Furthermore, the text introduces the innovative concept of a dynamic Balanced Scorecard powered by RAG, enabling real-time performance monitoring and informed decisions.

Further research could explore combining AI techniques for comprehensive strategic analysis and further develop explainable AI for trustworthy recommendations. AI models could be designed to evaluate resources and tailored for specific industries. Further, once a company has formulated and agreed a new strategic course, it could build an intelligent recommendation model that "cascades down" the operational implications of the strategy, thereby ensuring all line managers can use it to understand and buy into the overall organizational strategy. Such a model could also be used to understand the impact of deviations from the expected path, so increasing the quality and speed of implementation efforts.

Additionally, research could investigate making AI tools accessible for smaller companies, and integrating Environmental, Sustainability and Governance factors into decision-making. Addressing other concerns like mitigating bias in AI models, analyzing job impacts, and ensuring responsible data use could be useful as well. Finally, case studies of successful AI implementations, comparisons with traditional tools, and long-term impact studies would add real life application to the approach proposed.

Disclosure of Interests.. The author works at Globant.

References

1. Kampik, T.: Large Process Models: Business Process Management in the Age of Generative AI. Arxiv (2023)

2. Lewis, P.: Retrieval-Augmented Generation for Knowledge-Intensive NLP Tasks. Arxiv (2020)
3. Globant Press Release. https://www.globant.com/it/news/globant-fornitore-strategico-di-servizi-di-intelligenza-artificiale-allianz-bank-italia. Accessed 13 Apr 2024
4. Caldwell, N.: https://www.natwestgroup.com/news-and-insights/news-room/press-releases/innovation-and-partnerships/2023/nov/natwest-and-ibm-collaborateon-generative-ai-initiative-to-enhanc.html. Accessed 13 Apr 2024
5. Shahul, E.S.: RAGAS: Automated Evaluation of Retrieval Augmented Generation. Arxiv (2023)
6. Brynjolfsson, E.: Generative AI at work. SSRN Electron. J. (2023)
7. Shakeri, M.: AR search engine: semantic information retrieval for augmented reality domain. Sustainability (2022)
8. Barney, J.: The resource-based theory of the firm. Organ. Sci. (1996)
9. Warren, K.: Improving strategic management with the fundamental principles of system dynamics. Syst. Dyn. Rev. (2005)
10. Pavlovich, T., Dron, E.: Data quality and digital twins in decision support systems of oil and gas companies. In: Proceedings of the 8th Scientific Conference on Information Technologies for Intelligent Decision Making Support (2020)
11. Kaplan, R., Norton, D.: Putting the balance scorecard to work. Harv. Bus. Rev. (1993)
12. Kaplan, R., Norton, D.: Having trouble with your strategy? Then map it. Harv. Bus. Rev. (2000)
13. Jarrahi, M.H.: Artificial intelligence and the future of work: human-AI symbiosis in organizational decision making. Bus. Horiz. **61**(4), 577–586 (2018)
14. Liao, Q., Shing, M., Zhang, Y., Bellamy, R.: Introduction to explainable AI. In: CHI EA Conference Paper (2020)
15. Dosi, A., Jason Bell, J., Mirzayev, E., Vanneste, B.: Generative Artificial Intelligence and Evaluating Strategic Decisions. https://papers.ssrn.com/sol3/papers.cfm?abstract_id=4714776. Accessed 13 Apr 2024
16. Marr, B.: Generative AI in Practice, p. 216. Wiley (2024)
17. Mollick, E.: Co-Intelligence, Living and Working with AI, p. 168. Penguin Random House (2024)

Sustainable Optimization of Construction Project Portfolio Management Through a Data-Driven Approach

Marisa A. Lostumbo[1,2](✉), Marc Bara Iniesta[2], and Laura Calvet[3]

[1] Computer Science Department, Universitat Oberta de Catalunya, Barcelona, Spain
mlostumbo@uoc.edu

[2] EAE Business School, Barcelona, Spain
marc.bara@campus.eae.es

[3] Telecommunication and System Engineering Department, Universitat Autónoma de Barcelona, Sabadell, Spain
laura.calvet.linan@uab.cat

Abstract. Effective management of construction project portfolios demands informed decisions driven by data and mathematical models, aiming to enhance decision-making and address complex decision problems. This article introduces an innovative approach that prioritizes sustainability criteria selection, utilizing data-driven insights and mathematical optimization to optimize project portfolio management. Through multi-objective optimization, our study targets improved resource efficiency, time reduction, and minimized environmental impact, offering portfolio managers a powerful tool for decision-making.

Keywords: Project Portfolio Management · Decision-Making · Data-Driven · Sustainability · Optimization Algorithm

1 Introduction

In Project Portfolio Management (PPM) resources are inevitably limited (primarily budget, but also time, and resource availability, among others), and at the organizational level, this information is relevant for decision-making, as these constraints cannot be easily modified [15]. In project-based industries like construction, effective PPM is crucial [4]. PPM involves selecting projects that align with strategic objectives while considering resource constraints. As companies aim to execute the right projects at the right time, they face the challenge of managing multiple projects concurrently within limited resources. This underscores the growing demand for tailored PPM solutions to meet industry-specific needs [6].

In construction companies, with more projects than available resources, prioritization errors can lead to resource misallocation, causing delays and loss of potential benefits [5]. Correct prioritization ensures optimal resource utilization and timely project execution, maximizing organization benefits Figure 1.

A. A. Juan et al. (Eds.): DSA ISC 2024, LNCS 14778, pp. 60–71, 2025.
https://doi.org/10.1007/978-3-031-78238-1_6

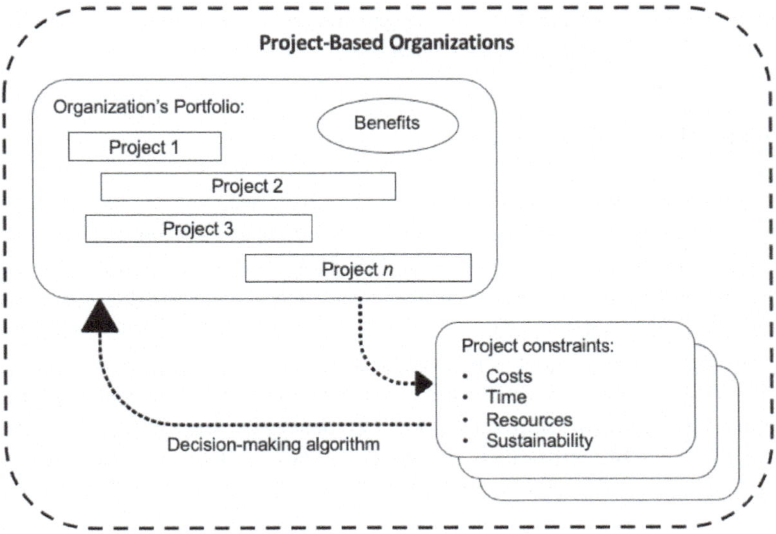

Fig. 1. Project portfolio in project-based organizations

In addition, the construction sector is responsible for approximately 25% of terrestrial water use, and therefore, improving environmental impact is becoming an increasingly critical objective for the industry [8]. Although the state of the art indicates numerous discussions surrounding portfolio theory, there currently exists a lack of frameworks integrating sustainability into the management of construction project portfolios [19]. In recent years, researchers have explored multi-objective optimization for multi-project scheduling, as [18] highlighted, with some studies specifically addressing sustainability concerns. For instance, [12] explores the Carbon Penalty. In contrast, our study introduces a Penalty for each day of delay in Water Consumption, calculated based on the estimated water usage for each activity duration within the project.

Moreover, the construction industry heavily relies on information, and project success demands efficient access and automated data management [1]. Accurate information resource management is crucial for productivity, on-time project execution, and staying within budget constraints [17]. Data sharing during project phases facilitates decision-making, yet each project lifecycle stage manages its data separately [9]. Besides, the effective implementation of data-driven models in construction project management still encounters several obstacles, including the integration of sustainability [3].

In this context, we explore how a data-driven approach can enhance decision-making, integrating sustainability considerations. A multi-objective model has been proposed, in which the primary objective function seeks to minimize the project time while considering resource constraints, while the secondary objective function is centered on maximizing environmental sustainability impact by minimizing water consumption during the construction phase. A solving approach is introduced and illustrated with a small example.

2 Theoretical Framework

The Project Management Institute's (PMI) 2017 edition of the Standard for Portfolio Management emphasizes the importance of acknowledging and integrating sustainability principles into PPM. However, it does not provide clear guidance on how to accomplish this integration effectively. Additionally, existing literature on PPM lacks comprehensive frameworks for incorporating sustainability strategies into project and portfolio decision-making processes. There is a clear need for further research and practical guidelines to assist organizations in enhancing the integration of sustainability practices. Improving PPM methodologies to align with sustainability objectives can lead to economic advantages for organizations while also contributing to environmental and social benefits [2,20].

In the construction industry, a sustainable approach extends beyond commercial boundaries. It involves considering environmental and social impacts. Construction companies need to prioritize their projects without compromising ecological integrity. Project assessments aligned with sustainability goals are essential. However, selecting appropriate projects amidst conflicting objectives and requirements represents challenges for the sector. PPM provides a solution by harmonizing criteria and promoting environmentally responsible decisions [11]. Sustainability should be included in construction projects from the early stages of the project and must be monitored throughout the project life [14].

More recently, authors have addressed international perspectives that unify the themes of strategic management, decision theory, and data science. Most decision-making models take advantage of the collection and processing of relevant data using appropriate analytics aimed at effective decision-making [21].

The large scale and complexity of the construction industry generate vast and diverse datasets, challenging traditional database management. Innovation in construction project management now relies on emerging information technology for real-time data management and decision support [13]. Integrating a data-driven approach into existing construction project management practices involves adopting advanced analytics tools to analyze project data. This enables informed decision-making, resource optimization, and risk mitigation, ultimately improving project performance and outcomes. By leveraging data-driven methodologies, companies can streamline workflows and enhance efficiency in construction projects [24].

Managing multiple projects within a portfolio with resource constraints entails scheduling activities across simultaneous projects, while considering precedence and resource limitations. Known as the Resource-Constrained Multi-Project Scheduling Problem (RCMPSP), it involves complex problem-solving, reconciling conflicting objectives like minimizing project duration and resource usage, and integrating uncertainty and dynamic changes. Researchers employ various methods, including heuristic algorithms, mathematical programming models, and the fusion of simulation and optimization techniques, to tackle these challenges effectively. Moreover, researchers have proposed various problem features, such as integrating multiple objectives, accommodating different resource types, and allowing flexible project structures, to better emulate real-world scenarios and improve the RCMPSP's applicability [7]. While some authors present comprehensive models for simultaneous project portfolio selection and scheduling, maximizing portfolio benefits and minimizing resource fluctuations [22], in the construction industry few others begin to incorporate sustainability objectives into the problem [12,16].

3 Solving Approach

Our approach integrates a multi-objective optimization algorithm to tackle the complexities of construction project portfolio management, prioritizing environmental sustainability. It hinges on merging historical data from portfolio projects with algorithms to effectively guide decision-making.

First, historical data from a construction project portfolio is collected, which includes project timelines, resource utilization, estimated durations, and environmental impact metrics, specifically the water consumption indicator during the construction phase.

Next, we propose a multi-objective optimization algorithm (see Algorithm 1) incorporating key objectives: economic efficiency, environmental sustainability, and resource optimization. Dynamically balancing these objectives, the algorithm considers interdependencies and trade-offs inherent in managing construction project portfolios. The algorithm is a novel adaptation of the NSGA-II Algorithm (Non-dominated Sorting Genetic Algorithm II) [10], implemented in Python. NSGA-II has been chosen for its specific design to address multi-objective optimization problems, which are characterized by several conflicting objectives requiring simultaneous optimization. Its fundamental aim is to identify solutions that are not dominated by any others in the objective space, ensuring no single solution is superior across all objectives. This capability makes NSGA-II exceptionally suited for addressing optimization challenges involving multiple criteria or goals, thereby making it an ideal selection for our research.

Our approach begins by representing each individual in the population as a list of activity priorities across the portfolio. These priorities, generated through the algorithm's evolutionary process, form the foundation for constructing a schedule that adheres to both precedence and resource constraints.

The first step of our algorithm focuses on creating a resource-compliant portfolio schedule given all activity priorities. To achieve this, the algorithm initiates with a topological sort of activities to ensure precedence constraints are met. However, recognizing that this schedule may not be feasible under resource constraints, we employ a priority-based activity delay mechanism. Activities are strategically delayed based on their assigned priorities to resolve resource conflicts, ultimately ceding a schedule that strikes a balance between precedence constraints and resource availability.

Algorithm 1. Pseudocode of the proposed Algorithm

Initialize population with random individuals
for each individual in the population **do**
    Find Precedence and Resource Feasible Solution
    Evaluate Objectives for the Individual
end for
while termination condition not met **do**
    Selection
    Crossover and Mutation
    Evaluate Fitness of Offspring
    Integrate Offspring into Population
end while
Identify Pareto-optimal front from the final population

Moving forward, the algorithm evaluates the fitness of each individual based on the resulting resource-compliant schedule. Two key objectives guide this evaluation: minimizing the total penalty for late delivery and reducing total water consumption during the construction phase. By assessing these objectives, the algorithm effectively navigates trade-offs between project delays, resource allocation efficiency, and environmental impact.

Through an iterative process of selection, crossover, and mutation, the NSGA-II algorithm guides the population towards a Pareto-optimal front. This process ensures the exploration of diverse solutions that offer optimal trade-offs between competing objectives. Furthermore, maintaining diversity within the solution population is crucial for preventing stagnation and ensuring the robustness of the optimization process. To address this, our algorithm incorporates specific features to prevent stagnation, such as a restart mechanism and dynamic population size adjustment. These features enable continuous solution space exploration, leading to more effective and sustainable scheduling outcomes.

4 Validation and Numerical Experiment

Our study has built a dataset specifically tailored to the RCMPSP problem in the construction sector (see Table 1). The dataset and the Python code implementing the algorithm described in this paper can be found on GitHub at the following link: https://bit.ly/ConstructionProjectsPortfolio.

The dataset consists of three distinct construction projects, each characterized by a specific start date, a delivery deadline, a daily penalty rate for late delivery (expressed in euros), and a daily water consumption rate (in cubic meters) during the construction phase.

The algorithm described in the previous section has been implemented in Python and the computational experiment was conducted on a high-performance computing environment featuring an Intel Core i9 10940X processor, 32 GB of DDR4-2666 RAM, and leveraging all 28 available threads. It was configured with

Table 1. Description of Construction Projects

Project	Start Date	Deadline	Penalty (€/day)	Water Consumption (m^3/day)
P1	18-09-2024	13-08-2025	500	250
P2	25-09-2024	15-12-2026	700	350
P3	02-10-2024	11-05-2028	900	300

specific settings to address the RCMPSP problem in the construction sector. The parameters shown in Table 2 were utilized for the experiments.

Table 2. Genetic algorithm parameter settings.

Parameter Name	Description	Value
Mutation Probability of Attributes	This defines the likelihood of mutation for each attribute in an individual, ensuring a balance between genetic diversity and stability	0.2
Number of Generations	The algorithm is executed for 40 generations, providing ample opportunity for the evolution and maturation of solutions	40
Crossover Probability	A high probability of crossover between two individuals, encouraging a diverse combination of genetic materials for varied solutions	0.7
Individual Mutation Probability	This parameter dictates the chance of an individual undergoing mutation, introducing necessary variability into the population	0.1
Population Size	A total of 100 individuals are included in the population, offering a broad genetic pool while maintaining computational efficiency	100
Offspring Size	Each generation produces 150 offspring, promoting a rich array of potential solutions for subsequent generations	150

The genetic algorithm employs adaptive parameters to dynamically adjust its behavior based on the observed performance over predefined segments. These adjustments aim to prevent stagnation and promote the exploration of new genetic spaces. The adaptive parameters are shown in Table 3.

Table 3. Adaptive genetic algorithm parameter settings.

Parameter Name	Explanation	Value
Segment Generations	The number of generations considered a segment, serving as a period for evaluating performance improvements	4
Stagnant Segments Threshold	Identifies stagnation if no improvement is observed after two consecutive segments, triggering adaptive mechanisms	2
Mutation Increase Rate	Increases the mutation rate by 50% upon stagnation to introduce more variability into the population	0.5
Population Increase Rate	Increases population and offspring sizes by 50% upon stagnation to explore a broader genetic space	0.5
Maximum Population Size	Caps the population size to prevent unmanageable growth, maintaining computational efficiency	500
Maximum Offspring Size	Similarly caps offspring size to maintain computational efficiency while allowing for sufficient exploration	700

Figure 2 shows three graphs depicting the evolution of solutions in terms of water consumption, penalties, and individuals evaluated over the 40 generations. Notably, the algorithm reached the best solution found just after 14 generations, with stagnation detected twice thereafter, in generations 24 and 32, as evidenced by an increase in the population size. The total computation time for evaluating a total of 7,364 individuals over the 40 generations was 7,734.08 s. All the solutions found by the algorithm, for each individual, are feasible in terms of predecessors and resources. For each solution, "penalties" and "water consumption" (the fitness values to be optimized) are calculated. In the first 5 iterations, it is seen that the dispersion of the population is decreasing.

Figure 3 displays a scatter plot of fitness values, which illustrates the distribution of solutions within the general population, the Hall of Fame (the best solutions found throughout the execution), and those on the Pareto front (the set of non-dominated solutions). The best solution identified by our algorithm achieved a total penalty across the three projects of 88,900 euros and a total water consumption of 370,800 cubic meters. This solution represents the best balance between the economic and environmental objectives as set by our problem definition.

As depicted in the scatter plot, the general population is spread across a wide range of values, indicating diverse genetic exploration. The Hall of Fame solutions cluster around the lower end of penalties and water consumption, indicating the algorithm's effectiveness in honing in on high-quality solutions. The top five solutions were utilized, representing the most outstanding results identified during

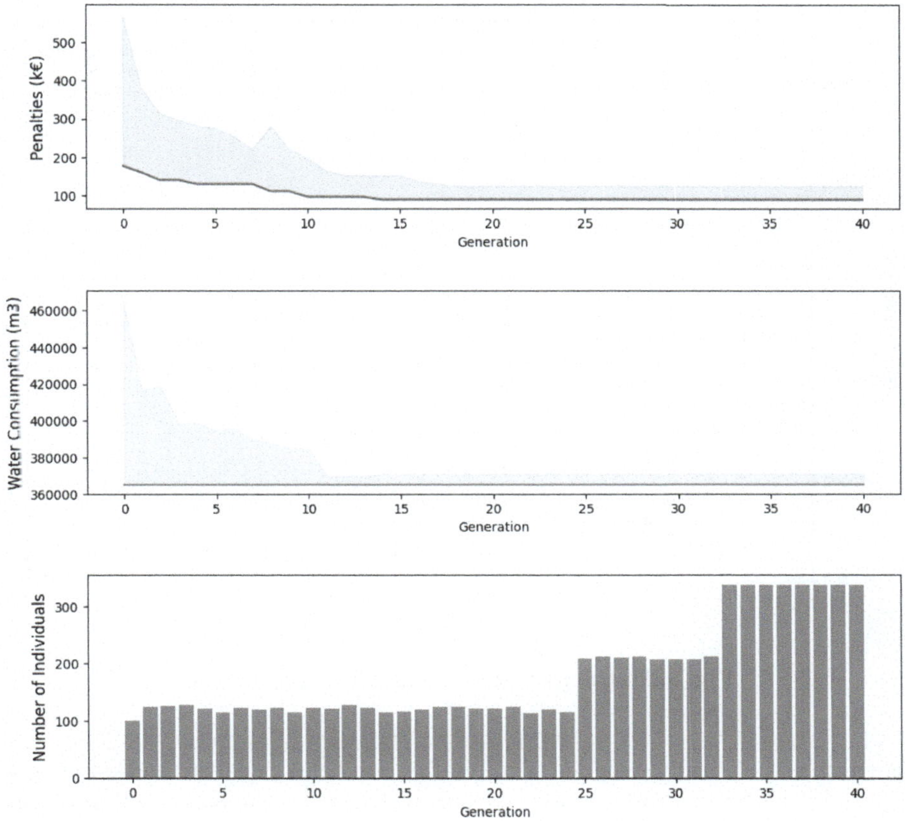

Fig. 2. Evolution of performance results and population size over generations

the genetic algorithm's execution. Moreover, the Pareto front solutions, marked distinctly, provide a boundary that defines the best trade-offs achieved between the two objectives. These solutions are particularly important as they offer various options for decision-makers who may prioritize one objective over the other to varying degrees. The scatter plot demonstrates that the algorithm successfully navigated the complex solution space to find a set of viable solutions along the Pareto front, which indicates a well-converged and effective optimization process. The spread of solutions along the front suggests that our algorithm can offer a variety of choices to stakeholders, depending on their specific preferences for penalty costs versus water consumption in construction project management. These results underscore the robustness of our adapted NSGA-II algorithm in handling multi-objective optimization problems within the construction sector, aligning economic penalties with sustainable water usage considerations.

Our study incorporates benchmarking methods with external datasets from MPLIB1 [23], a reputable source within the Resource-Constrained Multi-Project Scheduling Problem (RCMPSP) domain. It is important to highlight that our

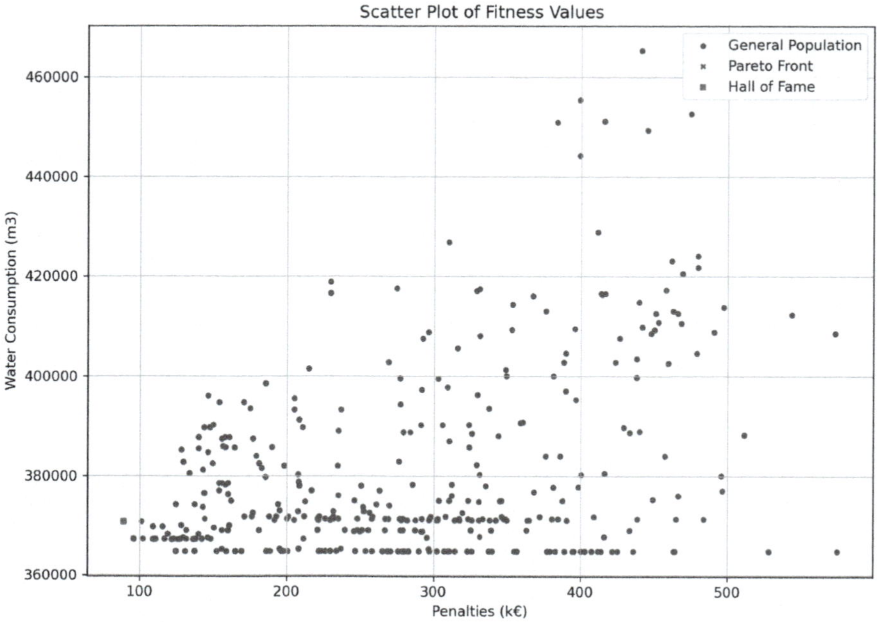

Fig. 3. Distribution of solutions within population

algorithm is specifically designed to minimize both delay penalties and water consumption, the latter of which occurs exclusively during the construction phase. However, the MPLIB1 datasets used for our benchmarking do not include phase-specific data. The datasets available in the literature are structured around project activities and do not differentiate between phases, thus limiting our assessment to delay penalties and precluding the evaluation of water consumption. Consequently, our benchmarking efforts have been primarily focused on the optimization of delay penalties.

Among the various benchmarking parameters, such as Total Portfolio Makespan (TPM) and Average Project Makespan (APM), APM is the key metric for gauging our performance in minimizing delay penalties. By aiming to uniformly reduce the duration of each project in the portfolio, we directly optimize APM. This methodology allows for a meaningful comparison of our algorithm against established metrics in the literature.

We selected a subset of 15 instances from the thousands available within the MPLIB1 datasets. This selection was made randomly to provide a representative cross-section of the data without introducing any bias towards specific instance types or complexities. The decision to limit the number of instances for our analysis was driven by the scope of our study. The primary aim was not to exhaustively test every instance within the MPLIB1 datasets but to perform a verification exercise that would validate the competitive nature of our algorithm.

Our benchmarking analysis has revealed that our NSGA-II algorithm performs competitively within the landscape of existing solutions (see Table 4 for a set composed of 15 projects). When compared against the 11 best-reported results in the literature, our algorithm (referenced as ML in the tables) consistently ranked within the top tier, achieving positions such as tie, 5th, 6th, or 7th. Although not leading the rankings, this performance is notable for several reasons.

Table 4. Comparison with the best results in the literature for the Average Project Makespan (APM) (MPLIB1 Set1) (ML: our algorithm).

Inst.	K06	H07	G08	C09	S15	V15	W15-S	W15-I	P16	V20	B23	ML	Rank
0	250.5	323.8	319.0	276.5	262.5	301.3	245.3	250.3	247.0	245.2	241.2	256.3	7th
50	511.8	628.8	611.0	560.5	530.8	679.2	496.0	667.2	499.5	488.3	498.2	537.3	7th
104	187.8	187.8	187.8	187.8	187.8	187.8	187.8	187.8	187.8	187.8	187.8	187.8	Tie
210	263.8	276.8	267.0	264.8	264.3	310.2	248.7	273.2	254.8	249.5	248.2	264.3	6th
228	138.6	138.6	138.6	138.6	138.6	138.6	138.6	138.6	138.6	138.6	138.6	138.6	Tie
329	908.8	1143.7	900.0	913.2	870.7	1199.2	875.2	1199.0	845.2	833.3	847.0	896.3	6th
349	706.8	932.2	907.5	766.7	714.5	917.8	697.0	923.8	701.3	694.8	673.5	713.8	6th
420	324.8	422.5	411.3	365.3	409.2	472.5	306.0	383.0	306.5	296.8	303.7	324.7	5th
442	290.2	346.0	336.3	332.2	355.2	414.2	270.7	378.8	271.7	260.3	263.2	282.3	5th
535	304.7	346.5	339.7	313.8	310.8	353.2	294.7	352.0	298.5	296.0	294.2	312.2	7th
613	537.0	741.8	746.2	599.5	540.8	775.2	520.7	654.5	520.0	505.3	497.2	540.2	6th
765	828.8	1084.0	1037.0	902.8	799.5	1131.0	800.0	1120.0	799.8	781.6	776.6	881.0	7th
808	501.6	573.3	541.1	521.1	502.5	632.7	479.3	550.1	484.6	471.1	469.1	505.1	7th
817	224.5	224.5	224.5	224.5	224.5	224.5	224.5	224.5	224.5	224.5	224.5	224.5	Tie
770	217.7	217.7	217.7	217.7	217.7	217.7	217.7	217.7	217.7	217.7	217.7	217.7	Tie

Firstly, our algorithm was not exclusively engineered to outperform on traditional scheduling metrics alone but was designed with a broader, more holistic aim. The focus of our NSGA-II algorithm is on sustainability-a critical aspect of modern project management that is not addressed by the current benchmark datasets. Our approach accounts for project phases within a portfolio, recognizing that different phases have varying impacts on resources, particularly water consumption during the construction phase. Secondly, our algorithm considers the unique deadlines and penalty rates of individual projects within a portfolio, which introduces a level of complexity not captured by the benchmark datasets. The fact that our algorithm remains competitive under these constraints is a testament to its robustness and effectiveness.

The results confirm that our algorithm stands strong among well-established methods, while also bringing the added value of sustainability considerations into the project scheduling domain.

5 Conclusions

This paper underscores the importance of data-driven decision-making in construction project portfolio management. By applying a data-driven decision-making approach, significant improvements in operational efficiency and environmental sustainability can be achieved. Our study has demonstrated that utilizing mathematical models and data analytics enables the identification of patterns, resource optimization, cost and time minimization, and environmental impact reduction. At the same time, we acknowledge that there are challenges to overcome in the practical implementation of these decision-making models. For instance, ensuring that stakeholders are adequately trained to understand and effectively utilize these tools. Additionally, considerations regarding data quality and availability, along with the need for continuous model updating to adapt to changing environmental conditions, are essential. A comprehensive approach is required to address both technical aspects and organizational and cultural challenges to achieve optimal outcomes.

References

1. Adekunle, P., Aigbavboa, C., Akinradewo, O., Oke, A., Aghimien, D.: Construction information management: benefits to the construction industry. Sustainability **14**(18), 11366 (2022)
2. Aghajani, M., Ruge, G., Jugdev, K.: An integrative review of project portfolio management literature: thematic findings on sustainability mindset, assessment, and integration. Project Manag. J. 87569728231172668 (2023)
3. Alzubi, Y., Khatatbeh, A.: Impact of modern information technologies on construction management. In: AIP Conference Proceedings, vol. 2847. AIP Publishing (2023)
4. Arsanjani, M.A., Ershadi, M.: Avenues to improving the effectiveness of project portfolio management in the construction industry. Int. J. Manag. Sci. Eng. Manag. **17**(4), 259–268 (2022)
5. Bakar, A.H.A., Yusof, M.N.: Project portfolio management and portfolio performance in construction industry: a conceptual framework. Res. J. Fish. Hydrobiol. **11**(3), 131–136 (2016)
6. Bilgin, G., Dikmen, I., Birgonul, M.T., Ozorhon, B.: A decision support system for project portfolio management in construction companies. Int. J. Inf. Technol. Decis. Mak. **22**(02), 705–735 (2023)
7. Bredael, D., Vanhoucke, M.: Multi-project scheduling: a benchmark analysis of metaheuristic algorithms on various optimisation criteria and due dates. Eur. J. Oper. Res. **308**(1), 54–75 (2023)
8. Calle Müller, C., Pradhananga, P., ElZomor, M.: Pathways to decarbonization, circular construction, and sustainability in the built environment. Int. J. Sustain. High. Educ. (2024)
9. Clement Desgourdes, J.R.: The role of big data analytics for decision-making in projects: uses and challenges. Enterprise Inf. Syst. 2317153 (2024)
10. Deb, K., Pratap, A., Agarwal, S., Meyarivan, T.: A fast and elitist multiobjective genetic algorithm: NSGA-II. IEEE Trans. Evol. Comput. **6**(2), 182–197 (2002)

11. Ershadi, M., Jefferies, M., Davis, P., Mojtahedi, M.: Incorporating environmental sustainability in project portfolio management by construction contractors. Int. J. Struct. Civil Eng. Res. **10**(3), 5 (2021)
12. Ghoroqi, M., Ghoddousi, P., Makui, A., Shirzadi Javid, A.A., Talebi, S.: An integrated model for multi-mode resource-constrained multi-project scheduling problems considering supply management with sustainable approach in the construction industry under uncertainty using evidence theory and optimization algorithms. Buildings **13**(8), 2023 (2023)
13. Huang, Y., Shi, Q., Zuo, J., Pena-Mora, F., Chen, J.: Research status and challenges of data-driven construction project management in the big data context. Adv. Civil Eng. **2021**, 1–19 (2021)
14. Kiani Mavi, R., Gengatharen, D., Kiani Mavi, N., Hughes, R., Campbell, A., Yates, R.: Sustainability in construction projects: a systematic literature review. Sustainability **13**(4), 1932 (2021)
15. Liesiö, J., Salo, A., Keisler, J.M., Morton, A.: Portfolio decision analysis: recent developments and future prospects. Eur. J. Oper. Res. **293**(3), 811–825 (2021)
16. Nabipoor Afruzi, E., Aghaie, A., Najafi, A.A.: Robust optimization for the resource-constrained multi-project scheduling problem with uncertain activity durations. Scientia Iranica **27**(1), 361–376 (2020)
17. Nalanda, W., Jadhav, S.S.: Analysis of resource management practices in construction industry-a systematic literature review. J. Adv. Zool. **45** (2024)
18. Sánchez, M.G., Lalla-Ruiz, E., Gil, A.F., Castro, C., Voß, S.: Resource-constrained multi-project scheduling problem: a survey. Eur. J. Oper. Res. **309**(3), 958–976 (2023)
19. Siew, R.Y.J.: Integrating sustainability into construction project portfolio management. KSCE J. Civ. Eng. **20**, 101–108 (2016)
20. Silvius, G., Marnewick, C.: Interlinking sustainability in organizational strategy, project portfolio management and project management a conceptual framework. Procedia Comput. Sci. **196**, 938–947 (2022)
21. Sinha, B.K., Bagchi, S.B.: Strategic Management, Decision Theory, and Decision Science. Springer, Cham (2021)
22. Tavakolan, M., Chokan, F., Dadashi Haji, M.: Simultaneous project portfolio selection and scheduling from contractor perspective. Int. J. Constr. Manag. **24**(3), 298–313 (2024)
23. Van Eynde, R., Vanhoucke, M.: Resource-constrained multi-project scheduling: benchmark datasets and decoupled scheduling. J. Sched. **23**, 301–325 (2020)
24. Wu, L.: Enhancing Data-driven applications in construction. Ph.D. thesis, University of Alberta (2021)

Artificial Intelligence in Marketing Strategic Decisions via Product Portfolio Optimization

Patricia Rodriguez-Garcia[1](✉), David Lopez-Lopez[2], and Jon Ander Martin Solano[3]

[1] Department of Computer Science, Universitat Oberta de Catalunya, 08018 Barcelona, Spain
patroga@uoc.edu

[2] Department of Marketing, ESADE Business School, 08172 Sant Cugat, Spain
david.lopez1@esade.edu

[3] Department of Computer Science, Universitat Oberta de Catalunya, 08018 Barcelona, Spain
jmartinsola@uoc.edu

Abstract. This article presents theoretical and numerical insights from employing artificial intelligence algorithms to strategic marketing decisions, focusing specifically on the optimization of product portfolio allocation. Given its critical importance in revenue generation, customer satisfaction and market positioning, it is crucial that marketing directors make appropriate and data-driven decisions on the implications of different strategies regarding the allocation of inventory and advertising spending, among others. Through computational analysis, utilizing the Non-dominated Sorting Genetic Algorithm II, we explore how varying allocation strategies impact key performance metrics. Beyond merely providing optimal portfolio allocation, the algorithm allows for quantifying trade-offs among objectives, shedding light on their implications for competitive dynamics, revenue generation, and inventory costs. By solving multi-objective optimization problems inherent in product portfolio management, this article offers comprehensive insights into the role of AI-driven approaches in enhancing strategic decisions that shape competitiveness and customer value in today's marketing landscape.

Keywords: Strategic decisions · Artificial Intelligence · AI-driven Strategic Decision-making · Marketing Strategy · MOPO · NSGA-II

1 Introduction

In the era of digitization and data abundance, the intersection of marketing and Artificial Intelligence (AI) has emerged as a transformative force, reshaping the landscape of strategic decision-making [2]. Among the different facets of marketing strategy, the optimization of product portfolio stands as a pivotal topic

A. A. Juan et al. (Eds.): DSA ISC 2024, LNCS 14778, pp. 72–83, 2025.
https://doi.org/10.1007/978-3-031-78238-1_7

with profound implications for revenue generation, market positioning, and customer satisfaction. The product portfolio refers to the totality of all products and/or services that a company offers to the market [20]. It involves allocating resources, specially inventory and advertisement investments, wisely to maximize revenues, strategically positioning the company within target markets, managing risks through diversification, optimizing revenue generation by leveraging product strengths, and reinforcing customer relationships by offering a range of solutions.

This article seeks to explore how the application of AI algorithms, such as the Non-dominated Sorting Genetic Algorithm II (NSGA-II) [12], could amplify the strategic-decision of marketing directors in the domain of product portfolio by enhancing the efficacy and precision of product portfolio optimization [3,6]. The primary objectives of this study can be outlined as follows: (i) solve multi-objective optimization problem: traditional approaches to product portfolio optimization often focus on single-dimensional objectives, such as maximizing revenue or minimizing costs, however, the reality of marketing strategy encompasses a multitude of objectives; (ii) development of an intelligent algorithm based on the NSGA-II technique to identify Pareto-optimal solutions and generate diverse and well-balanced product portfolios; (iii) performance evaluation and validation of the algorithm to assess its efficacy and robustness by quantitatively analyzing key performance metrics such as convergence speed, solution quality, and computational efficiency; and (iv) provide practical implications and managerial insights for marketing directors in their strategic decision-making. Through a combination of theoretical analysis, algorithmic development, empirical experimentation, and practical insights, this article endeavors to contribute to the field of AI-driven marketing strategy.

2 Literature Review: AI for Marketing Strategic Decisions

In recent years, the study of AI into strategic decision-making has seen a remarkable growth. Some of the notable contributions on the application of AI to strategic decisions are as follow: in business strategy improvement with predictive analytics [7]; in strategic execution with machine learning and key performance indicators [30,31]; in effective implementation of decision support systems to guide strategic decision- making [5]; in new business opportunities and capacity for innovation and competitive advantage with customer experience improvement [15]; in enhancing customer experience by providing better personalization and quality of service [27]; in marketing analytics and AI [2,32]; in creation of corporate ecosystems [28]; and in providing insights related to a brand competitors behaviors and marketing strategies [1].

AI can assist marketing directors in several areas [15,16,22,37]; such as (i) predictive analytics: machine learning algorithms can analyze historical sales data to identify patterns, trends, and seasonality in product demand; (ii) customer segmentation: by clustering customers based on purchasing behavior, demographics, and preferences, machine learning algorithms can help identify distinct market segments with unique needs and preferences and then tailor the product assortment for each segment to maximize sales revenue and customer satisfaction; and also in (iii) optimization: machine learning optimization algorithms, such as genetic algorithms or reinforcement learning, can explore the vast solution space of possible product combinations to find the optimal portfolio that maximizes sales revenue while minimizing inventory costs.

A comprehensive review of the state of the art of data science in product portfolio is given in [21,36]. Traditional approaches to product portfolio optimization often rely on historical data and expert judgment, which can be limited in scope and accuracy. In contrast, AI-driven techniques offer the ability to analyze vast datasets and identify patterns, trends, and customer preferences with greater precision. In the literature, there are several articles that deal with the product portfolio optimization where [34] forecast, using a special form of regression, demand trends of individual product features from sold product configurations over time. [18] extend these approaches with an automatic optimisation process for model selection, parameter setting, and initial state estimation. [33] elaborated on machine learning algorithms, including clustering, classification, and regression, effectiveness in segmenting markets, forecasting demand, and identifying opportunities for product differentiation. Natural language processing (NLP) techniques enable sentiment analysis of customer feedback, providing valuable insights into product perception and sentiment [13]. Additionally, according to [20] reinforcement learning algorithms can optimize pricing and promotion strategies by continuously learning from customer interactions.

Finally, regarding construction of investment portfolios by using the methodology introduced by [19], investors consider 2 factors: expected return and variance in return. Variance measures the variability in realized return around the expected return, giving equal weight to realizations below the expected and above the expected return. The essence of portfolio optimization is to pursue the conflicting goals of maximizing return and minimizing risk, and these goals should be optimized simultaneously. Part of the appeal of the Markowitz model is it can be solved by efficient quadratic programming methods. [25] transformed the multi-objective portfolio selection model into a single-objective programming model using fuzzy normalization and uniform design method. Meanwhile, some researchers studied meta-heuristic algorithms [12,23], which can provide near optimal solutions even for large-sized problem within a reasonable time.

3 Problem Formulation: Mathematical Model for MOPO

One of the most strategic decision that marketing managers are tasked to refers to product portfolio optimization. At least annually, when they prepare budget and objectives for the coming year, they face a complex strategic problem due to the need to balance multiple objectives simultaneously. One of the primary objectives is to maximize sales revenues, this involves identifying products with high demand and optimizing their presence in the portfolio. Other objective is to minimize the risk. On the other hand they face constraints, which among others we highlighted, the inventory costs, the advertisement investment and the customer satisfaction. Keeping excessive inventory ties up capital and incurs storage costs while stock-outs can lead to lost sales opportunities and dissatisfied customers. Therefore, marketing managers need to find the right balance between maintaining sufficient inventory levels to meet demand and minimizing carrying costs associated with excess inventory. Particularly for the product portfolio optimization is crucial to balance product assortment in order to cater a wider range of customer preferences and increase market penetration. However, focusing solely on revenue maximization may lead to overlooking other important factors such as customer satisfaction or efficiently investment of advertising budget.

As a result, marketing managers face a Multi-Objective Problem Optimization (MOPO) which refers to the process of simultaneously optimizing multiple conflicting objectives [9]. MOPO aims to explore the Pareto front, by iteratively search for solutions that are Pareto optimal [4,8] and which represents the trade-offs between the different objectives. By exploring the Pareto front, marketing directors can identify a range of feasible solutions and select the one that best aligns with their strategic goals and constraints [14].

3.1 Mathematical Formulation of the Marketing Director Problem

The mathematical formulation of the MOPO for product portfolio in marketing can vary depending on the specific decision variables, objectives and constraints involved in the problem. However, a general formulation involves the following factors:

- Objective function (F): In MOPO, there are multiple objective functions to optimize simultaneously, $F = (f_1, f_2)$, where F represents the vector of objective functions, f_1 is to maximize sales revenue, and f_2 is to minimize risk.
- Decision variables (X and Y): Represent the actions that marketing directors can control or influence, including the allocation of resources for inventory and advertising spending. $X = (x_1, x_2, ..., x_m)$ and $Y = (y_1, y_2, ..., y_m)$, where x_i is the allocation of inventory to each product and y_i is the allocation of advertising budget to each product.
- Constraints (G): Define the limitations or restrictions that must be satisfied. $G = (g_1, g_2, ..., g_k)$. G represents the vector of constraints, where g_1 represents the inventory constraint, ensuring that the total inventory across all products does not exceed the available capacity or desired levels; g_2 represents the advertising investment constraint, limiting the total advertising investment to stay within budget; g_3 represents the budget limitation, ensuring that the total cost of products, advertising, and any other expenses should not exceed the available budget.
- Optimization goals: specifies the overall objectives of the MOPO, including finding Pareto-optimal solutions, maximizing overall utility, minimizing overall cost, etc.

$$\text{Maximize Revenue:} \quad \text{Revenues} = \sum_i R_i \times x_i \tag{1}$$

$$\text{Maximize Satisfaction:} \quad \text{Satisfaction} = \sum_i S_i \times x_i \tag{2}$$

subject to:

$$\text{Budget Constraint:} \quad \sum_i C_i \times x_i + \sum_i A_i \times y_i \leq B \tag{3}$$

$$\text{Inventory Constraint:} \quad \sum_i x_i \leq \text{Inventory Capacity} \tag{4}$$

$$\text{Advertising Budget Constraint:} \quad \sum_i y_i \leq \text{Advertising Budget} \tag{5}$$

where:
x_i be the decision variable of the quantity of product i to sell.
y_i be the decision variable of the advertising investment for product i.
R_i be the revenue generated by product i.
S_i be the satisfaction score associated with product i.
C_i be the cost associated with producing product i.
A_i be the cost associated with advertising product i.
B be the total budget available.

Solving this formulation involves finding the set of Pareto-optimal solutions, where no solution can be improved in one objective without degrading at least one other objective, thus representing a trade-off between competing objectives. Various optimization algorithms, such as genetic algorithms, particle swarm optimization, or simulated annealing, can be used to explore the solution space and identify Pareto-optimal solutions.

4 Non-Dominated Sorting Genetic Algorithm II (NSGA-II)

The NSGA-II is a powerful multi-objective optimization method that can be effectively applied to MOPO product portfolio [12]. As explained in [10,11,24] introduced the GA and the notion of inspired by Darwin's theory. It uses genetic operators and Pareto dominance to search for a set of solutions that are non-dominated, providing a trade-off between conflicting objectives [26]. According to [17], NSGA can handle complex interactions and nonlinear relationships be tween objectives. [35] worked in advanced GA design including the application of local search methods to improve solutions and speeding up the convergence process.

The algorithm contains the following stages:

1. Definition of the problem parameters: (i) assets: list of assets (products) in the portfolio; (ii) returns: array of expected returns (revenues) for each asset; (iii) inventory, advertising spending, and customer satisfaction: arrays representing scores for different objectives for each asset; (iv) total investment: total budget available for investment in the portfolio. (v) target return: target return to achieve; (vi) penalty: penalty applied if the target return is not achieved.

2. Genetic Algorithm (GA) parameters: (i) population size and n generations: size of the population and number of generations for the genetic algorithm, (ii) restart interval: number of generations before restarting the algorithm, (iii) mutation rate, crossover rate and gene mutation rate: parameters controlling mutation and crossover probabilities. (iv) generations and HoF size: parameters for the MOPO, (v) cov matrix: covariance matrix representing the risk associated with each asset.
3. Evaluation function ('evaluate'): evaluates the fitness of each individual in the population based on multiple objectives, including expected return, risk, inventory and advertising spending or customer satisfaction. Applies penalties if the target return is not achieved.
4. DEAP (Distributed Evolutionary Algorithms in Python) initialization: initializes the DEAP framework for MOPO (framework for building and experimenting with evolutionary algorithms (EAs) in Python; for NSGA is 'eaMuPlusLambda') by defining the fitness function and the individual chromosome structure and registration of genetic operators. Individuals are ranked based on their dominance relationships, identifying non-dominated individuals that are not worse than any other individual in the population.
5. Initialization function ('initChromosome'): randomly initializes the chromosome (portfolio allocation) for an individual. Ensures that at least one asset is selected for investment. A population of potential solutions, known as individuals, is generated randomly. Each individual represents a potential portfolio of projects with varying combinations of investments. A crowding distance is calculated for each individual to quantify its density within the non-dominated solutions. This distance measure encourages diversity among the non-dominated solutions and helps maintain a well-distributed Pareto front.
6. Genetic operators registration: registers genetic operators such as crossover, mutation, and selection. Individuals are selected to form the next generation population, and the selected individuals undergo genetic operators (crossover and mutation) to create offspring. The process aims to combine the characteristics of promising solutions to produce new and potentially better solutions. The population is updated by replacing the old generation with the newly created offspring, maintaining a diverse set of solutions that represent the Pareto front.
7. Statistics and Hall of Fame (HoF) setup: sets up statistics collection and HoF to store best solutions.
8. Evolutionary algorithm execution: runs the evolutionary algorithm with a restart mechanism. Uses NSGA-II to evolve the population and then collects statistics and updates the Hall of Fame. The algorithm continues to evolve the population through generations until a termination criterion is met, such as a maximum number of generations or convergence to a sufficiently optimal Pareto front.
9. Results visualization: plots the evolution of best fitness values over generations, prints the optimal portfolio allocation and other relevant metrics such as expected return, risk, inventory and advertising spending or customer sat-

isfaction and plots the Pareto front and prints the best solutions along the Pareto front.

5 Experimental Setup

In order to test and to validate our NSGA-II algorithm we use a solved problem from the Portfolio Optimization book. Particularly, we use publicly available data from [19,29]. The database contains two instances of financial information including the increase in price, considering dividends, over a twelve-year period and also included the change each year in the Standard and Poor's 500 stock index. Based on the twelve years of data, there was the standard statistical package to calculate a covariance matrix for three stock as well as the correlation matrix. The first instance contains information of three stocks: A (AT&T from telecom industry), B (General Motors from the automotive industry) and C (USX Corporation from the steel industry). The second instance contains information of three stocks not identified: A, B and C. For the first instance, the solution recommends about 53.0% of the portfolio in ATT, about 36.0% in GMC and over 11.0% in USX. For the second instance, the solution recommends about 41.5% of the portfolio in asset A, about 5.0% in asset B and over 53.5% in asset C. In both instances the expected return is 15.0% with a variance of 0.02. Moreover, in order to incorporate the remaining variables of the problem we employed a random data generation approach for the inventory, advertisement and customer satisfaction components.

The following figures show the results of applying the NGSA-II algorithm to the MOPO of product portfolio, representing the optimal portfolio allocation and associated metrics (Figs. 1, 2, 3 and 4).

1. Optimal portfolio allocation: by applying the NSGA we arrive to the optimal allocation of the portfolio across different assets (products) which is expressed as a percentage of the total investment. In experiment 1: Asset A: 52.9%, Asset B: 35.8%, Asset C: 11.2%; and in experiment 2: Asset A: 49.7%, Asset

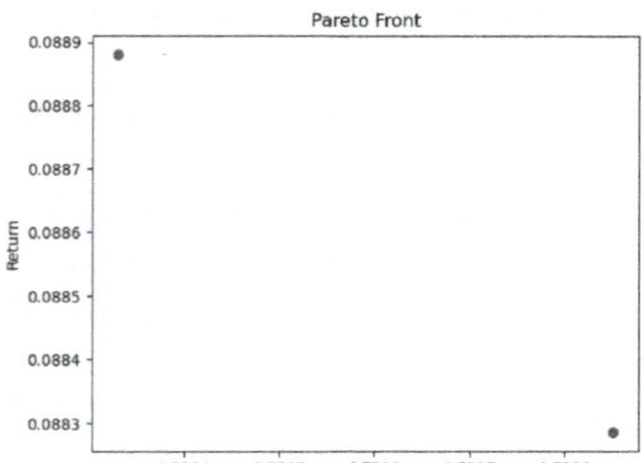

Fig. 1. Experiment 1 Pareto

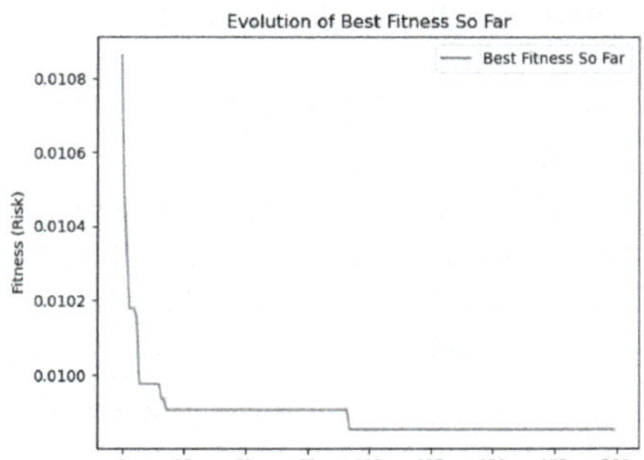

Fig. 2. Experiment 1 Best Fitness

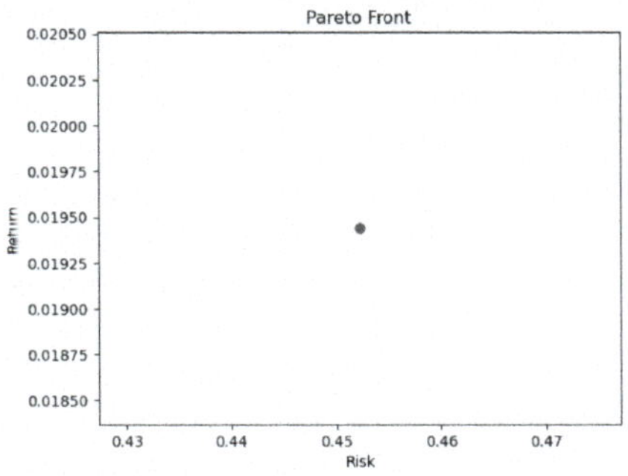

Fig. 3. Experiment 2 Pareto

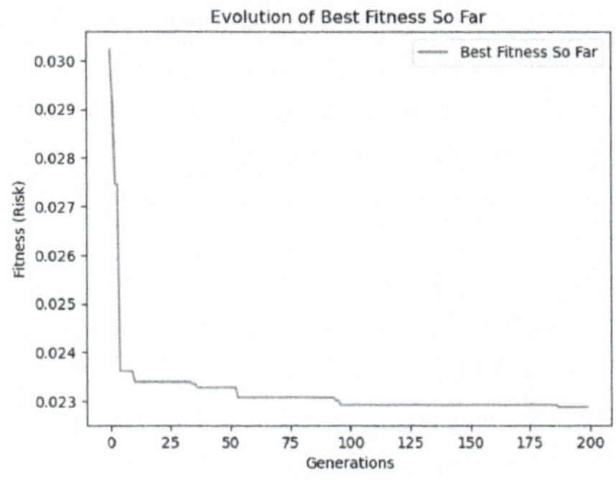

Fig. 4. Experiment 2 Best Fitness

B: 0.0%, Asset C: 50.3%. In terms of efficiency of the NSGA, comparing the results of NSGA with those of the solution: in experiment 1, the optimal allocation obtained suggests a nuanced distribution of investments across assets with a higher weight in asset A while notably while in experiment 2, Asset B receives no allocation, signaling its marginal contribution to achieving the portfolio's objectives, and the other assets achieved stakes near 50%.

2. Expected return: as this is a target in both cases the return is 1.15. For calculating the Pareto frontier there will be combination of target return with is associated risk for the unconstrained curve and then a set of constrained curves with a combination of results for the advertisement, inventory and customer satisfaction scores. This is illustrated in Fig. 5.
3. Portfolio risk: in the experiment 1, the portfolio risk is approximately 0.0232 even while slightly to the best solution of 0.0224, it remains within an acceptable range. In experiment 2 the best result is 0.0228 not far distant from the objective 0.0232. This highlights the importance of comprehensively assessing risk exposure when crafting portfolio strategies
4. Advertising spending, customer satisfaction and inventory expenses Scores: these scores represent the performance of the portfolio with respect to additional objectives beyond return and risk. In the provided results, each score is set to 0.5, indicating a neutral performance in these aspects. Consistent with previous findings, advertising spending, customer satisfaction and inventory expenses scores exhibit neutral performance which reaffirms the need for a holistic evaluation framework encompassing diverse objectives.
5. Runtime: indicates the time taken for the optimization process to complete; in the provided results, the runtime is approximately 245.16 s in experiment 1 and 210.63 s in experiment 2. This runtime demonstrates efficiency on the optimization process, with a notably shorter runtime compared to previous iterations. This underscores the scalability and applicability of AI-driven optimization techniques in real-world marketing contexts.
6. Best solutions along the Pareto front: lists the best solutions found along the Pareto front, which represents optimal combination of assets with the trade-offs between risk-return. Across the presented results, all solutions exhibit

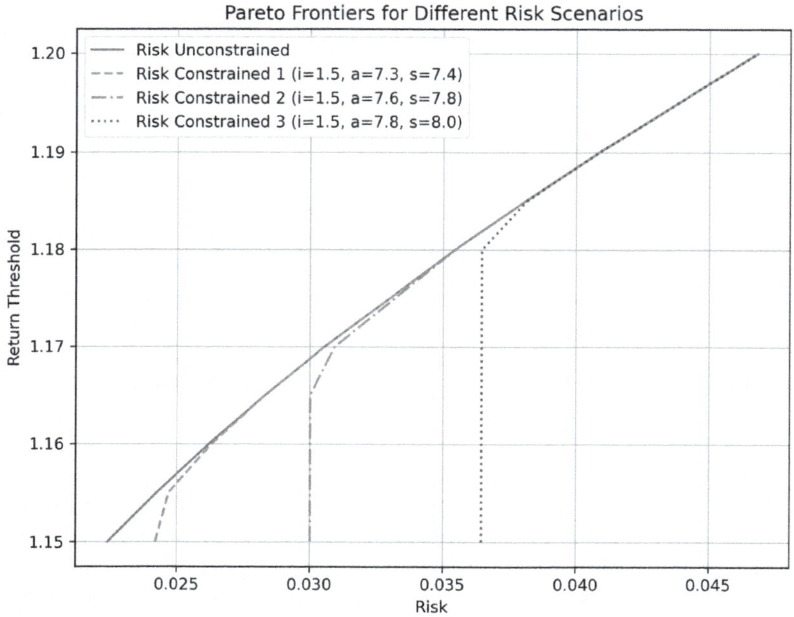

Fig. 5. Pareto frontiers: trade-offs approach

closely aligned metrics, underscoring their Pareto optimally. This coherence in values underscores the efficiency and effectiveness of the identified solutions in striking a balance between risk and return.

7. Pareto frontiers trade-offs: delineate the optimal solutions discovered across Pareto frontiers, illustrating the trade-offs between different objectives. In the provided results, the metrics exhibit comparable values across all solutions. Figure 5, from experiment 1, represents the unconstrained Pareto-frontier alongside three additional frontiers, each reflecting levels of advertisement, inventory and customer satisfaction. It can be seen how an escalation in the level of customer satisfaction and advertising correlates with a diminished return for equivalent risk levels. The noteworthy aspect of the intelligent algorithm lies in its ability to quantify the involved trade-offs. As illustrated in the lower left quadrant of the figure, notable disparities emerge between the unconstrained frontier and subsequent frontiers as constraints on advertising, inventory, or customer satisfaction are tightened. Conversely, in the upper right quadrant, all frontiers converge towards a singular optimal solution, highlighting the universality of the best approach.

6 Conclusion and Future Work

This article offers a practical approach of applying AI in strategic decisions, in this case for marketing directors, specially for those that imply a multi-objective need. NSGA-II offers a good solution in terms of robustness as well as in computational results for the product portfolio optimization and show that by leveraging AI-driven to MOPO techniques, marketing directors can navigate the complex landscape of trade-offs and craft robust portfolio strategies that align with organizational objectives. Although the literature review shows a growing and extensive contribution on the use of AI in connection with strategic decision and product portfolio optimization, the results indicate that the practical approach is little explored by the academy and still holds open questions and challenges. Moving forward, we conclude that further research and experimentation hold four main areas: (i) the potential to drive enhanced strategic-decisions making under uncertainty, (ii) contrasting to real-world data, (iii) integrating data from emerging technologies such as Internet of Things (IoT) that will lead to dynamic results, (iv) and finally ethical and fairness considerations and challenges associated with the use of AI in marketing decisions, emphasizing the importance of responsible AI practices, transparency, and privacy protection.

Disclosure of Interests. The authors have no competing interests to declare that are relevant to the content of this article.

References

1. Arora, S., Sahney, S., Parida, R.R.: Drivers of showrooming behaviour: insights from integrated perspectives. Int. J. Retail Distrib. Manag. **50**(3), 398–413 (2022)
2. Bella, K.M.J.: A study on marketing analytics and artificial intelligence **XV Issue 55**, 53–57 (2024)
3. Borges, A.F., Laurindo, F.J., Spínola, M.M., Gonçalves, R.F., Mattos, C.A.: The strategic use of artificial intelligence in the digital era: systematic literature review and future research directions. Int. J. Inf. Manag. **57**, 102225 (2021)
4. Bullah, S., van Zyl, T.L.: A learnheuristic approach to a constrained multi-objective portfolio optimisation problem. arXiv preprint arXiv:2304.06675 (2023)
5. Cannavacciuolo, L., Iandoli, L., Ponsiglione, C., Zollo, G.: Knowledge elicitation and mapping in the design of a decision support system for the evaluation of suppliers' competencies. Vine - J. Inf. Knowl. Manag. Syst. **45**(4), 530–550 (2015)
6. Davenport, T., Guha, A., Grewal, D., Bressgott, T.: How artificial intelligence will change the future of marketing. J. Acad. Mark. Sci. **48**, 24–42 (2020)
7. Demirkan, H., Delen, D.: Leveraging the capabilities of service-oriented decision support systems: putting analytics and big data in cloud. Decis. Support Syst. **55**(1), 412–421 (2013)
8. He, Q., He, Z., Duan, S., Zhong, Y.: Multi-objective interval portfolio optimization modeling and solving for margin trading. Swarm Evol. Comput. **75**, 101141 (2022)
9. Henderson, H.: Creating alternative futures. Media Resources Center, Iowa State University (1978)
10. Juan, A.A., Keenan, P., Martí, R., McGarraghy, S., Panadero, J., Carroll, P., Oliva, D.: A review of the role of heuristics in stochastic optimisation: from metaheuristics to learnheuristics. Ann. Oper. Res. **320**(2), 831–861 (2023)
11. Juan, A.A., Kelton, W.D., Currie, C.S., Faulin, J.: Simheuristics applications: dealing with uncertainty in logistics, transportation, and other supply chain areas. In: 2018 Winter Simulation Conference (WSC), pp. 3048–3059. IEEE (2018)

12. Kalayci, C.B., Polat, O., Akbay, M.A.: An efficient hybrid metaheuristic algorithm for cardinality constrained portfolio optimization. Swarm Evol. Comput. **54**, 100662 (2020)
13. Khotinskaya, G., Chernikova, L., Likhacheva, O., Setchenkova, L., Slepneva, T.: Digital transformation as a driver for business ecosystem development. In: E3S Web of Conferences, vol. 403, p. 08017. EDP Sciences (2023)
14. Kolm, P.N., Tütüncü, R., Fabozzi, F.J.: 60 years of portfolio optimization: practical challenges and current trends. Eur. J. Oper. Res. **234**(2), 356–371 (2014)
15. Kumar, D.T.S.: Data mining based marketing decision support system using hybrid machine learning algorithm. J. Artif. Intell. Capsule Netw. **2**(3), 185–193 (2020)
16. Li, S.: The development of a hybrid intelligent system for developing marketing strategy. Decis. Support Syst. **27**(4), 395–409 (2000)
17. Loke, Z.X., Goh, S.L., Kendall, G., Abdullah, S., Sabar, N.R.: Portfolio optimisation problem: a taxonomic review of solution methodologies. IEEE Access (2023)
18. Ma, J., Kim, H.M.: Continuous preference trend mining for optimal product design with multiple profit cycles. J. Mech. Des. **136**(6), 061002 (2014)
19. Markowitz, H.M., Markowitz, H.M.: Portfolio Selection: Efficient Diversification of Investments. J. Wiley, Hoboken (1967)
20. Mehlstäubl, J., Braun, F., Denk, M., Kraul, R., Paetzold, K.: Using machine learning for product portfolio management: a methodical approach to predict values of product attributes for multi-variant product portfolios. Proc. Des. Soc. **2**, 1659–1668 (2022)
21. Mehlstäubl, J., Braun, F., Paetzold, K.: Data mining in product portfolio and variety management–literature review on use cases and research potentials. In: 2021 IEEE Technology & Engineering Management Conference-Europe (TEMSCON-EUR), pp. 1–6. IEEE (2021)
22. Mustak, M., Salminen, J., Plé, L., Wirtz, J.: Artificial intelligence in marketing: topic modeling, scientometric analysis, and research agenda. J. Bus. Res. **124**, 389–404 (2021)
23. Pai, G.V., Michel, T.: Metaheuristic multi-objective optimization of constrained futures portfolios for effective risk management. Swarm Evol. Comput. **19**, 1–14 (2014)
24. Panadero, J., Juan, A.A., Freixes, A., Grifoll, M., Serrat, C., Dehghanimohamamdabadi, M.: An agile simheuristic for the stochastic team task assignment and orienteering problem: applications to unmanned aerial vehicles. In: 2020 Winter Simulation Conference (WSC), pp. 1324–1335. IEEE (2020)
25. Pouya, A.R., Solimanpur, M., Rezaee, M.J.: Solving multi-objective portfolio optimization problem using invasive weed optimization. Swarm Evol. Comput. **28**, 42–57 (2016)
26. Qu, B., Zhou, Q., Xiao, J., Liang, J., Suganthan, P.N., et al.: Large-scale portfolio optimization using multiobjective evolutionary algorithms and preselection methods. Math. Probl. Eng. **2017** (2017)
27. Rana, J., Gaur, L., Singh, G., Awan, U., Rasheed, M.I.: Reinforcing customer journey through artificial intelligence: a review and research agenda. Int. J. Emerg. Mark. **17**(7), 1738–1758 (2022)
28. Rodriguez-Garcia, P., Li, Y., Lopez-Lopez, D., Juan, A.A.: Strategic decision making in smart home ecosystems: a review on the use of artificial intelligence and internet of things. Internet Things 100772 (2023)
29. Schmidt, C.P., Gould, F.J., Schmidt, C.: Study Guide [to] Introductory Management Science, Gary. D. Eppen, Floyd J. Gould, Charles P. Schmidt. Prentice Hall, Hoboken (1991)

30. Schrage, M., Kiron, D.: Improving strategic execution with machine learning. MIT Sloan Management Review (2018). pp. 0_1–7
31. Schrage, M., Kiron, D.: Understanding the real keys to effective kpi's. MIT initiative on the digital economy research biref, pp. 1–3 (2018)
32. Shahid, M.Z., Li, G.: Impacto of artificial intelligence in marketing: a perspective of marketing professionals in Pakistan. Glob. J. Manag. Bus. Res. **19**(2), 27–33 (2019)
33. Tang, J., Liu, G., Pan, Q.: A review on representative swarm intelligence algorithms for solving optimization problems: applications and trends. IEEE/CAA J. Autom. Sin. **8**(10), 1627–1643 (2021)
34. Tucker, C.S., Kim, H.M.: Data-driven decision tree classification for product portfolio design optimization, p. 041004 (2009)
35. Ünal, H.T., Başçiftçi, F.: Using evolutionary algorithms for the scheduling of aircrew on airborne early warning and control system. Defence Sci. J. **70**(3) (2020)
36. Verma, S., Sharma, R., Deb, S., Maitra, D.: Artificial intelligence in marketing: systematic review and future research direction. Int. J. Inf. Manag. Data Insights **1**(1), 100002 (2021)
37. Vlačić, B., Corbo, L., e Silva, S.C., Dabić, M.: The evolving role of artificial intelligence in marketing: a review and research agenda. J. Bus. Res. **128**, 187–203 (2021)

Challenges of Artificial Intelligence in Strategic Decisions: Talent Management

Sergio Carol[1(✉)] and Patricia Rodriguez-Garcia[2]

[1] Director de Talento, Occident, Catalonia, Spain
sergio.carol@gco.com
[2] Department of Computer Science, Universitat Oberta de Catalunya, 08018 Barcelona, Spain
patroga@uoc.edu

Abstract. Is about changing and dynamic management, which requires constant revision and updating.

Organizational talent management involves areas such as recruitment and selection, talent development and retention, performance evaluation, strategic workforce planning, executive talent assessment, and management of promotions, among others. In each of these areas, there are specific challenges that entail multiple decision-making processes whose outcomes can be very significant and even compromise the sustainability of the company. Some of these challenges include identifying optimal candidates, reducing biases in the selection process, personalizing talent development, and ensuring objectivity in performance evaluation, for example.AI holds promise in automating candidate filtering, predictive turnover risks, facilitating AI-assisted assessment tools, and recommending growth opportunities. However, integrating AI faces significant obstacles, such as biases in data, data privacy and security concerns, and the need for integration with existing systems. This article emphasizes the importance of a careful and considered approach to integrate AI in talent management, ensuring that technology serves as an enabler for achieving organizational goals ethically and sustainably.

Undoubtedly, by addressing biases, AI can create more fair, prosperous, and efficient organizations, reducing inequalities, and contributing to economic growth in Society.

Keywords: Strategic Decision-Making · Artificial Intelligence · Talent Management

1 Introduction

Talent management is undoubtedly one of the most relevant strategic decision for organizations. Optimizing the management of this resource has a direct impact on the profitability, sustainability and competitiveness of companies [3, 9]. Having the right talent is what allows organizations to differentiate themselves and adapt to the changing environment; it is the factor that drives innovation, creativity, and the quality of the products and services offered. The organization's objectives vary according to market circumstances, customer needs, and technological evolution, and the necessary talent must adapt to the

A. A. Juan et al. (Eds.): DSA ISC 2024, LNCS 14778, pp. 84–91, 2025.
https://doi.org/10.1007/978-3-031-78238-1_8

new competencies and skills demanded; as a result talent management requires dynamic decision-making with constant revision and updating.

To address the gaps in existing literature and provide a more focused perspective, this paper aims to delineate the transformative impact of artificial intelligence (AI) integration in talent management while elucidating the key challenges therein. Drawing on seminal works by Cai et al. [1] and Gonzalez et al. [3], it is evident that AI presents novel opportunities and challenges in optimizing talent management processes. Despite its potential, the implementation of AI in talent management necessitates careful consideration of various factors, including fairness, transparency, data security, and integration with existing practices.

To date, we have seen artificial intelligence (AI) applications in talent management for matching candidates to positions (LinkedIn), the design of personalized training itineraries (IBM), Google's Project Oxygen (which uses data analysis to identify the qualities of the best managers), or applications like HireVue for analyzing video interviews, evaluating not only what candidates say but how they say it, to predict their suitability for a position [6].

However talent management faces an unprecedented transformation with the incorporation of AI into its processes [8]. This evolution promises to revolutionize the way organizations identify, attract, and retain their most valuable employees. However, despite its potential to improve efficiency and effectiveness, the implementation of AI in talent management presents several challenges that must be carefully addressed. One of the main challenges is ensuring fairness and transparency in decision-making processes. The possibility of inherent biases in the algorithms and data used can lead to unfair decisions, negatively affecting diversity and inclusion within the organization. Additionally, interpreting decisions made by AI systems can be complex, requiring additional effort to explain and justify these decisions to relevant stakeholders. Data security and employee privacy also emerge as significant concerns. Moreover, talent management through AI involves handling large volumes of personal and sensitive data, which demands robust protection measures to prevent privacy breaches and ensure compliance with current regulations. Another important challenge is the integration of AI-based systems with existing talent management practices. Resistance to change by employees and the need to redefine roles and responsibilities are obstacles that must be overcome to fully leverage their benefits. Finally, skills development and continuous training are outlined as key elements for the successful adoption of AI in talent management. Preparing leaders and employees to work effectively with emerging technologies is fundamental to fostering a culture of innovation and adaptability in the digital age [5].

These challenges underline the importance of a careful and considered approach to the implementation of AI in this field, ensuring that technology is a facilitator for reducing biases of any kind, contributing to forming more just, prosperous, and efficient organizations, reducing inequalities, and bringing economic growth to society.

To provide a clearer research focus, this paper articulates the following key questions: (i) how can organizations effectively navigate the complexities of AI integration in talent management?, and (ii) what strategies can mitigate biases and ensure ethical decision-making in AI-driven talent processes?.

Structured into distinct sections, this article will first provide a comprehensive review of current literature to establish the research gap and delineate the scope of AI integration in talent management. Subsequently, it will delve into the challenges posed by AI adoption in talent management, supported by empirical evidence and relevant case studies. Finally, the paper will propose strategies and recommendations for addressing these challenges and maximizing the benefits of AI in talent management, drawing on insights from recent research.

2 Challenges and Opportunities of AI in Talent Management

The talent management process in organizations essentially comprises the following stages: (i) recruitment and selection, (ii) talent development and retention, (iii) professional and executive evaluation and (iv) compensation and promotion.

In all of these areas, opportunities arise where decision-making based on or assisted by AI can contribute to making them faster, more efficient, and with higher success rates [6, 7]. However, there are significant challenges to overcome, as these decisions involve knowing, analyzing, and categorizing, for example, information like cultural values, emotional management, relational competencies. One of the main strategic questions to solve is: "how can talent managers ensure accuracy in selecting a new team leader or executive considering all these aspects, in addition to purely objective ones?".

Tables 1a, 1b and 1c collects, for each of these sub-processes, the most significant challenges, as well as the possible contributions that AI can offer to overcome them together with the main contributions from the current scientific community.

3 Challenges that AI Must Overcome and Strategies to Do so

Overcoming the challenges described so far is not easy and will necessarily involve overcoming significant difficulties such as biases of all kinds in the data, poor calibration of the data among different evaluators, consistency and homogeneity of the data (since it is collected by different systems, in different organizations and over extended periods of time, it is often highly compromised), the objective assessment of the sought-after managerial qualities and their measurement, the predictive identification of managerial potential, or the difficult assessment of ethical, cultural, or soft skills [10].

To navigate these challenges effectively, organizations can leverage a spectrum of AI-driven solutions. Implementation of anti-bias algorithms for instance, offer a robust mechanism to mitigate biases in data and decision-making processes; strengthening of data security ensures the integrity and confidentiality of sensitive information, bolstering trust in AI-driven talent management systems and sentiment analysis tools provide invaluable insights into employee sentiments and organizational dynamics, enabling proactive interventions to foster a positive work environment.

Integrated performance management platforms facilitate seamless data aggregation and analysis, empowering organizations to derive actionable insights for talent optimization; dynamic predictive models leverage AI algorithms to forecast future talent needs

Table 1a. Challenges and Contributions of AI in Talent Management Processes.

Field	Challenge	AI contributions	Authors
Recruitment and Selection	Identification of optimal candidates in large volumes of applications	Automation of candidate filtering through machine learning algorithms	Cai, L., Ji, Y., Wijekoon, C., & Yuan, Y. (2023). Decision-Making on Selection of Talent Management Methods in the Era of Digitalization; and Nocker, M., & Sena, V. (2019). Big data and human resources management: The rise of talent analytics
	Reduction of biases in the selection process	Predictive analysis to identify candidates with higher potential for long-term success	
	Customization of the recruitment process for different roles	AI-assisted interview tools to objectively evaluate skills and competencies	
Talent Retention	Identification of personalized training and development needs	Adaptive learning platforms for personalized training	Kravariti, F., & Johnston, K. (2020). Talent management: a critical literature review and research agenda for public sector human resource management; and Nocker, M., & Sena, V. (2019). Big data and human resources management: The rise of talent analytics
	Succession management and career planning	Predictive models to identify turnover risks and dissatisfaction factors	
	Recommendation systems for opportunities of career growth and development	Prediction and prevention of staff turnover	

and identify emerging leadership potential; and AI-assisted leadership development platforms offer personalized learning experiences tailored to individual competencies and career aspirations.

Additionally, behavioral analysis tools enable organizations to discern nuanced behavioral patterns and preferences, facilitating targeted talent development initiatives; and predictive analysis for assessing the impact of promotions enable informed decision-making, ensuring alignment with organizational objectives and individual career trajectories [3].

Table 1b. Challenges and Contributions of AI in Talent Management Processes.

Field	Challenge	AI contributions	Authors
Performance Evaluation	Objectivity and precision in performance evaluation	Algorithms to identify trends and patterns in performance	Iles, P., Preece, D., & Chuai, X. (2010). Talent management as a management fashion in HRD: Towards a research agenda; and
	Identification of areas for improvement and strengths	Integration of multiple sources of feedback	Lee, J., & Kim, S. O.: People Analytics Case Study: Performance Prediction Model and Effective Decision-making Timing; and Sivathanu, B., & Pillai, R. (2020). Technology and talent analytics for talent manage-ment–a game changer for organizational performance
	AI-assisted performance management systems for personalized recommendations	Customization of development plans based on performance	

4 Case Study Proposal: The Challenge of Executive Assessment

We propose here a specific case study where AI can foster strategic-decisions in talent management, with the aim of clearly outlining the existing challenges in a real organization.

Context: Financial company, 150 years of history, multinational, strong growth in the last 20 years through acquisitions (which entails the integration of very diverse historical ways of working, different processes and systems), average age of 48.2 years, average seniority of 16 years with little voluntary turnover (<4%), 3,600 employees in Spain and 3,800 distributed in 53 countries, with an annual wage cost of 1.5 billion euros.

Available Information: (1) Individual: Results of annual professional evaluations, 360-degree executive assessment, Insight Discovery and Basadur tests (not available in all cases), quantitative business results from previous years, complete remuneration information, promotion history, and career path. (2) Corporate: Results reports, various organizational charts, financial statements, and documents that help understand the history, culture, and values of the Organization (relevant topics when determining the leadership styles that best fit each situation).

Table 1c. Challenges and Contributions of AI in Talent Management Processes.

Field	Challenge	AI contributions	Authors
Executive Talent Evaluation	Objective evaluation of leadership and executive skills	Text analysis tools to evaluate feedback from multiple sources	Cappelli, P., Tavis, A., Burrell, L., Barton, D., Carey, D., & Charan, R. (2018). The new rules of talent management; and Zhang, X., Zhao, Y., Tang, X., Zhu, H., & Xiong, H. (2020). Developing fairness rules for talent intelligence management system
	Development of leadership competencies adapted to the future	Predictive tools to identify potential leaders within the organization	
	Recommendation systems for personalized leadership development programs	Measurement of leadership impact on organizational performance	
Promotion Management	Fair and objective identification of candidates for promotion	Predictive analysis to determine the impact of promotions on team morale	Gonzalez, M. F., Capman, J. F., Oswald, F. L., Theys, E. R., & Tomczak, D. L. (2019). "Where's the IO?" Artificial intelligence and machine learning in talent management systems; and Wiblen, S., & Marler, J. H. (2021). Digitalised talent management and automated tal-ent decisions: the implications for HR professionals
	Alignment of promotions with the strategic needs of the organization	Evaluation of employee readiness and potential for higher roles	
	Aligning employee aspirations with organizational opportunities	AI systems for a comprehensive evaluation of performance and promotion potential	
	Maintaining motivation among non-promoted employees	AI-driven competency assessments to identify development needs before promotion	

Objective: With this available information and in this organizational context, the goal is to find a tool that allows for a better objectification of executive talent and its degree of alignment with the culture and values of the organization and its future needs. Implicitly, this goal involves knowing specifically which competencies will be necessary in the short and medium term and establishing procedures to improve the quantity and quality of the available data.

Uses: Having this strategic decision-support tool should facilitate better success rates in promotion and appointment processes for certain key positions or projects. By facilitating more informed strategic decision-making, it aims to bolster success rates while mitigating errors and unnecessary turnover. Furthermore, it promises to accelerate organizational agility and responsiveness to dynamic market demands.

5 Conclusion

In this article, a review of the various talent management processes in large organizations has been conducted, identifying the challenges posed by decision-making in each one and the opportunities to improve these strategic decisions through the application of AI. While AI holds the promise of revolutionizing recruitment, talent development, performance evaluation, executive assessment, and promotion management, it also requires careful navigation to mitigate inherent biases, ensure data security, and integrate with existing practices effectively. The proposed case study highlights the complexities faced by organizations in assessing executive talent and aligning it with organizational culture and future needs. By leveraging AI-driven tools and strategies, organizations can better objectify executive talent, improve decision-making processes, and accelerate internal processes, ultimately leading to increased productivity, reduced turnover, and improved business outcomes.

From the several futures lines of work the article identifies higher potential in: (i) exploring advanced data analytics techniques, such as natural language processing (NLP) and sentiment analysis, to extract deeper insights from employee feedback, performance reviews, and other unstructured data sources; (ii) investigating strategies for fostering effective collaboration between humans and AI systems in talent management processes to provide explainable AI-driven recommendations, and incorporating human oversight to ensure the ethical use of AI; and (iii) expanding the use of predictive analytics models to forecast future talent needs, identify emerging skills gaps, and proactively address talent shortages to stay ahead of workforce trends and make data-driven decisions in talent planning and development.

Disclosure of Interests. The authors have no competing interests to declare that are relevant to the content of this article.

References

1. Cai, L., Ji, Y., Wijekoon, C., Yuan, Y.: Decision-making on selection of talent management methods in the era of digitalization. Systems **11**(9), 450 (2023)
2. Cappelli, P., Tavis, A., Burrell, L., Barton, D., Carey, D., Charan, R.: The new rules of talent management. Harv. Bus. Rev.
3. Gonzalez, M.F., Capman, J.F., Oswald, F.L., Theys, E.R., Tomczak, D.L.: Where's the IO?" Artificial intelligence and machine learning in talent management systems. Pers. Assess. Decis. **5**(3), 5 (2019)
4. Iles, P., Preece, D., Chuai, X.: Talent management as a management fashion in HRD: towards a research agenda. Hum. Resour. Dev. Int. **13**(2), 125–145 (2010)
5. Kravariti, F., Johnston, K.: Talent management: a critical literature review and research agenda for public sector human resource management. Public Manag. Rev. **22**(1), 75–95 (2020)
6. Lee, J., Kim, S.O.: People analytics case study: performance prediction model and effective decision-making timing. Dongguk Bus. Res. Inst. (2022)

7. Nocker, M., Sena, V.: Big data and human resources management: the rise of talent analytics. Soc. Sci. **8**(10), 273 (2019)
8. Sivathanu, B., Pillai, R.: Technology and talent analytics for talent management–a game changer for organizational performance. Int. J. Organ. Anal. **28**(2), 457–473 (2020)
9. Wiblen, S., Marler, J.H.: Digitalised talent management and automated talent decisions: the implications for HR professionals. Int. J. Hum. Resour. Manag. **32**(12), 2592–2621 (2021)
10. Zhang, X., Zhao, Y., Tang, X., Zhu, H., Xiong, H.: Developing fairness rules for talent intelligence management system. In: 53rd Annual Hawaii International Conference on System Sciences, pp. 5882–5891. IEEE Computer Society (2020)

Tourist Routes Recommender System Supported by Artificial Intelligence

Cristina González-Navasa[1](✉), José Andrés Moreno Pérez[2], and Julio Brito[2]

[1] I3MA, San Cristóbal de La Laguna, Spain
alu0100457599@ull.edu.es

[2] Universidad de La Laguna, IUDR and DIIS, San Cristóbal de La Laguna, Spain
{jamoreno,jbrito}@ull.edu.es

Abstract. This work presents the development of a personalised intelligent tourist route recommendation system based on artificial intelligence and soft computing methodologies. The system is made up of different modules and each one includes intelligent techniques and tools, leverages customer data, information on points of touristic interest, and user feedback to offer personalised recommendations. Specifically, by applying clustering techniques, customers are segmented into similar groups, allowing for more precise targeting. Customer segmentation facilitates profile analysis to capture the characteristics and preferences of tourists. Using content-based and filtering recommendation methods, the system further personalises POI selections. Moreover, the integration of metaheuristic algorithms, such as GRASP, addresses challenges in route planning and takes into account user restrictions such as location, preferred attractions, and available time. Additionally, the integration of visualisers and mobile interfaces represents an important impact on the success of recommendation systems on platforms to enhance user satisfaction and the efficacy of recommendations. This comprehensive approach not only improves the user experience, but also contributes to the success and adoption of recommendation systems in the tourism area.

Keywords: Recommender system · Tourist trip planning · Metaheuristics

1 Introduction

Within the tourism framework, the issue of designing personalised tourist itineraries arises [11,27]. In the literature, these problems are defined as Tourist Trip Design Problems (TTDP). To best represent the real context, it is formulated as an orientation problem (OP), [9,22,24] and its variants.

This work deals with the development of a personalised intelligent system for recommending tourist routes within the Atlantis Technology Tourist Services Platform[1]. The platform is private and integrates information about customers,

[1] https://www.atlantistecnologia.com/.

A. A. Juan et al. (Eds.): DSA ISC 2024, LNCS 14778, pp. 92–106, 2025.
https://doi.org/10.1007/978-3-031-78238-1_9

available tourist options, and ratings provided by users. In the future, the intelligent recommender will be offered as a product to customers or will be integrated in other touristic services like CanariasViaja[2].

This innovative system is methodologically based on artificial intelligence and soft computing techniques to generate route recommendations based on tourist profiles [1,20], more specifically, those related to recommendation systems and metaheuristic optimisation. Metaheuristics constitute the appropriate intelligent tool for route optimisation, considering that proposals must comply with tourist restrictions, the locations of attractions to visit, and available time at the destination, among others [12,25]. At the destinations, we can find a very wide range of points of interest to visit and activities to do. Henceforth, we shall refer to both as Points of Interest (POIs).

The system is made up of different modules and each one includes intelligent techniques and tools. Clustering techniques are used for customer segmentation and profile analysis to capture the characteristics and preferences of tourists. Numerical weights are also assigned to each service or activity in two ways, based on objective criteria considering the characteristics or popularity of POIs and reflecting the interest or level of satisfaction of the tourist, as shown by their interaction implicitly or explicitly.

Furthermore, recommendation methods based on tourist characteristics and preferences, as well as categorisation of activities and offered services, are used. For optimal routes, optimisation techniques, metaheuristics, and soft computing methodologies are used to provide viable high-quality itineraries adapted to the preferences and limitations of tourists based on the previous personalised recommendations.

All system modules are integrated to share information (see Fig. 1). The profile module, responsible for segmenting individuals, must be integrated with other system modules to provide profile data to recommendations and route design modules. The recommendation module is the centre of the system. On the one hand, it directly connects with the profile manager, receiving user data, and generating profiles as starting points for offering personalised recommendations. On the other hand, it provides personalised recommendations based on user preferences and suggests points of interest or activities that align with the preferences of tourists.

Finally, the route design module receives profile information and recommendations from previous modules to design tourist routes that are tailored to user preferences. All of this is integrated into the customer interaction tool of the platform to display recommendations and associated information and capture the reactions and satisfaction of tourists to provide feedback to the system.

The main innovation of this work lies in the structure of the proposed system, as shown in Fig. 1, together with numerous adjustments and advancements in the intelligent components to leverage the outcomes of one module for the subsequent component. These improvements facilitate their integration within the system and contribute to the overall success of its practical application.

[2] https://www.canariasviaja.com/.

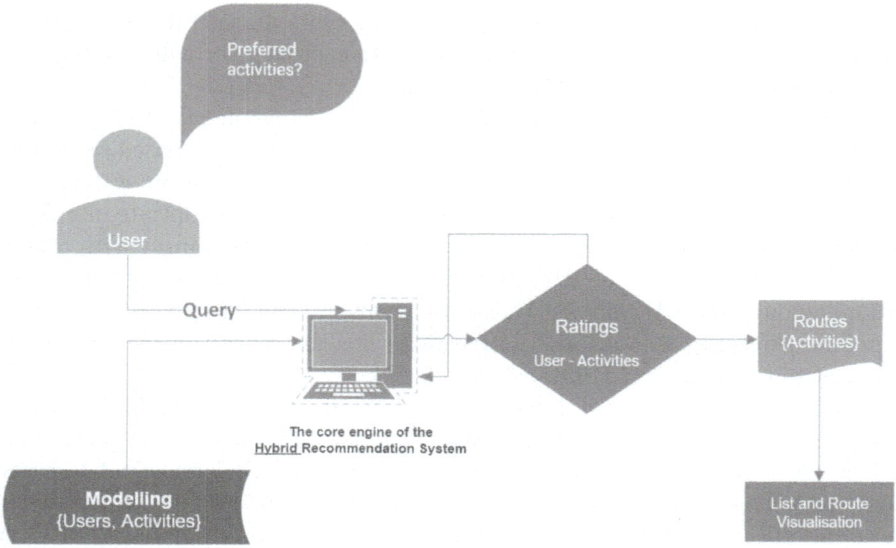

Fig. 1. Recommender System components.

The rest of the paper is organised as follows. After the general introduction of the system, which also contains a brief overview of the related literature and an explanation of the integration of the modules, in Sect. 2, we present the IA technologies and algorithms used in detail. We present the Profile Manager module in Sect. 2.1, for client segmentation. In Sect. 2.2, the methodology of the Recommendation Generator is described as a hybrid approach. The next Subsect. 2.3, introduces the proposed metaheuristic, Greedy Randomized Adaptive Search Procedure (GRASP), and also presents a formal description of the optimisation problem that we used to find an optimal solution. Finally, Sect. 3 contains the conclusions and directions for future work.

2 Technologies and Algorithms

The main technology used is the Python programming language, with the profile management, recommendation, and route design modules implemented using this language. A customised version of the GRASP algorithm is implemented in the route design module for route optimisation.

The libraries used include Pandas and Scikit-learn for data manipulation and processing and for employing segmentation and classification algorithms in profile creation, respectively.

Additionally, SQL is used to extract information from the databases. User data, including sociodemographic information, travel characteristics, and preferences, are collected. These data are then processed to obtain client-level information.

2.1 Profile Manager: Segmentation Algorithm

The segmentation algorithm used in the Profile Manager module aims to group tourists into distinct categories based on their shared characteristics and preferences. This process involves analysing various factors such as demographics, travel behaviour, and interests to identify meaningful clusters within the tourist population.

The algorithm uses a combination of clustering techniques, such as k-means, to divide tourists into homogeneous groups, although other classification techniques can be used. These techniques allow the algorithm to identify similarities and differences among tourists and assign them to the appropriate segments. The k-means algorithm is used because of its simplicity and effectiveness in grouping users into profiles based on their preferences and demographic characteristics.

Once the segmentation process is complete, each tourist is associated with a specific segment that represents their profile. This segmented approach allows for more personalised recommendations and route designs tailored to the preferences of each group of travellers.

In general, the segmentation algorithm has a fundamental impact on understanding the diverse needs and preferences of tourists, facilitating targeted marketing strategies, and improving the overall tourism experience.

Algorithm 1. Profile Segmentation with k-means

```
from sklearn.cluster import KMeans
kmeans = KMeans(n_clusters)
user_clusters = kmeans.fit_predict(user_data)
```

The k-means algorithm is a widely used unsupervised learning technique for clustering data into distinct groups. In the context of profile segmentation, k-means can be applied to partition users into groups based on similarities in their characteristics and preferences [2,23].

A critical aspect in the application of the k-means algorithm is the determination of the optimal number of clusters. The elbow method is a commonly used technique for selecting the optimal number of clusters. This method involves running the k-means algorithm for a range of values and calculating the sum of squared distances of each point to the nearest centroid (also known as cluster inertia). Then, a plot of inertia is drawn against the number of clusters. The point at which the curve starts to flatten out is known as the *elbow*. This point indicates the optimal number of clusters, as adding more clusters does not provide a significant improvement in inertia.

Once the optimal number of clusters has been determined, the k-means algorithm is applied to group customers into homogeneous segments. Each cluster represents a group of customers with similar characteristics, allowing for a better understanding of the needs and preferences of different segments of the customer base.

The profiles module is an essential part of the tourist route recommendation system. Its main function is to analyse the information available from users and segment the profiles into specific groups based on their characteristics and travel preferences. This module is crucial to provide relevant and personalised tourist route recommendations. Key technical functions used for profile management include the gathering and analysis of user information to understand their preferences and needs and even collecting historical information to understand the previous preferences of the user [1].

Data encompass a wide range of information. From the user perspective, it includes demographic details such as age, gender, nationality, whether the user is travelling alone or accompanied, as well as previous purchase history and preferences (adventure, culture, nature, etc.). Simultaneously, data related to points of interest comprises geolocation (latitude and longitude), availability, types of attractions, ratings, and categories. These comprehensive datasets are utilised to continually update user profiles with automated feedback mechanisms, ensuring they accurately reflect user preferences and visit histories.

2.2 Recommendation Generator

The recommendation generator module is an essential part of the tourist route recommendation system. Its main objective is to provide useful suggestions to users about various products and services, and personalised recommendations for tourist routes and activities at the destination [6].

In this component, we will apply techniques to extract knowledge to infer the preferences of tourists with respect to destination points of interest. Based on tourist profiles obtained from survey data, some demographic information, and available activity information, we generate a recommendation for a set of attractions to visit. Subsequently, the route planning component uses this set to generate personalised travel plans by applying an optimisation algorithm (see Fig. 2).

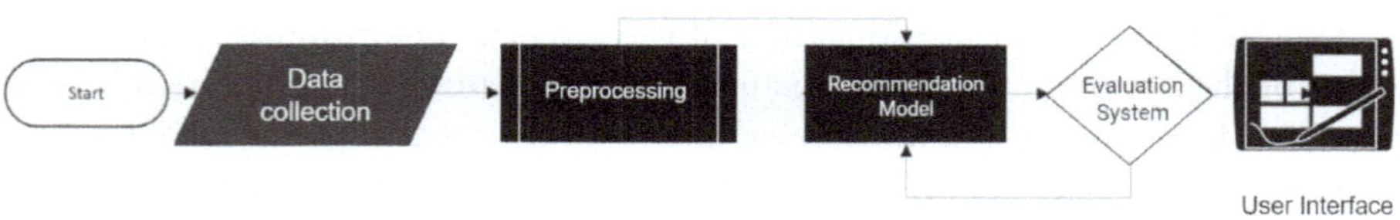

Fig. 2. Basic Framework of Recommender Systems

In well-established tourist destinations, we can find a very wide range of points of interest to visit and activities to do. From a usability standpoint, it is also not feasible to have an interface that asks and evaluates each of the tourist attractions. Additionally, the limited knowledge of the destination that tourists have makes it difficult to establish preferences and select the subset of places to visit that maximises their satisfaction. Therefore, it is much more likely that the input data from tourists will be limited to establishing their interests in a particular typology or characteristics of the visit to be made.

We propose a content-based method [19] to recommend points of interest and activities. From the total number of points of interest to visit and activities to engage in, this technique estimates preferences and recommends a subset of points that are expected to maximise the preferences of the tourists. Furthermore, a hybrid approach [4] that combines this content-based method and collaborative filtering [7] allows for taking into account user preferences and similarities between tourists. [8]. The proposed method consists of the following:

The development of the user profile involves taking advantage of information on user behaviour, sociodemographic profile, such as that obtained from a Tourism Expenditure Survey[3], as well as the general characteristics of their trip to the Canary Islands.

Clustering techniques are used to determine the characteristics of tourists, their interests, and preferences regarding the places and activities available. An example of a group or profile obtained could be Spanish women between 35 and 45 years old, travelling as a couple, interested in sun and beach, cultural activities, and hiking. These profiles are then used to derive a weighted characteristic vector based on minimal demographic information available for a user and explicit ratings characterising the preferred activities obtained from the profile.

Categorisation of points of interest involves conducting a content analysis to determine the characteristics of each activity and assigning weights to these categories. This produces a feature vector with associated weights for each point of interest, for example: Excursion to Masca (30% beach, 60% hiking, 10% cultural). This categorisation is achieved using natural language processing techniques to determine the categories of each activity from existing textual descriptions and classification methods that facilitate the weighting of these categories.

Determining the activities to recommend involves calculating the similarity between the weighted characteristics of the tourist and the categorised activities. The cosine similarity measure is applied, which is widely used in the literature. The values of this similarity measure express the degree of preference of tourists for points of interest.

Based on the explanation above, the system creates user profiles utilising information gathered from the Tourism Expenditure Survey to understand user behaviour. These profiles are then grouped to determine interests and preferences, specifically identifying which variables influence the choice of particular activities by tourists.

In addition, a content analysis of the activities is performed. Activity data is analysed to determine the defining characteristics of each activity, and they are classified based on these characteristics and the percentage of belonging to each category.

Once a user profile is identified, either explicitly or implicitly, the system can recommend activities by calculating the similarity of the user profile to the

[3] https://www.gobiernodecanarias.org/istac/.

classified activities. This allows for personalised recommendations based on the user's preferences and interests.

2.3 Route Generator: Metaheuristics

The main purpose of the tourist route generator is to provide high-quality and personalised itineraries to users. The generator considers user profile data including expressed preferences previously offered personalised recommendations and geographical, temporal, or other constraints.

For optimal operation of the itinerary planner, each point of interest should have a predefined average visit time and the ability to incorporate travel times between points by using Google Maps API, for example.

The itinerary generator uses metaheuristics for route construction. It applies widely used search and optimisation strategies to solve optimisation problems, such as the TTDP and its variants. These techniques are especially useful when addressing complex problems where finding the optimal solution is computationally expensive or even impossible within a reasonable timeframe.

The OP and its variants are commonly used to model route design problems such as the TTDP. This is because it allows flexibility in planning, modelling the needs of tourists to plan their daily activities, considering time and constraints, maximising the value of the tourist experience, and adapting to different preferences and travel styles of tourists. These characteristics make OP a quality tool for addressing the challenges associated with tourist trip planning, where it is not necessary to visit all available points of interest, but allows for a selection of the most notable ones for tourists. The mentioned characteristics makes difference with respect to other models like VRP among others.

The OP model is considered the most reliable because the majority of tourists are more likely to plan trips by searching for a daily route. By simplifying TTDP as OP, the problem can be expanded to consider a specific aspect of route planning that may be relevant to the tourist experience [26]. Another frequently used model is the extension to the Team Orienteering Problem (TOP), which involves finding multiple optimal routes over a few days while preventing intersecting itineraries. [24] It is also common to find models with time constraints. For example, considering specified opening hours for visiting points of interest, would be modelled as OP with Time Windows (OPTW), [17]. Taking into account the travel time between points of interest, it would be modelled as time dependent OP (TDOP) categorised by time of travel or choice of transport mode [10,14].

These models may also have multiple constraints to meet, which add complexity; in these cases, they are often referred to in the literature as Multi-constraints TTDP (MCTTDP) [18] It may contain additional realistic constraints that reflect features like personal interests, itineraries for heterogeneous groups, clustering the set of POIs to avoid very similar activities, etc. [5,15,16]. Other TTDP models that incorporate multiple objectives are categorised as multi-objective TTDP (MOTTDP), the objectives that are most commonly tackled include reducing the travel time between various POIs and increasing profits [13].

We also note that multi-objective optimisation is challenging due to requiring significantly more function evaluations.

Using the available information, we design the route that best satisfies the tourist and adapts to the constraints of the destination and the limitations of the tourist. To achieve this, a mathematical optimisation model is used to formalise the description of possible alternatives, along with their corresponding constraints and the optimisation objective.

The formal description of the TTDP modelled as a simple OP with a given set of POIs is detailed below. The goal is to determine a path from the accommodation or starting point, limited in duration or length by a value, that visits some of the POIs, in order to maximise the total collected score. We provide a mathematical formulation of the basic OP. An excellent review of OP and its applications and variants, including the team orienteering problem (TOP), the OP with time windows (OPTW), and the TOPTW, is given in [26].

Let $I = \{1, 2, \ldots, n\}$ denote the set of POIs and $I^0 = I \cup \{0\}$ also include the point from which each route starts and ends, denoted by 0 (usually the tourist accommodation) [3]. Indices i and j represent points, the accommodation or a POI. Each route is determined by the routing variables x_{ij}, for all indices $i, j \in I^0$. The value of x_{ij} is 1 if the route goes from point i to point j and is 0 otherwise. To formalise the model, two additional series of variables are used. The selection variables y_i for every index $i \in I$. The value of y_i is 1 if POI i is visited on the route and $y_i = 0$ otherwise. The routing variables u_i, for each index $i \in I$, give the position of the selected POIs on the route.

The OP is formulated using the following notation:

Parameters:
- s_i is the score value of POI i; $i = 1, 2, \ldots, n$.
- t_{ij} is the travel time from point i to point j; $i, j = 0, 1, \ldots, n$.
- r_i is the visit time of POI i; $i = 1, 2, \ldots, n$.
- T_{max} is the limit time of the route, including travel and visit times.

Variables:
- Route variables $x_{ij} \in \{0, 1\}$; $i, j = 0, 1, \ldots, n$.
- Selection variables $y_i \in \{0, 1\}$; $i = 1, 2, \ldots, n$.
- Position variable $u_i \in \{1, 2, \ldots n\}$; $i = 1, 2, \ldots, n$.

This OP is formulated as an Integer Linear Programming Problem (ILP) as follows:

Maximise

$$\sum_{i=1}^{n} s_i y_i \tag{1}$$

Subject to:

$$\sum_{j=1}^{n} x_{0j} = \sum_{i=1}^{n} x_{i0} = 1 \quad (2)$$

$$y_i \leq 1 \qquad \forall i \in I \quad (3)$$

$$\sum_{j=1}^{n} x_{ij} = \sum_{j=1}^{n} x_{ji} = y_i \qquad \forall i \in I \quad (4)$$

$$\sum_{i=1}^{n} r_i y_i + \sum_{i=0}^{n} \sum_{j=0}^{n} t_{i,j} x_{ij} \leq T_{max} \quad (5)$$

$$1 \leq u_i \leq n \qquad \forall i \in I \quad (6)$$

$$u_i - u_j + 1 \leq (n-1)(1 - x_{ij}) \qquad \forall i, j \in I^0 \quad (7)$$

$$x_{ij} \in \{0, 1\} \qquad \forall i, j \in I^0 \quad (8)$$

$$y_i \in \{0, 1\} \qquad \forall i \in I \quad (9)$$

$$u_i \text{ integer} \qquad \forall i \in I \quad (10)$$

Equation (1) represents the objective function which is to maximise the total score collected. Constraints (2) guarantee that each route starts and ends at accommodation or starting point 0. The set of constraints (3) establishes that every POI is visited at most once. Constraints (4) ensure that, if a POI is visited in the given tour, it is preceded and followed by exactly one other POI in the tour. Constraints (5) ensure the limited time for the tour. Constraints (6) and (7) avoid subtours. Finally, (8), (9) and (10) establish the conditions of the variables.

Optimisation techniques allow for addressing the task of finding a high-quality solution to the problem with appropriate computational effort, therefore, approximate procedures that find nearly optimal solutions are the only reasonable way to approach these problems. Metaheuristics are the most suitable intelligent tools for this purpose.

Specifically, the GRASP approach is used to find high-quality routes [21].

Algorithm 2. GRASP

function GRASP(*MNitera*):

readInput()
initiate *BestSolution*;
for $k = 1, ..., MNitera$ **do**
(a) ConstructedSolution = GRASPConstructPhase();
(b) *ImprovedSolution* = LocalSearch(*ConstructedSolution*);
(c) UpdateSolution(*ImprovedSolution*, *BestSolution*);
return *BestSolution*;

end GRASP

GRASP is a metaheuristic algorithm that iteratively constructs solutions by combining greedy construction with randomised improvement. It starts by generating a feasible solution using a greedy algorithm and then applies a local search procedure to improve it. GRASP iterates this process for a set number of iterations or until a termination condition is met, selecting the best solution found throughout the iterations as the final result.

In the framework for designing tourist routes, GRASP can efficiently explore the solution space to discover high-quality itineraries that meet the user preferences and limitations criteria. GRASP uses candidates as a potential new component in the partial solutions that are evaluated during the iterative process of constructing solutions, and the Restricted Candidate List (RCL) to progressively iteratively build routes. This approach maintains a balance between exploration and exploitation to create optimal routes by choosing a subset of promising candidates at each construction step.

GRASP functions in a step-by-step manner, progressively constructing potential solutions by combining greedy selection and randomisation. During each step, the algorithm picks a potential solution (candidate) using a greedy strategy that steers the selection process towards favourable choices. However, GRASP introduces randomness by utilising a RCL to prevent early convergence and encourage exploration in the exploration process.

The RCL plays a vital role in the GRASP approach. Rather than directly choosing the top candidate solution at every stage, GRASP generates a list of potential candidates by evaluating a subset of the most favourable choices according to specific criteria or quality measures. This subset, defined by the RCL, aims to strike a balance between utilising the best-known solutions and investigating other potential options. The selection among the candidates in the RCL is done in a random manner, adding diversity to the search process and enabling the exploration of various areas within the solution space.

Construction Phase: During the construction phase, the algorithm iteratively selects destinations for the route, considering user preferences and the distance between destinations. The randomised component introduces diversification, preventing the algorithm from getting stuck in local optima.

Algorithm 3. GRASP Construction Phase

function GRASPConstructPhase():

Take an initial empty *PartialSolution*;
while there are candidate POIs to add to *PartialSolution*:
(a) construct RCL with the best candidates that can be added to the PartialSolution;
(b) choose at random a candidate from RCL;
(c) update the *PartialSolution* by adding the selected candidate into the solution;
return *PartialSolution*;

end GRASPConstructPhase

Local Search: After the construction phase, a local search is applied to improve the generated route. This may involve swapping nearby destinations to reduce the total distance or including additional destinations if possible within temporal and geographical constraints.

Algorithm 4. Local Search

function LocalSearch(*Solution*):

  repeat:
    (a) find the best *Neighbour* of *Solution*;
    (b) **if** *Neighbour* is better than *Solution* **do**:
      Solution = *Neighbour*;
  until no better *Neighbour* is found;
  return *Solution*;

end LocalSearch.

To evaluate the performance of the algorithm, an objective function is used, considering the total distance of the route, user preferences, the tourist score, and other relevant factors to assess the quality of the generated route.

In addition, user feedback, such as ratings and comments on recommended routes, can be utilised to enhance route evaluation and adjust future recommendations.

Recommendation systems and route generators are two applications with the common goal of providing personalised solutions based on user needs and preferences. However, their relationship becomes more interesting when considering the GRASP approach.

In the context of a recommender system, GRASP can be used to select the optimal combination of elements that maximise user satisfaction based on their preferences and constraints. In contrast, a route generator can use GRASP to find the optimal route between various points. Although recommender systems and route generators may seem different on the surface, both can benefit from the GRASP approach to provide personalised and optimised solutions to users. This demonstrates the power and versatility of modern artificial intelligence and data science techniques.

2.4 Interactive Recommendation Visualiser

Visualisers and interfaces play a crucial role in the context of modern recommendation systems and on mobile devices, and their influence is of paramount importance to both the user experience and the effectiveness of the recommendations provided. These elements serve as key facilitators for understanding recommendations by providing visual representations that convey information more effectively than by mere textual or numerical means. Appropriately arranged graphs, images and tables can significantly contribute to understanding the underlying reasons for recommendations and their correlation with individual preferences.

Mobile interfaces involve a more intuitive interaction with the recommendation system, as the inclusion of features such as advanced search, filters, and comment options gives users a greater level of control over recommendations, leading to a more personalised experience. This level of personalisation is crucial, as the relevance of recommendations is intrinsically linked to the information provided by users and their ability to adjust their preferences as needed. Furthermore, these mobile interfaces cater to the limitations of mobile devices, including smaller screens and touch interaction modalities. Appropriate and responsive design for mobile devices ensures a smooth and efficient user experience, an essential element for user retention and continued engagement with the application or platform. Thus, comfort and usability on mobile devices emerge as fundamental pillars to maintain active user participation.

Visualisers and mobile interfaces have a fundamental impact in promoting recommendation personalisation, user context adaptation, and facilitating the exploration and discovery of new elements. Equally important is its ability to collect feedback data, which is invaluable for refining the accuracy of recommendations and enhancing overall service quality.

When designing visualisers and mobile interfaces, considerations such as personalisation, should be taken into account to allow users to customise their viewing and recommendation preferences. The context is also essential as it shows recommendations relevant to the current situation of the user. Exploration should be facilitated to enable the discovery of new elements. Moreover, data collection is crucial as it collects user feedback data that can be used to improve the accuracy of recommendations and the overall quality of the service.

In this way, the viewers themselves can promote the personalisation of recommendations based on interests, history, or location, among other factors. Additionally, mobile interfaces can collect user feedback data noninvasively and in compliance with data protection laws, utilising surveys and ratings, or comments analysis.

Visualisers and mobile interfaces have a fundamental impact on the success of recommendation systems. They can aid in improving recommendation accuracy, user satisfaction and overall service quality.

3 Conclusions

The project addresses a real-world application. The final applicability in real circumstances is highly dependent on the availability of quality data and efficient accessibility to the corresponding databases. With the available data, we delve into the management of tourist profiles by establishing useful clusters and configuring the structure of the profiles module that captures user preferences and limitations. The profiles module is responsible for structuring and organising this information, enabling information updates and adaptations to any changes that may occur.

The profile module is important for customising tourist route recommendations in our system. Through the collection, processing and segmentation of user data, we can offer more relevant and appealing travel experiences. Continuous adaptation ensures that our recommendations remain accurate and relevant over time.

The recommender module is another key component of the tourist route recommendation system. After the corresponding study, it was concluded that the most appropriate option is a hybrid system that combines the advantages offered by each model, constituting a collaborative recommender based on both knowledge and content. Its ability to understand user preferences and provide precise and relevant recommendations is essential to enhance user experience and increase customer satisfaction.

The route design module is essential to offer users attractive and personalised tourist routes. As a route generation algorithm, the suitability of choosing metaheuristic systems capable of generating efficient and high-quality routes with adequate resource usage was demonstrated. Given the essentially heterogeneous characteristics of points of interest and uncertainty, metaheuristics can be tailored to the particular circumstances and operating context.

Using the GRASP approach allows for finding high-quality routes that adapt to preferences and limitations of the users. Specifically, the most appropriate strategy is to combine a GRASP construction system with the incorporation of a post-optimisation system. The constructive phase, which is greedy, random and adaptive, efficiently produces various high-quality initial routes. The post-optimisation phase, based on the implementation of destructive and constructive operations, allows for adaptation to the user changes post-selection, including unforeseen points of interest or the discarding of some selected for the route, as well as specific changes in the visiting order and unforeseen delays.

Furthermore, continuous user feedback contributes to the constant improvement of the system. The adequate and effective implementation of visualisers and mobile interfaces represents a critical component for the success of recommendation systems on platforms or mobile devices, increasing user satisfaction and the effectiveness of the recommendations offered.

Further research can be conducted following multiple directions. Focus on taking into account real-world characteristics and their integration into the system, constraints, and different objectives to target, considering the possibility of incorporating multiple functions into the objective. Exploring the impact on real-time decision-making, and being able to replanning the trip. Extending the MCOP model to incorporate the mentioned time windows or traffic considerations would further improve the planning process, making it more practical and realistic. Moreover, the capabilities of other methods, such as machine learning, can be investigated to enhance the prediction and improve the accuracy of the recommendation.

References

1. Abbasi-Moud, Z., Vahdat-Nejad, H., Sadri, J.: Tourism recommendation system based on semantic clustering and sentiment analysis. Expert Syst. Appl. **167**, 114324 (2021). https://doi.org/10.1016/j.eswa.2020.114324

2. Blömer, J., Lammersen, C., Schmidt, M., Sohler, C.: Theoretical analysis of the k-means algorithm – a survey. In: Kliemann, L., Sanders, P. (eds.) Algorithm Engineering. LNCS, vol. 9220, pp. 81–116. Springer, Cham (2016). https://doi.org/10.1007/978-3-319-49487-6_3
3. Brito, J., Expiosito-Marquez, A., Moreno, J.A.: A fuzzy GRASP algorithm for solving a tourist trip design problem. In: 2017 IEEE International Conference on Fuzzy Systems (FUZZ-IEEE). IEEE (2017). https://doi.org/10.1109/fuzz-ieee.2017.8015656
4. Burke, R.: Hybrid web recommender systems. In: Brusilovsky, P., Kobsa, A., Nejdl, W. (eds.) The Adaptive Web. LNCS, vol. 4321, pp. 377–408. Springer, Heidelberg (2007). https://doi.org/10.1007/978-3-540-72079-9_12
5. Choachaicharoenkul, S., Coit, D., Wattanapongsakorn, N.: Multi-objective trip planning with solution ranking based on user preference and restaurant selection. IEEE Access **10**, 10688–10705 (2022). https://doi.org/10.1109/access.2022.3144855
6. De Biasio, A., Montagna, A., Aiolli, F., Navarin, N.: A systematic review of value-aware recommender systems. Expert Syst. Appl. **226**, 120131 (2023). https://doi.org/10.1016/j.eswa.2023.120131
7. Fkih, F.: Similarity measures for collaborative filtering-based recommender systems: review and experimental comparison. J. King Saud Univ. Comput. Inf. Sci. **34**(9), 7645–7669 (2022). https://doi.org/10.1016/j.jksuci.2021.09.014
8. Fonseca, B.B., Cornelio, O.M.: Sistemas de recomendación para la toma de deciciones. estado del arte. UNESUM - Ciencias. Revista Científica Multidisciplinaria **6**(1), 149–164 (2022). https://doi.org/10.47230/unesum-ciencias.v6.n1.2022.289
9. Gavalas, D., Konstantopoulos, C., Mastakas, K., Pantziou, G.: A survey on algorithmic approaches for solving tourist trip design problems. J. Heuristics **20**, 291–328 (2014). https://doi.org/10.1007/s10732-014-9242-5
10. Gavalas, D., Konstantopoulos, C., Mastakas, K., Pantziou, G., Vathis, N.: Heuristics for the time dependent team orienteering problem: application to tourist route planning. Comput. Oper. Res. **62**, 36–50 (2015). https://doi.org/10.1016/j.cor.2015.03.016
11. Ghobadi, F., Divsalar, A., Jandaghi, H., Barzegar Nozari, R.: An integrated recommender system for multi-day tourist itinerary. Appl. Soft Comput. **149**, 110942 (2023). https://doi.org/10.1016/j.asoc.2023.110942
12. Gunawan, A., Lau, H.C., Vansteenwegen, P.: Orienteering problem: a survey of recent variants, solution approaches and applications. Eur. J. Oper. Res. **255**(2), 315–332 (2016). https://doi.org/10.1016/j.ejor.2016.04.059
13. Han, B., Zheng, X., Guan, M., Sun, L., Zhang, Y.: Personalized route recommendation with hybrid Tabu search algorithm based on crowdsensing. Int. J. Intell. Syst. 3054888 (2023). https://doi.org/10.1155/2023/3054888
14. Khodadadian, M., Divsalar, A., Verbeeck, C., Gunawan, A., Vansteenwegen, P.: Time dependent orienteering problem with time windows and service time dependent profits. Comput. Oper. Res. **143**, 105794 (2022). https://doi.org/10.1016/j.cor.2022.105794
15. Li, J., Fu, P.: A label correcting algorithm for dynamic tourist trip planning. J. Softw. **7**(12) (2012). https://doi.org/10.4304/jsw.7.12.2899-2905
16. Lin, S.W., Yu, V.F.: A simulated annealing heuristic for the multiconstraint team orienteering problem with multiple time windows. Appl. Soft Comput. **37**, 632–642 (2015). https://doi.org/10.1016/j.asoc.2015.08.058

17. Lin, S.W., Yu, V.F.: Solving the team orienteering problem with time windows and mandatory visits by multi-start simulated annealing. Comput. Ind. Eng. **114**, 195–205 (2017). https://doi.org/10.1016/j.cie.2017.10.020
18. Paulavičius, R., Stripinis, L., Sutavičiutė, S., Kočegarov, D., Filatovas, E.: A novel greedy genetic algorithm-based personalized travel recommendation system. Expert Syst. Appl. **230**, 120580 (2023). https://doi.org/10.1016/j.eswa.2023.120580
19. Pérez-Almaguer, Y., Yera, R., Alzahrani, A.A., Martínez, L.: Content-based group recommender systems: a general taxonomy and further improvements. Expert Syst. Appl. **184**, 115444 (2021). https://doi.org/10.1016/j.eswa.2021.115444
20. Ricci, F., Rokach, L., Shapira, B.: Recommender systems: techniques, applications, and challenges. In: Ricci, F., Rokach, L., Shapira, B. (eds.) Recommender Systems Handbook, pp. 1–35. Springer, New York (2022). https://doi.org/10.1007/978-1-0716-2197-4_1
21. Ruiz-Meza, J., Brito, J., Montoya-Torres, J.R.: A GRASP-VND algorithm to solve the multi-objective fuzzy and sustainable tourist trip design problem for groups. Appl. Soft Comput. **131**, 109716 (2022). https://doi.org/10.1016/j.asoc.2022.109716
22. Ruiz-Meza, J., Montoya-Torres, J.R.: A systematic literature review for the tourist trip design problem: extensions, solution techniques and future research lines. Oper. Res. Perspect. **9**, 100–228 (2022). https://doi.org/10.1016/j.orp.2022.100228
23. Steinley, D.: K-means clustering: a half-century synthesis. Br. J. Math. Stat. Psychol. **59**(1), 1–34 (2006). https://doi.org/10.1348/000711005X48266
24. Vansteenwegen, P., Gunawan, A.: Orienteering Problems: Models and Algorithms for Vehicle Routing Problems with Profits. Springer, Cham (2019). https://doi.org/10.1007/978-3-030-29746-6
25. Vansteenwegen, P., Gunawan, A.: State-of-the-art solution techniques for OPTW and TOPTW. In: Orienteering Problems. EATOR, pp. 67–81. Springer, Cham (2019). https://doi.org/10.1007/978-3-030-29746-6_6
26. Vansteenwegen, P., Souffriau, W., Oudheusden, D.V.: The orienteering problem: a survey. Eur. J. Oper. Res. **209**(1), 1–10 (2011). https://doi.org/10.1016/j.ejor.2010.03.045
27. Yuan, Y., Zheng, W.: Your trip, your way: an adaptive tourism recommendation system. Appl. Soft Comput. **154**, 111330 (2024). https://doi.org/10.1016/j.asoc.2024.111330

Using Data Analytics, Machine Learning, and KPIs for Urban Planning and Design

Veronika Tsertsvadze[1(✉)], Majsa Ammouriova[2], Trinidad Fernandez[3], and Leon Kapetas[4]

[1] Universitat Politècnica de València, Ferrandiz-Carbonell, 03801 Alcoy, Spain
vtserts@upv.es
[2] German Jordanian University, Amman 11180, Jordan
maysa.ammouri@gju.edu.jo
[3] Fraunhofer Institute for Industrial Engineering IAO, Nobelstrasse 12, 70569 Stuttgart, Germany
trinidad.fernandez@iao.fraunhofer.de
[4] Resilient Cities Network, Korte Hoogstraat 31, 3011 GK Rotterdam, The Netherlands
lkapetas@resilientcitiesnetwork.org

Abstract. Worldwide initiatives aim to facilitate urban transitions toward climate neutrality objectives through diverse urban planning and design methodologies. This paper focuses on approaches that integrate empirically-grounded tools to guide stakeholders and local authorities in achieving climate neutrality targets. Within this context, the paper explores the utilization of data analytics, machine learning, and key performance indicators for informed decision-making in urban planning and design. The paper explores some key performance indicators, in the context of sustainable development goals, recorded for European countries and analyses them utilizing machine learning techniques. Accordingly, the main indicators are identified.

Keywords: Key performance indicators · Urban planning and design · sustainable cities · clustering

1 Introduction

The environment is significantly affected by cities' economic activities, resource consumption, and reliance on non-renewable energy. Cities consume 70% of global energy, contribute to 75% of greenhouse gas emissions, and accommodate over 50% of the world's population [3]. Projections suggest that this figure will rise to 70% by 2050 [11]. Sustainability goals guide urban transition toward climate neutrality [14], that is represented by sustainability development goals (SDGs) 11 and 13 (https://sdgs.un.org/goals). These SDS are related to sustainable transport and sustainable cities as well as to actions taken concerning the climate change. Amidst increasing urbanization and global uncertainty, cities

A. A. Juan et al. (Eds.): DSA ISC 2024, LNCS 14778, pp. 107–117, 2025.
https://doi.org/10.1007/978-3-031-78238-1_10

struggle to achieve sustainable environmental configurations. City governments around the world, including advanced nations, face these challenges, leading to higher rates of energy consumption, environmental damages such as degradation and pollution, inefficient infrastructure and resource management and planning, inadequate decision-making, and socio-economic disparities [2]. This transition is facilitated by various initiatives and projects aiming in redesigning urban areas facilitating the smooth climate neutral transition, and many aspects of urban design planning become relevant to this transition in the form of area utilization, transportation planning, energy building efficiency, waste and water management, community engagement, and the integration of smart technology [19]. Within the Mission on Climate-neutral and smart cities from the European Commission, projects to support the implementation of urban planning and design solutions for just, sustainable, resilient and climate-neutral cities by 2030 are developed. One of them projects is the UP2030 project, funded by the Horizon Europe Framework Programme (HORIZON), based on an innovative 5UP approach aiming at accelerating implementation, upscaling solutions, and addressing climate change challenges. This methodology consists of five interlinked phases: UP-Dating, UP-Skilling, UP-Grading, UP-Scaling, and UP-Taking. The objective of these phases is to stimulate cities, improve capabilities, prototype transformations, broaden solutions, and facilitate knowledge transfer for the advancement of sustainable urban development. UP2030 supports ten pilot cities in their urban transition journey towards climate neutrality, resilience, and spatial justice. The project measures proposed for the pilot themes will encourage robust co-creation activities and facilitate collaboration among multiple stakeholders.

In this context, key performance indicators (KPIs) are measured to evaluate the transition progress and provide insight for decision makers concerning the performance of the implemented solutions encompassing physical interventions, digital tools, products, and innovative policies in the pilot areas. For example, KPIs are used to evaluate alternative urban design plans, in context with sustainability goals [22].

The measured indicators supply a valuable asset for the assessment and predicting the evolution and impact of the planned measures and interventions. Data analytics and machine learning (ML) techniques are used to provide supporting tools to serve this purpose [16]. For example, forecasting techniques predict the KPI values based on historically recorded ones and, hence, provide an insight to the future trends. The ML techniques have shown their impact on the achieving sustainability development goals, which exceeds the cost of developing ML models [20]. Additionally, these KPIs could be used to describe the current state (As-Is state) and compared to other similar case. Further investigation could support finding a relation between KPI values and the pilot solutions.

This paper demonstrates the utilization of ML techniques in clustering countries based on their KPIs related to urban planning and design. This paper explores some related work in Sect. 2. Section 3 describes the UP2030 project. In Sects. 4 and 5, KPI values for European countries are described and the managerial insights related to KPI analysi are derived. The paper is concluded in Sect. 6.

2 Related Work

The transition to climate neutrality at the urban level is a multifaceted challenge that demands innovative solutions in city planning and design. Previous research has highlighted that the assessment of strategies that meet climate targets initially focuses on 100% renewable energy and progressively adopts a broader approach that encompasses power, heat, transport and industry [5]. This holistic perspective outlined by [21] is key to a complete energy transition, which requires detailed analysis and integration of models from different sectors, as well as promoting interdisciplinary collaboration. In this context, the present study is part of a growing body of literature that explores the role of advanced technologies, such as data analytics and ML, in promoting key learnings around policy development, planning strategies and socio-technical solutions towards social innovation and carbon neutrality in cities.

The in-depth analysis conducted by [4] and [10] facilitates the identification of patterns in energy consumption and mobility respectively, which informs more efficient and sustainable public policies. This approach has been particularly valuable in the evolution towards urban sustainability, allowing urban planners to anticipate needs and optimize resources with unprecedented accuracy, [9]. Thus, there is a clear push towards the development of predictive and prescriptive models to proactively influence future urban policies and projects, beyond merely describing current trends. In the context of sustainability, a variety of KPIs are defined to track the development and the progress of sustainability development goals. Related data to these KPIs are collected and documented in databases, such as "Our World in Data" that stores data from the UN and international organizations, (https://ourworldindata.org/sdgs).

The traditional approach to evaluating mobility policies described by [7] is based on the systematic collection and analysis of data, including real-time demographic and traffic information. Policy implementation requires coordination and continuous monitoring, adjusting as necessary to optimize results. The incorporation of ML and KPIs accelerates this process, enabling more efficient policy recommendations and resource optimization. KPIs, when integrated with ML, significantly improve the accuracy and efficiency of decision making, aligning actions with strategic objectives and minimizing biases, according to the work of [15].

3 The UP2030 Project and KPIs

The UP2030 project aims to drive the urban transition to climate neutrality, pro-poses a radical shift from current planning practices to strategic challenge-based innovative planning and testing approach (https://up2030-he.eu/about- project/, last access date on 9 Feb. 2024). Through the innovative 5UP methodology, it seeks to update policies, improve skills, and prototype solutions, aiming for broad and participatory implementation. The 5UP concept of the UP2030 project outlines a comprehensive strategy to transform urban environments towards climate neutrality. It involves a sequence of steps: UP-Dating policies and codes, UP-Skilling and building capacities among stakeholders involved in urban design, UP-Grading practices through prototyping, UP-Scaling solutions for broader impact, and UP-Taking through sharing knowledge towards the Mission and fostering Communities of Practice. This methodology represents an innovative and holistic approach to urban planning, ensuring that cities can achieve their sustainability goals through coordinated, cross-sectoral efforts.

According to [8], the project aims to revolutionize urban planning and design, focusing on policy updates, with the aim of improving decision making guided by KPIs to assess the effectiveness of urban mobility policies and progress towards climate neutrality goals. This strategic approach ensures that decisions are based on detailed data and analysis, enhancing more effective and sustainable planning and management processes in cities.

In order to improve policy effectiveness, it is crucial to adopt evidence-based policymaking, which requires comprehensive evaluation using robust data and methods [17]. Urban policy evaluation faces challenges such as scarce resources, gaps in expertise, and difficulties in defining and managing performance indicators and data [13]. Recent advances in data availability, especially open data, offer new perspectives for evidence-based policy making, potentially enriching urban mobility policy evaluations [16]. Additionally, ML techniques facilitates tools for trend analysis and prediction, benefiting urban planning and design [22]. The objective of this paper is to explore how these advances in data can be leveraged, providing insights and recommendations for urban mobility policy makers to foster evidence-based approaches.

4 European Countries and Selected KPIs

In the context of this paper, we consider European countries for the demonstration of KPI analysis and deriving managerial insights. Table 1 tabulates these countries. The Eurostat dataset was examined to retrieve recorded KPIs values for the European countries in 2016 and 2021 (https://ec.europa.eu/eurostat/web/main/data/database last accessed on 25 Feb. 2024), and 57 KPIs were selected reflecting various aspects in urban planning and design. Table 2 lists some of these KPIs. These KPIs describes the transportation planning, waste management, area utilization, and social aspects that are related to SDGs 11 and 13. Decision makers could utilize these KPIs in the comparison between alternative urban design and track the development of urban areas [12,22].

5 Empirical Results and Managerial Insights

The retrieved values for identified KPI values in Sect. 4 form the base for the analysis in this section. At first, the tabulated KPIs are examined to ensure that their values remain independent of population size or country area. Thus, rates are preferred over summations. For example, the total energy consumption has been divided by the population size to get the rate of energy consumption

Table 1. European countries and their IDs.

Country	ID	Country	ID
Albania	AL	Liechtenstein	LI
Austria	AT	Lithuania	LT
Belgium	BE	Luxembourg	LU
Bulgaria	BG	Latvia	LV
Switzerland	CH	Montenegro	ME
Cyprus	CY	North Macedonia	MK
Czechia	CZ	Malta	MT
Germany	DE	Netherlands	NL
Denmark	DK	Norway	NO
Estonia	EE	Poland	PL
Greece	EL	Portugal	PT
Spain	ES	Romania	RO
Finland	FI	Bosnia and Herzegovina	BH
France	FR	Serbia	RS
Croatia	HR	Sweden	SE
Hungary	HU	Slovenia	SI
Ireland	IE	Slovakia	SK
Iceland	IS	Turkiye	TR
Italy	IT	United Kingdom	UK

Table 2. Selected KPIs.

Ccde	KPI	Unit
CC	Contiribution to climate related expending	Million euro
CC_E	Control of corruption: estimate	Number
EE	Unemployment expenditure	Euro/inhabitant
VA_E	Voice and accountability, freedom of expression: estimate	Number
CP	Corruption Perceptions Index	Number
GDP_PC	GDP/capita	USD
EH	Expenditure on Sickness/healthcare and disability	Euro/inhabitant
PCECB	Population with confidence in EU institutions	Percentage
GE_PR	Governance effectiveness: percentile rank	Percentage
RC	Recycling of batteries and accumulators, cadmium content	Tonne
EHS	Housing and social exclusion expenditure	Euro/inhabitant
RL_E	Rule of low: estimate	Number
zeroEV	Total number of electricity, hydrogen and fuel cells passenger cars	Number

per inhabitant. Then, missing KPI values have been examined, since not all KPIs values are tabulated for all countries for all KPIs. To address this issue, K-nearest neighbors imputer [18], a ML technique, is employed to estimate the missing values based on the tabulated KPI values. This imputation relies on the values of neighboring KPIs, and the algorithm is implemented in Python (https://www.python.org/), and scikit-learn package is utilized (https://scikit-learn.org/stable/). Accordingly, the missing values have been replaced by their estimation.

Another ML technique is employed to cluster the considered European countries. The clustering algorithm is K-mean clustering [1]. For this purpose, the optimal number of clusters have been investigated and identified as four clusters using the elbow method [23]. Figure 1 demonstrates the implementation of elbow method to determine the optimal number of clusters. According to Fig. 1, the number of clusters is set to four.

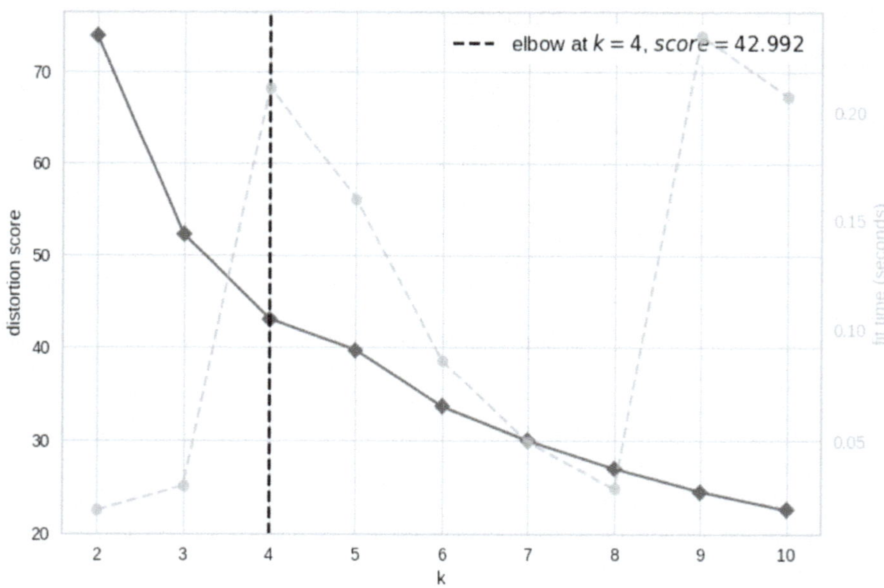

Fig. 1. Elbow method for the determination of the number of clusters.

The contribution of each of the 57 KPIs vary; thus, the importance of these KPIs in the clustering is evaluated, and the top ten important KPIs are shown in Fig. 2. These KPIs are listed in Table 2 as well. Two KPIs that stand out from the rest are the control of corruption in 2016 and the expenditures related to climate awareness in 2021; having the greatest impact on the clustering with respect to the other 57 KPIs. Both are ranked as the top ten impacting KPIs in clustering the countries in 2016 and 2021. It is noticed that most of the highly important KPIs in 2016 are important in 2021 clustering as well. During these five years, the world was influenced by the COVID-19 pandemic, starting in 2020. As a result, various actions were taken by policy makers to hinder the spread of the COVID-19 virus. These actions changed the way of life and priorities that are reflected on different measures. Additionally, these actions affected the evolution of KPIs, such as those related to gas emissions. The increased awareness in zero emission vehicles is reflected on the ranked KPIs; the increased impact of the number of zero emission vehicles. However, countries differ in their adaptation of this transportation solution.

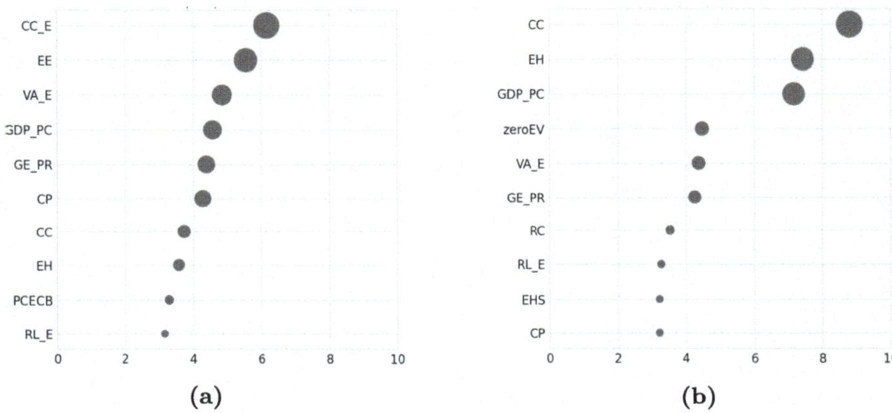

Fig. 2. The top ten important KPIs in clustering in (a) 2016 and (b) 2021.

Figures 3 and 4 display the distribution of clusters over the European countries in 2016 and 2021. It is noted that UK is not clustered in 2021 due data limitation at this year. According to the differences between the countries, they are

Fig. 3. The distribution of clusters.

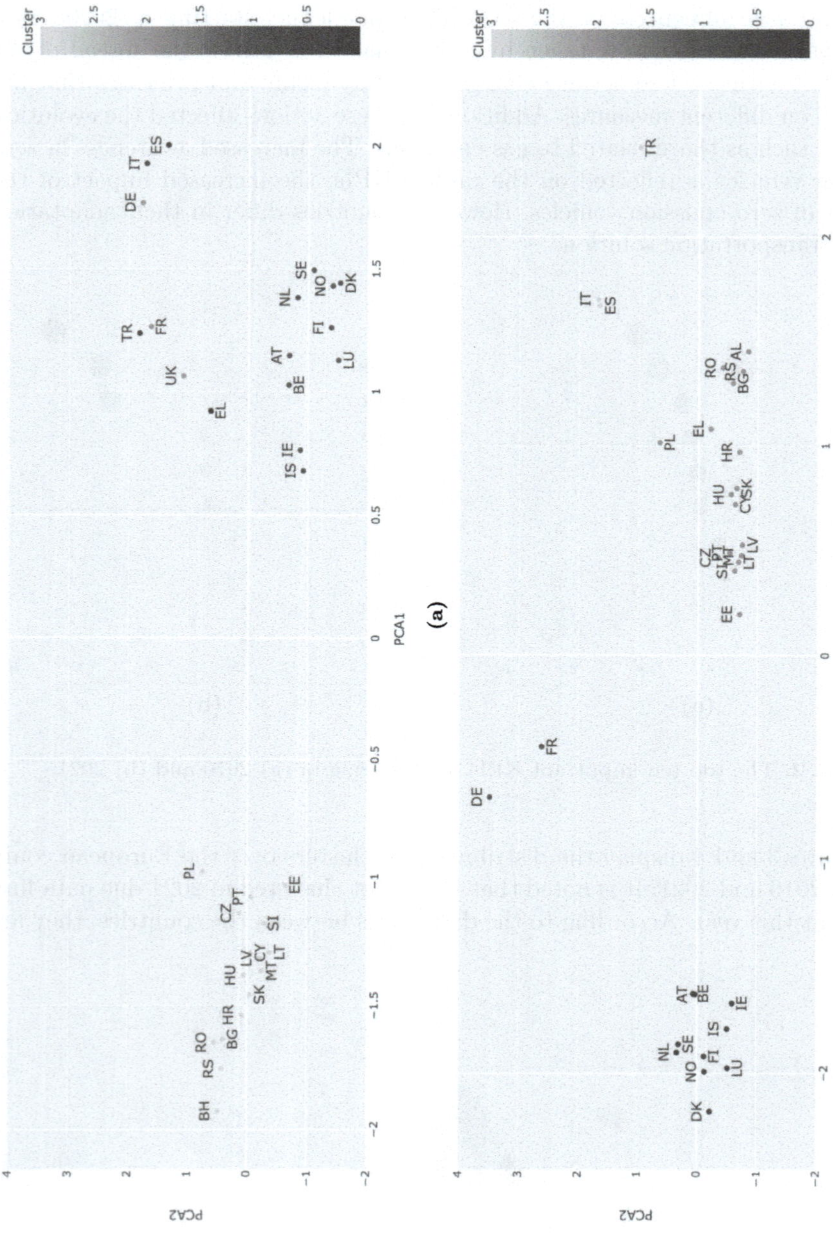

Fig. 4. Defined clusters with respect to two principal components in (a) 2016 and (b) 2021.

included into different sized clusters. For example, one of the clusters include ES, TR, and IT, while another cluster include more than five countries in 2016. Each cluster consists of countries that have similar KPI values. Countries remained in same clusters in 2016 and 2021, indicating similar actions and policies taken before and in 2021. Figure 4 demonstrates the distribution of countries with respect to two principal components. The principal components analysis is used to reduce the number of features, in this case KPIs, to represent the data [6]. Accordingly, the clusters are defined based on this distribution, and the values of these principal components are influenced by the most important KPIs. The similarity of clustering in both years indicates that changing the importance of KPIs has not affected the definition of clusters.

6 Conclusions

This paper focuses on KPIs and ML utilized for data analytics and evidence-based urban planning decision-making and design. In this context, many KPIs are defined for the monitoring of urban transition and implementation performance toward climate neutrality, resilience and spatial justice, where several initiatives, such as UP2030, are aiming to contribute. In urban planning and design, the importance of KPI definition as a steppingstone not only for the monitoring process, but also on how data can play a strategic role in shaping urban policy design. In this paper, a list of KPIs related to contributions towards climate neutrality and social innovation were collected for European countries.

The recorded KPIs values are analyzed using data analytics and ML techniques. Within the analysis, countries were clustered with respect to the recorded KPIs. Each cluster identifies countries with similarities. The defined clusters for years 2016 and 2021 show that similar countries undertook actions with similar impact on their KPIs, resulting on similar classifications in both years.

Additionally, the analysis showed how KPIs were affecting the clustering; the adaptation of zero emission vehicles, gross domestic product, expenditures related to healthcare and climate changes vary between countries and, accordingly, influence the identification of clusters. 70% of the top ten most impacting KPIs on clustering are shared between the years 2016 and 2021. These KPIs reflects different policies and concerns of European countries.

Future work details the analysis of the relation between KPIs and clusters. Additionally, the plan is to extend this work to include the analysis of KPIs related to smaller districts, such as cities. These districts for pilots for other cities or cities in the same or different country, as it is in the context of the UP2030 project. The analysis of KPI's evolution forms another extension to the utilization of ML in the KPI analysis and assessment of the impact of considered policies. These direction facilitates an input for decision making process and status evaluation of a country, city, or district.

Acknowledgment. This work has been funded by the European Commission project UP2030 (HORIZON-MISS-2021-CIT-02-01-101096405).

References

1. Baig, M.R., Govindan, G., Shrimali, V.R.: Data Science for Marketing Analytics: A Practical Guide to Forming a Killer Marketing Strategy Through Data Analysis with Python. Packt Publishing Ltd. (2021)
2. Bibri, S.E., Krogstie, J.: Environmentally data-driven smart sustainable cities: applied innovative solutions for energy efficiency, pollution reduction, and urban metabolism. Energy Inform. **3**, 1–59 (2020)
3. Bibri, S.E., Krogstie, J.: Smart eco-city strategies and solutions for sustainability: the cases of royal seaport, Stockholm, and western harbor, Malmö, Sweden. Urban Sci. **4**(1), 11 (2020)
4. Cascone, L., Sadiq, S., Ullah, S., Mirjalili, S., Siddiqui, H.U.R., Umer, M.: Predicting household electric power consumption using multi-step time series with convolutional LSTM. Big Data Res. **31**, 100360 (2023)
5. Christidis, P., Ulpiani, G., Stepniak, M., Vetters, N.: Research and innovation paving the way for climate neutrality in urban transport: analysis of 362 cities on their journey to zero emissions. Transp. Policy **148**, 107–123 (2024)
6. Hasan, B.M.S., Abdulazeez, A.M.: A review of principal component analysis algorithm for dimensionality reduction. J. Soft Comput. Data Min. **2**(1), 20–30 (2021)
7. Hristov, I., Chirico, A.: The role of sustainability key performance indicators (KPIS) in implementing sustainable strategies. Sustainability **11**(20), 5742 (2019)
8. Juan, A.A., Ammouriova, M., Tsertsvadze, V., Osorio, C., Fuster, N., Ahsini, Y.: Promoting energy efficiency and emissions reduction in urban areas with key performance indicators and data analytics. Energies **16**(20), 7195 (2023)
9. Liu, X., Dijk, M.: The role of data in sustainability assessment of urban mobility policies. Data Policy **4**, e2 (2022)
10. Majumdar, S., Subhani, M.M., Roullier, B., Anjum, A., Zhu, R.: Congestion prediction for smart sustainable cities using IoT and machine learning approaches. Sustain. Urban Areas **64**, 102500 (2021)
11. Nations, U.: World urbanization prospects. The 2014 revision, highlights **32** (2014)
12. Quijano, A., Hernández, J.L., Nouaille, P., Virtanen, M., Sánchez-Sarachu, B., Pardo-Bosch, F., Knieilng, J.: Towards sustainable and smart cities: replicable and KPI-driven evaluation framework. Buildings **12**(2), 233 (2022)
13. Ravensbergen, L., El-Geneidy, A.: Toward evidence-based urban planning: integrating quality assessments in literature reviews. J. Am. Plann. Assoc. **89**(3), 389–398 (2023)
14. Shtjefni, D., Ulpiani, G., Vetters, N., Koukoufikis, G., Bertoldi, P.: Governing climate neutrality transitions at the urban level: a European perspective. Cities **148**, 104883 (2024)
15. Shulajkovska, M., Smerkol, M., Dovgan, E., Gams, M.: A machine-learning approach to a mobility policy proposal. Heliyon **9**(10) (2023)
16. Soriano-Gonzalez, R., Perez-Bernabeu, E., Ahsini, Y., Carracedo, P., Camacho, A., Juan, A.A.: Analyzing key performance indicators for mobility logistics in smart and sustainable cities: a case study centered on Barcelona. Logistics **7**(4), 75 (2023)
17. Stéphane, J.: How can governments leverage policy evaluation to improve evidence informed policy making? (2020)
18. Triguero, I., García-Gil, D., Maillo, J., Luengo, J., García, S., Herrera, F.: Transforming big data into smart data: an insight on the use of the k-nearest neighbors algorithm to obtain quality data. Wiley Interdisc. Rev. Data Min. Knowl. Discov. **9**(2), e1289 (2019)

19. Uppal, D.: The green city dimension-striving towards a greener Europe-climate adaptation and energy transition driving change in (2021)
20. Vinuesa, R., et al.: The role of artificial intelligence in achieving the sustainable development goals. Nat. Commun. **11**(1), 1–10 (2020)
21. Wiese, F., Thema, J., Cordroch, L.: Strategies for climate neutrality. lessons from a meta-analysis of German energy scenarios. Renew. Sustain. Energy Transit. **2**, 100015 (2022)
22. Yang, L., van Dam, K.H., Zhang, L.: Developing goals and indicators for the design of sustainable and integrated transport infrastructure and urban spaces. Sustainability **12**(22), 9677 (2020)
23. Yuan, C., Yang, H.: Research on k-value selection method of k-means clustering algorithm. J **2**(2), 226–235 (2019)

Product Portfolio Rationalization via SKU Deletion: A Framework

Seán McGarraghy[1(✉)], Terry Harrison[2], Nevil Kirwan[1], and Peter Keenan[1]

[1] School of Business, University College Dublin, Belfield, Dublin 4, Ireland
{Sean.McGarraghy,Peter.Keenan}@ucd.ie

[2] Smeal College of Business, Pennsylvania State University, 459 Business Building, University Park, PA 16802, USA
tharrison@psu.edu

Abstract. We describe a product portfolio rationalization framework focused on the question of stock keeping unit (SKU) deletion. While innovation and adding a new SKU to a product portfolio is well researched, there does not seem to be a complementary literature on the obverse—and much more organizationally complex—process of product/stock unit discontinuation. Our work is motivated by our own observations, opinions expressed in conversations with practitioners, and the responses to an initial survey on SKU rationalization, which we summarize here.

We draw together results from a variety of research literatures and outline the scope of a research effort that would establish a process for SKU deletion. This effort includes: a definition of supply chain complexity which includes the effect of removal of products/stock units; describing supply chain cost as a function of (in part) this complexity; developing approaches to optimising performance measures (including cost) based on this complexity; formulating other research questions and hypotheses, including those arising from a survey of companies.

This paper makes progress towards a definition of supply chain complexity. It also considers a necessary condition for SKU deletion, namely, for a given SKU, estimating the lifetime and total sales over lifetime to aid a supply chain manager's decision on when to discontinue the SKU.

This use of data/evidence in decision making is at the heart of business analytics and so argues for the use of business analytics methods in the addressing of this problem.

Keywords: stock keeping unit · SKU · product · deletion · rationalization · supply chain · complexity · survey

1 Introduction

The management of a product portfolio is one of the most strategic decisions for a firm. It implicitly integrates a view of its competitive advantage while making tradeoffs between cost, quality, service and innovation. However, the bulk of the research in the area of supply chain (SC) product lifecycle management has

A. A. Juan et al. (Eds.): DSA ISC 2024, LNCS 14778, pp. 118–128, 2025.
https://doi.org/10.1007/978-3-031-78238-1_11

primarily sought to address the nature of product development and expanding the product portfolio, rather than how existing components, products or lines are discontinued. As a result, there is a considerable body of work focusing on innovation and the introduction of products, but very little on product/stock unit rationalization via product deletion. To address this missing piece of the product portfolio literature, we seek to establish a research theme around the topic of stock keeping unit (SKU) deletion. A preliminary necessary condition for rationalization of a given SKU is to be able to estimate the SKU life and total sales over this lifetime. While there are other complexities to the deletion decision, understanding the lifecycle of the product is a critical consideration.

Here, we outline the issues. In the next section, we review past work from a broad range of literatures which appears to be relevant to this theme. The succeeding Sect. 3 describes an initial survey carried out on SKU rationalization. This survey's aim was to help frame and develop research questions and hypotheses based on the responses to the survey. These questions center on:

- how to measure supply chain complexity;
- how to relate SKU complexity to supply chain cost;
- how to connect SKU portfolio decisions to supply chain performance;
- how to determine the best product/service portfolio for the organization.

In subsequent sections we develop a research framework/agenda motivated by responses to a survey of practitioners, other interactions with industry practitioners, and our own thinking on the problem. This research framework identifies the need for a comprehensive definition of supply chain complexity, which includes—for each product/component—the impact of deleting that product from the portfolio of offerings. This definition can then be used in a model of supply chain cost (or profit), or indeed other measures of performance, and methods of business analytics can be brought to bear on the problem.

The ultimate aim of this agenda is to develop mathematical models for determination of an optimal portfolio mix (according to a suitable performance measure) for a given organization in a given supply chain at a given point in time. We argue that this optimization exercise must incorporate supply chain complexity as a fundamental organizing principle, and will necessarily be data driven, using concepts and methods of business analytics. While elusive to define, it is clear that many supply chain practitioners and academics recognize that increased complexity, at least of certain kinds, leads to overall and possibly significant cost increases.

As an attempt at defining this agenda, we propose the research questions:

- what criteria do organizations currently use when deciding which products to discontinue?
- what criteria should organizations use for this decision?
- what processes do firms adopt in "rationalizing" their product portfolios?

From the survey results, we framed hypotheses:

- there is a broad range of approaches used, from detailed procedures, to *ad hoc* decision making;

- the detailed procedures are more used by companies which perceive the process as important;
- the *ad hoc* decision making is more used by companies which perceive the process as unimportant.

2 Prior Work

The span of relevant literature for this work is broad, including areas in supply chain, marketing, management, economics and business analytics; hence, in this paper, we perform a selective review, highlighting especially relevant papers.

De Groote [4] identifies two complementary problems: the marketing choice of the breadth of the product line, and the manufacturing choice of the flexibility of the production process. One measure of complexity of a product line is product variety, which may incorporate such terms as the number of products and the relative values of their demand rate, setup time, and unit cost of production [5]. Product variety has implications not only for production facilities but for the associated supply chain; research suggests that firms which match supply chain structure to the product variety they offer outperform firms which fail to match such choices [10]. An analytical approach to supply chain management requires systematic measurement and control of performance measures relating to suppliers, delivery performance, customer service, and inventory and logistics costs [6]. Flexible supply chains are needed to achieve high levels of performance: flexibility requires that management focus on improving information flows in the supply chain and leveraging new process technologies [14]. Upstream complexity, internal manufacturing complexity, and downstream complexity all have a negative impact on manufacturing plant performance; and supply chain characteristics that drive dynamic complexity have a greater impact on performance than those that drive only detail complexity [2]. Supply chains may be viewed as complex systems: [1] summarizes complexity properties in four interdependent categories: structural, dynamic, behavioral and decision making; and provides five mechanisms for dealing with supply chain complexity. Another approach to complexity considers disruption risk and impact on a large number of SKUs [11].

Product variety therefore poses an important business problem, and the problem differs for different products [13]. Efficient management of retail and wholesale inventories is also crucial [9]. The potential of business analytics in dealing with product variety is shown by the Franz Edelman Prize-winning paper [15] discussing the use of analytics in Hewlett Packard (HP), specifically a revenue coverage optimization tool used to manage product variety. Using such tools to manage product complexity allowed HP to increase its profits by more than $500M in 4 years, streamline product offerings, improve execution, achieve faster delivery, reduce overhead, and increase customer satisfaction and market share.

3 Survey

In this section we describe and discuss a survey that was conducted to provide insight into the processes that firms currently use to manage SKU portfolios. We are especially interested in how they determine when to remove a SKU from the active set of product offerings. We first describe our survey methods and then discuss the results and their implications.

3.1 Methodology

Survey Instrument. The instrument used to assess current practices and attitudes was a mail survey questionnaire, consisting of twelve freeform text questions and one question on a seven-point Likert measurement scale anchored by "1 = completely unimportant, 4 = important, 7 = critically important".

This draft questionnaire was pre-tested to check its content validity, clarity and terminology and modified based on the commentaries.

Sample. We mailed the survey to 55 industrial members of the Penn State Smeal College of Business Center for Supply Chain Research. Two follow-up mailings were executed to encourage participation. We received 17 completed surveys for a response rate of 30.9%. The majority of the responding firms were large, multinational manufacturing companies.

We asked each company to forward the survey to a person who was well acquainted with the internal processes that lead to SKU deletion. This step was taken in order to improve the quality and quantity of responses as well as to reduce the impact of potential inaccurate recall, hindsight bias and subconscious attempts to maintain self-esteem [8].

3.2 Results and Conclusions

Given the textual nature of the survey and the small number of responses, we cannot perform standard statistical analyses to test hypotheses regarding the SKU deletion practices of the responding firms.

However, we note that for the single question (Q2) that was answered on a seven point scale the mean was 4.9, indicating "very important". Figure 1 shows a boxplot of the Q2 responses: they range from 2 to 7 with a slight left skew.

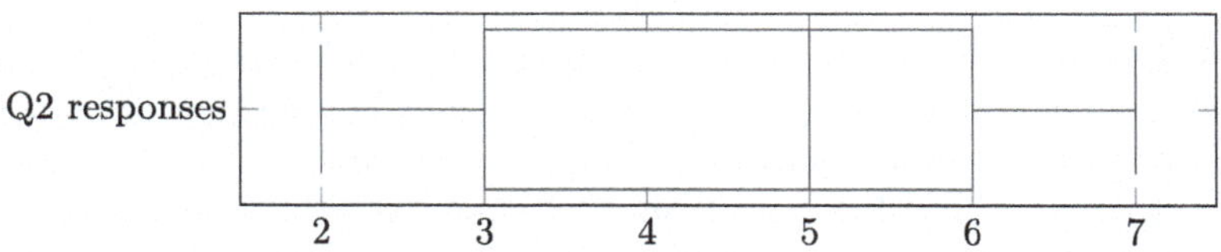

Fig. 1. Box and whisker plot of the responses to survey question 2.

The written responses, discussed below, indicate a strong sentiment among the group that SKU management in general, and SKU deletion in particular, are important issues.

With the small number of completed questionnaires, we have treated the written responses as "interviews". The level of detail varied considerably in the responses, ranging from answers of a few words to extended discussions which included multiple pages of submitted supplemental materials. We summarize the content of these responses with the following observations. Note that 5–7 indicate potential for the use of concepts and methods of business analytics.

1. *SKU deletion is an important, ongoing issue for most firms.* This is expressed, either explicitly or implicitly, repeatedly throughout the responses.
2. *The process for identifying SKUs for deletion is ill-defined for most firms.* While a few firms had objective methods for removing SKUs for deletion, most processes were *ad hoc.*
3. *The typical process has a firm-dependent cutoff for revenue or profitability.* The majority of firms that described a SKU deletion process used a criterion that identified candidates by gross annual dollar volume or profitability.
4. *Most firms use processes that bring together a variety of groups within the organization.* These groups vary from firm to firm but typically include manufacturing, supply, sales, marketing and logistics. The decision may rest with the collective group or a recommendation is made for SKU deletion to a senior executive.
5. *Most firms have a variety of measures to determine the impact of a SKU on profitability, cost, service delivery or manufacturing complexity; however, these vary widely from firm to firm.*
6. *A variety of data is used to inform the SKU deletion decision.* These items include: product volume, gross profit, incremental profit, fully allocated costs, impacts on storage and handling, "manufacturing complexity", "strategic value", market share, inventory impacts, and stage of the product life cycle. Overall, the data is diverse and often difficult to define and assess.
7. *Similar measures are used for SKU "addition" as for SKU deletion.*
8. *The overall process can be ill-defined and exacerbated by difficulty in the shared, and often unclear, ownership across various groups within the firm and driven by competing performance metrics.*

4 Research Framework

We outline a research framework and agenda for addressing the question of SKU deletion. The basis for posing and answering the question of managing a portfolio of products, with particular emphasis on SKU deletion, is maximizing overall profitability, broadly defined. To measure the impacts of deleting a SKU requires more precise, and possibly new, approaches to understanding supply chain dynamics. In particular, new and complex metrics are needed to determine this impact by measures that relate changes in the structure of the supply chain to impacts on cost, quality, speed and flexibility. A preliminary necessary step is to be able to estimate with reasonable accuracy, for a given SKU, the SKU lifetime and total sales over lifetime, to allow the supply chain manager to make an informed decision, in good time, on when to cease ordering from his/her supplier(s) and so discontinue the SKU. Initially, we focus on the following questions.

- *Definition(s) of Product/SKU/Supply Chain Complexity.* A key measure identified by many of the survey respondents was that of "Supply Chain Complexity". There appeared to be general agreement that the vague notion of increased complexity leads to increased cost. However, even defining a measure for supply chain complexity is elusive. It encompasses the spatial arrangement of facilities and processes; the temporal flows of goods, services, information and funds; and the mix of sourcing, manufacturing, service delivery and distribution activities. Additionally, there does not exist a well-defined, widely accepted measure for complexity.
- *Determine Supply Chain Cost as a function of Supply Chain Complexity.* Once a metric for supply chain complexity is defined, the next step is to quantify the way complexity affects supply chain cost. These relationships are multi-factored, highly interdependent and possibly nonlinear and discrete in their interaction.
- *Optimize Supply Chain Complexity (using a performance measure based on Supply Chain Complexity, e.g., partly based on Supply Chain cost).* The ultimate goal of this research agenda is to improve supply chain performance via the decision to remove selected SKUs from the product portfolio. However, this requires more than a mathematical formulation and optimization, since the removal of SKUs is a process that has effects across many parts of the firm. Ultimately, one must offer an organizational process that develops consensus and shared decision making among groups with conflicting incentives and objectives.

4.1 Towards a Definition of Supply Chain Complexity

The factors contributing to Supply Chain Complexity are many and varied, with complicated interactions. Thus a complete yet usable definition of Supply Chain Complexity may not be readily attainable, at least initially. It is clear, however, that the contributing factors include the following:

- complexity level of
  1. product/SKU
  2. number of echelons in the supply chain
  3. characteristics of processes within each echelon, such as manufacturing time, equipment requirements, etc.
  4. the level of interaction between echelons;
- variety/complexity/cost
  1. the number of products involved in a particular step
  2. the number of sub-steps involved with a particular item
  3. the kinds of equipment, people and process demands.

To be of use in decision making, any overall measure of complexity must ultimately be reducible to a dollar cost. Furthermore, a distinction needs to be drawn between complexity in the *design* of the supply chain and complexity in the *running* of the supply chain.

4.2 Product/SKU Complexity

A relatively easy measure to begin assessing complexity is simply the number of SKUs carried in the product portfolio or the number of product components involved in the manufacture or delivery of a SKU. It is commonly believed that great numbers of interactions lead to more complexity and, ultimately, greater manufacturing cost. However, this measure is an imperfect one, as we have received reports of firms that have decreased their SKU count and yet manufacturing costs did not correspondingly decrease. There are other complicating factors, some of which we discuss below.

4.3 Supply Chain Complexity

Complexity can arise in (at least) three contexts:

- internally within an organization;
- in interactions with other organizations (including customer(s));
- as a result of the external environment;

and can arise for (at least) three reasons:

- uncertainty;
- level of detail/complication of Product/Service delivered to end customer (*e.g.*, number of components/modules, intricacy of individual components/processes);
- organization.

It is clear that the contributors to complexity are many and varied: too many to consider in one paper. In order to characterize complexity, we must decide which of these (or other) contributors to include in our scope and which not to. For example, gaming effects are specifically excluded.

From the business viewpoint (see also [12]), we posit:

- SKU rationalisation is founded on financial performance for each SKU. Specifically, Revenue, Gross Margin or Contribution margin hurdles are used to determine candidates for deletion. However, often costs allocated to each SKU are not totally variable. Real and tangible savings from SKU reduction derive from reducing the costs associated with crewing, utilization and maintenance of manufacturing and distribution assets: typically production lines, warehouses or whole factories.
- Supply Chain Complexity is typically determined by the routings of products through a manufacturing asset network, and the inefficiency of this routing. Inefficiency can be driven by factors including setup and changeover times, bottlenecks, lead times, buffer stock, SKU-asset specific production rates and routing options.
- Thus, SKU complexity is not driven by the SKU alone, but also by the assets and the SKU routing through these assets. Deleting SKUs is only meaningful and impactful if assets or routings are deleted as part of the SKU deleting.

- As a result, SKU deletion needs to be asset and routing focused and based on the business case of deleting an asset and the associated SKUs and not deleting SKUs with no impact on retained assets.
- Taking this further, analysing financial performance by asset in consideration of the SKUs on the asset will provide direction to determine which assets contribute least to the financial performance of the business.

This viewpoint indicates a further role for: methods of business analytics in identifying potential assets to remove to make a significant impact on cost; and methods of operational research, such as routing modeling and optimization. For example, system dynamics and simulation approaches are proposed in [3].

Any useful complexity modeling effort must:

- prioritize which influences to include, and in what order when successively enriching the model;
- measure and quantify effects, perhaps using such analytics techniques as regression;
- restrict included influences to those for which we have enough data to successfully capture;
- create functional forms for the influences on complexity, or—failing that—at least identify which influences
  1. increase complexity,
  2. decrease complexity,
  3. do not affect complexity.

With this in mind, we tentatively provide Table 1, listing influences on, or causes of, complexity, which are included in the proposed research agenda on SKU deletion. The influences are put in descending order of importance, with the constraint that there must exist sufficient data to capture them to a reasonable level of accuracy. For each influence, we indicate whether a functional form exists for it, what is its effect on complexity (increase/decrease/neutral) and which aspect of the supply chain does it affect: the design or the running (managing).

4.4 SKU Deletion and Cost

While the general belief is that decreasing the number of SKUs in a product portfolio leads to lessened supply chain complexity and therefore, decreased overall costs, this may not always be the case. We have anecdotal evidence from conversations with a large multi-national manufacturer that at the conclusion of a large SKU deletion effort, resulting in thousands of deleted SKUs, costs were reduced only slightly.

Possible causes of this effect may include:

- a reduction in complexity of one stage (*e.g.*, Make) may lead to an increase in complexity of other stages (*e.g.*, Source or Delivery);
- although some products or lines which used a component may no longer be produced, others that used it may still be produced; in this case, the component used to replace a deleted SKU may be more expensive to make or source.

Table 1. Influences on supply chain complexity which are proposed for inclusion in this modeling effort.

Influence	*Functional Form* (Y/N)	*Effect* (+/−/0)	*Design/Run*
Stochastic demand	Y	+	Run
Stochastic lead times	Y	+	Run
Uncertainty in resources/skills	N	+	Both
Lack of knowledge about trends	N	+	Run
Difficulty forecasting	N	+	Both
Bullwhip effect	Y	+	Run
Complexity of SKU design	Partly	+	Both
Complexity of production process	Partly	+	Both
Managing returns	N	+	Both
Shorter lifecycles	Y	+	Both (?)
Outsourcing/offshoring	Y (?)	+	Both
New technology	N	+	Both
Differing demands	N	?	Run
Changing resource requirements	N	+	Both
Business process issues	N	?	Both
Organizational structure issues	N	?	Both
Different strategies	N	?	Both
Uncoordinated actions	N	+	Run
Information issues	N	+	Both
Incompatible IT systems	N	+	Run
Legal changes	N	?	Both
Environmental changes	N	?	Both

5 Estimating SKU Total Lifetime Demand

As an example of the potential use of business analytics approaches to address this problem of product rationalization, we summarize here (a subset of) our results from work on estimating SKU total lifetime demand. Knowledge or a good estimate of total lifetime demand will be a major factor in the decision of when is the right time to rationalize a SKU.

Many SKUs have finite life times, for example, in consumer electronics. We consider two datasets: one of sales of personal computers and one of sales of phone accessories. The decision on whether to discontinue a SKU becomes relevant after the peak of sales (as determined by a heuristic that looks at historical weekly sales) has been passed.

At a given post-peak week in the lifetime of a SKU, three methods are used to estimate the total lifetime sales and the total life in weeks (or other relevant time period) of the SKU:

- A linear fit to the data from the peak onwards; this assumes a triangular distribution of data, which has been found to be reasonable for many SKUs, particularly those we investigate [7];
- Using a Holt model that is, exponential smoothing allowing for a trend;
- Regarding the sale of an item as analogous to a death in survival analysis, we fit a Weibull distribution to the data.

When multiple methods of these three make reasonable predictions we build an ensemble prediction from them. These approaches, particularly the Weibull distribution approach, show promise for estimating total lifetime sales and total life. Details will be provided in a forthcoming paper.

6 Conclusions

We have argued that managing a product portfolio is one of the most important efforts for a business; yet, although there is a much work on innovation and the introduction of products, there is little on product/stock unit rationalization via product deletion. To address this, we have proposed a research theme around SKU deletion.

We have described an initial survey on SKU rationalization, which aimed to help frame and develop research questions and hypotheses centering on: measuring supply chain complexity; relating SKU complexity to supply chain cost; connecting SKU portfolio decisions to supply chain performance; determining the best product/service portfolio for the organization.

We have developed a research framework motivated by responses to this survey, as well as other conversations with practitioners, and our own thinking. This framework identifies the need for a comprehensive definition of supply chain complexity, which includes, for each SKU, the impact of deleting that SKU from the portfolio, and whether this can mean the rationalization of assets such as assembly lines or factories. This definition can then be used to model supply chain cost measures of performance, and methods of business analytics can be brought to bear on the problem.

A necessary condition for rationalization of a given SKU is the ability to estimate the SKU life and total sales over lifetime. We have summarized work in progress on this topic.

We foresee the need for use of various analytics approaches, including simulation, modeling of product lifecycle curve, clustering of SKUs by similarity of lifetime, product lifecycle curve and other measures, and the resulting classification of a new SKU meaning that data on similar past SKUs may help in guiding predictions on the new SKU.

Future work will involve elaborating on these definitions and the inputs to a general model or models incorporating these ideas. This will require the input of many researchers.

We acknowledge that there may be business decisions *e.g.*, marketing mix where the contribution of business analytics may be limited: the business needs to know what it wants in order to properly formulate objectives.

References

1. Abbasi, M., Varga, L.: Steering supply chains from a complex systems perspective. Eur. J. Manag. Stud. **27**(1), 5–38 (2022). https://doi.org/10.1108/EJMS-04-2021-0030
2. Bozarth, C.C., Warsing, D.P., Flynn, B.B., Flynn, E.J.: The impact of supply chain complexity on manufacturing plant performance. J. Oper. Manag. **27**, 78–93 (2009)
3. Golrizgashti, S., Hosseini, S., Zhu, Q., Sarkis, J.: Evaluating supply chain dynamics in the presence of product deletion. Int. J. Prod. Econ. **255**, 108722 (2023). https://doi.org/10.1016/j.ijpe.2022.108722
4. de Groote, X.: Flexibility and marketing/manufacturing coordination. Int. J. Prod. Econ. **36**, 153–167 (1994)
5. de Groote, X.: Flexibility and product variety in lot-sizing models. Eur. J. Oper. Res. **75**, 264–274 (1994)
6. Gunasekaran, A., Patel, C., Tirtiroglu, E.: Performance measures and metrics in a supply chain environment. Int. J. Oper. Prod. Manag. **21**(1/2), 71–82 (2001)
7. Hu, K., Acimovic, J., Erize, F., Thomas, D.J., Mieghem, J.A.V.: Forecasting new product life cycle curves: practical approach and empirical analysis. Manuf. Serv. Oper. Manag. **21**(1), 66 (2019)
8. Kumar, N., Stern, L., Anderson, J.: Conducting interorganizational research using key informants. Acad. Manag. J. **36**(6), 1633–1651 (1993)
9. Marzolf, M.J., Miller, J.W., Peinkofer, S.T.: Retail & wholesale inventories: a literature review and path forward. J. Bus. Logist. **45**(1), e12367 (2024). https://doi.org/10.1111/jbl.12367
10. Randall, T., Ulrich, K.: Product variety, supply chain structure, and firm performance: analysis of the U.S. bicycle industry. Manag. Sci. **47**(12), 1588–1604 (2001)
11. Simchi-Levi, D., et al.: Identifying risks and mitigating disruptions in the automotive supply chain. Interfaces **45**(5), 375–390 (2015). https://doi.org/10.1287/inte.2015.0804
12. Staskiewicz, A.M., Hvam, L., Haug, A., Mortensen, N.H.: The role of SKU management in product variety reduction projects, pp. 37–60. Springer, Cham (2023). https://doi.org/10.1007/978-3-031-09782-9_2
13. Sweeney, K., Evers, P.T., Duan, Y., Windle, R.: Trade-offs between operational performance and sales in a retail environment: the simultaneous impact of product variety on inventory levels, product availability, and sales. J. Bus. Logist. **44**, 463–479 (2023). https://doi.org/10.1111/jbl.12334
14. Vachon, S., Klassen, R.D.: An exploratory investigation of the effects of supply chain complexity on delivery performance. IEEE Trans. Eng. Manag. **49**(3), 218–230 (2002)
15. Ward, J., et al.: HP transforms product portfolio management with operations research. Interfaces **40**(1), 17–32 (2010)

Decision Science for Personalized Medicine: Data-Driven and Model-Driven Methods

Pier Giorgio Villani(✉)

Department of Emergency and Critical Care Medicine, Hospital of Cremona, Cremona, Italy
piergiorgio.villani@asst-cremona.it

Abstract. The history of medicine is a path of progress and discoveries, characterized by a constant search for knowledge for human care and well -being. Nowadays medical practice is based on three principles: knowledge (model), experience and statistics (evidence-based medicine, EBM). Mathematics is increasingly entering medicine with two distinct approaches: namely, data-driven and model-driven. However, this must not limit professional freedom or distort the traditional pathophysiological modelling approach that is the only one accepted by doctors, who must remain the protagonists of the therapeutic decisions. This paper wants to show how model-driven approaches offer more guarantees, compared to data-driven ones, to maintain a pathophysiological model (accepted by clinicians) and therefore bring an important contribution to the knowledge for human care. On the other side, only data-driven approaches can discover variables involved in a complex system. So, the integration of these two approaches is the future. In addition, only the collaboration between different skills and the construction of a common language can bring innovative solutions for the management of health care and patient treatment.

Keywords: decision science · medicine · healthcare

1 Introduction

Nowadays, medical knowledge rests on three pillars: experience, pathophysiology and statistics (Evidence Based Medicine) [1]. Unlike what happened in the past, the current objective of medicine is to know in order to cure. To produce the necessary knowledge, medicine tries to use the scientific method (i.e. an iterative and never-ending refinement of conceptual models through experience), and if it cannot do so as is done in physics and hard sciences, it is not by a choice of physicians, but for a fundamental reason: the human body system is extremely complex and non-deterministic. Furthermore, even if it were deterministic, each patient is different from any other. For these reasons, it is not possible to exactly know the consequences of a single medical act on a patient in terms of risks and

A. A. Juan et al. (Eds.): DSA ISC 2024, LNCS 14778, pp. 129–137, 2025.
https://doi.org/10.1007/978-3-031-78238-1_12

benefits, except in strictly statistical terms, using few variables and standardizing the system under study.

With the increase in computing capabilities and the development of increasingly efficient algorithms, in this era we are witnessing a new revolution; mathematics enters medicine, providing the ability to predict the effects of a treatment more and more accurately. In this short paper we want to analyze the different mathematical models used in the clinical field of the data science to conclude that data-driven algorithms do not allow doctors to understand how the solution is generated and therefore they cannot be trusted. On the other side, model-driven algorithms, understood and accepted by doctors, do not allow to explore all the variables involved in a complex system. I think that the integration of these approaches is the solution. The search for correlations with artificial intelligence (AI) methods and the construction of a model starting from those correlations may be the future of mathematics in the medical field.

2 Historical Development of Medicine

The therapeutic act was originally carried out autonomously. The sick person applied medicines to himself or carried out certain practices that were the result of personal experience or orally handed down.

The first great revolution in medicine occurred with the birth and development of the first great civilizations. As the complexity of the community system increased, there was a need to define roles and, consequently, the figures necessary for keeping them stable. In this context, the figure of the Healer was born, the one who presides over the well-being of the community, that is, who dispenses and passes down the practices necessary for this purpose. Each Healer applied medicines on the basis of his own experience after acquiring them from his predecessors [2].

The second great revolution occurred in ancient Greece, where Hippocrates and Galen played a critical role in the development of modern medicine. Galen, with his studies on human anatomy and physiology, laid the foundations on which to build scientific thought. Hippocrates, considered the father of modern medicine, emphasized rational diagnosis and rational treatment of diseases. By these two figures, medicine was organized around a decidedly rational, rigorous and empirical methodology [3]. From the analysis of the symptoms, the doctor had to trace the internal causes of the pathology, building an overall and coherent theoretical framework, from which the choice of therapy then derived. In other words, the physician had to create a pathophysiological model. In the centuries that followed, the idea of a pathophysiological model established itself in medical thought, enriching and refining itself more and more with the knowledge acquired. Today one of the pillars of medicine is precisely this idea of a pathophysiological model.

The third great revolution occurred in the nineteenth century with statistics. Today, statistics plays a critical role in all phases of research and clinical practice, helping to guide informed medical decisions, to improve the quality of care and to advance scientific knowledge [4].

3 Decision Science in Medicine

The applications of decision science in medicine can be classified into two broad categories: optimization of healthcare systems and support to clinical practice. In the former case the objective is to find the best use of available resources to optimize some specified outcomes, by improving the processes that govern a complex system, such as the healthcare system. In the latter case the objective is to provide decision support to physicians in all phases of a clinical pathway: diagnosis, prognosis and therapy or treatment.

3.1 Decision Science and Optimization of Healthcare Systems

Operations research plays a fundamental role in supporting management decisions in any field, including healthcare. It allows optimizing the use of available resources and providing efficient, accessible and sustainable processes to the advantage of patients, sanitary operators and administrators. Its use is easier when the decision problems are well-defined and data are certain. Over the years, hundreds of studies (theses, articles, books) have been produced that address different aspects of optimization in healthcare, by proposing various algorithms and a rich set of optimization and decision support methods. Applications of operations research can be found in the management of personnel as well as of resources [6].

Nevertheless, operations research is not always easily accepted by healthcare professionals, because they can see optimization as a limitation to their profession and a kind of undesired control over their daily activities. The use of optimization models and methods to establish performance indicators or to search for weak points and inefficiencies in healthcare systems contribute to increase this negative perception and attitude [12]. Perhaps, this is the main reason why, at the moment, in spite of the vast related literature, operations research is still poorly represented in many healthcare systems and hospitals.

3.2 Decision Science and Support in Clinical Practice

In modern medicine the use of decision support systems is mandatory. For instance, it is estimated that there exist about 10,000 known human diseases, but doctors are able to remember only a small fraction of them at any given time. A Johns Hopkins Medicine study estimated that up to 40,500 critically ill U.S. hospital patients die each year from an unknown medical condition that may have caused or contributed to their deaths [8].

Ever larger archives are emerging that contain an enormous amount of data, not just medical ones. Repositories can be subject-specific, accepting specific types of structured data, or generalist, accepting multiple types of data. Users should select the repository suited to their field of study. The PLOS Medicine journal has created a list of databases divided by topic, guaranteeing a high standard in data acquisition and processing [15]. In the biomedical field, the Physionet repository [16] contains the most complete and used database of biomedical signals from patients hospitalized in ICU, the MIMIC IV. The Medical Information Mart for Intensive Care (MIMIC)-IV database provided critical care data for over 40,000 patients admitted to intensive care units at the Beth Israel Deaconess Medical Center (BIDMC) [17]. Furthermore, there are numerous repositories in many research or university institutes or in research groups: for example, The Cancer Imaging Archive (TCIA) which contains several radiological images, CDC Wonder, from the US Centers for Disease Control and Prevention and many others.

Today, this approach to treatment, which guides modern medicine, clashes with the evidence that the number of variables that characterize any patient has increased significantly (personalized medicine). The complexity of such a model cannot be explored with traditional methods: managing such a large number of parameters is far beyond the capabilities of any clinical doctor. Therefore we need a more personalized model that considers each patient as a whole, thus integrating all variables and parameters that characterize him (genetics, lifestyle, environment, etc.).

From the above it is clear that decision science can play a critical role in clinical practice, by helping physicians to improve both their diagnosis and their construction of a personalized pathophysiological model of each patient.

4 From Data to Models

To evaluate the usefulness of decision science methods in helping physicians in their daily clinical practice, it is necessary to analyze the way a physician "thinks" and what skills he uses in evaluating a patient to define a diagnosis and therefore a therapy.

As recalled in Sect. 2, after Hippocrates, medicine is based on a form of knowledge that seeks to relate causes and effects, as it happens with the study of natural phenomena, rather than simply trying to produce the desired effects (healing of the patient) without knowing the reason ("magical/religious" approach).

In this context, the physician, starting from signs and symptoms (effects of the pathology), searches for a cause (pathophysiological model) that explains them all (diagnosis). Successively, he defines which tests to carry out to better specify the model. Finally, a prognosis and a therapy are defined on a statistical basis: the chosen therapy is the one from which the majority of patients with the same pathology enrolled in multicomputer studies have obtained benefit.

This is *EBM* (Evidence-Based Medicine) which is based on three pillars: doctor experience, pathophysiological models and statistical analyses. This approach to treatment, developed over centuries, is, currently, the best way to treat each patient [1].

The interaction between independent and dependent variables is very complex, in general. The physician, despite knowing the qualitative model very well, cannot manage its complexity; hence, he typically corrects each independent variable separately by evaluating the overall outcome a posteriori.

Over time, the comparison between a pathophysiological model and the real data measured and observed from a patient allows the physician to iteratively refine and adjust the model itself. In this way, the propagation of errors due to the inaccuracy of the initial model or the lack of precision of the measurements carried out is amortized and the continued use of incorrect models is also avoided. In this reasoned "trial and error" procedure [7] the necessary experiments are made on the patient himself. A mathematical model could provide useful information a priori, suggesting appropriate actions and corrections, i.e. planning the path before undertaking it.

4.1 Data-Driven Methods

At a first glance, deep learning and machine learning algorithms may look like ideal tools for leveraging digital data for supporting decisions. Large repositories, now public and easy to access, allow for the application of these methods.

- Supervised learning (SL) is a paradigm in machine learning where input objects (for example, a vector of predictor variables) and a desired output value (also known as human-labelled supervisory signal) train a model. For instance, this method has recently been used to recognize and diagnose COVID pneumonia by analysing chest radiology images. Although perfectly working on the training set, when tested on samples extracted from the reference database, the results turned out to be unreliable for clinical purposes. A detailed analysis of the neural network revealed a spurious correlation with the framework of the image instead of the chest image [13].
- Unsupervised learning is a method in machine learning where, in contrast to supervised learning, algorithms learn patterns exclusively from unlabelled data. This method is used for understand the clinical characteristics of ICU patients. An enormous amount of data is collected in ICUs. Dividing patients into homogeneous groups allows us to define common characteristics for a possible different therapeutic approach or to find correlations between data [19]
- Reinforcement learning (RL) is an interdisciplinary area of machine learning and optimal control concerned with how an intelligent agent ought to take actions in a dynamic environment in order to maximize the cumulative reward. This method was used to determine the optimal treatment in terms of intravenous fluids or vasopressors administration for ICU patients with sepsis.

However, it should be remarked that, beyond failing to be "intelligence" in any meaningful sense of the word, artificial intelligence, and specifically machine learning, consists of pattern recognition on a statistical basis.

Unfortunately, statistical pattern recognition may fail in a profoundly non-human way in situations that a human doctor can immediately recognize.

Current artificial intelligence models in intensive care medicine have been largely generated using retrospective data, without external validation or perspective evaluation. Reproducibility of solutions is not guaranteed and there are no clear protocols to examine it thoroughly. Adhesion to signalling standards and the risks of distortion is not optimal: 6 out of 81 non-randomized studies had been tested in a real clinical environment and 72% of them showed high distortion risks [20].

Beyond the limits of machine learning illustrated by this example, all methods based on statistics suffer from the well-known "devil-in-the-variance" effect. Although it is useful to have reliable information about average correlations between causes and effects, statistics does not provide any hint about how to locate a specific patient in the distribution of all possible cases. Few or no patients are "average patients"; there is a conceptual gap between statistics necessarily based on large numbers and the need to provide the best possible treatment to each specific patient.

Hence, summing up, the main advantage of statistical methods or data-based methods consists in discovering hidden correlations even when they are hard to conjecture on the basis of physiology. On the other side, the main drawback is considering patients by categories with little or no control over the parameters involved.

A medical doctor does not need "solutions" that he can not trust. If he does not understand how the results are obtained, he will not use the method. A medical doctor needs to have maximum possible control over the course of each patient, he needs to have a clear understanding of the reason why a certain process occurs and therefore on how to stop it, if necessary. Furthermore, the responsibility for therapeutic decisions lies with the doctor and any error falls completely on his shoulders; hence, final decisions on whether to apply a therapy or not must remain up to him and cannot be assigned to any algorithm. As already said, the objective of medicine is to know in order to cure and not to cure without knowing; from this point of view, a blind confidence in data-driven statistical methods, looks like a regression from science to ancient magic and shamanism.

We will not truly understand the ethical issues until AI becomes mainstream. However, there are already some ethical dilemmas for the use of AI in intensive care units (ICU). The first one is certainly the leak of sensitive data. The collection and pre-processing phase, especially if carried out using external platforms, represents a real threat to patient privacy, today. De-identification and parallel-distributive computing could provide partial solutions to the problem. A second ethical issue is the safety of the AI model at the patient's bedside. At present, the safety and liability of the AI model cannot be blindly guaranteed. The decision-making process therefore falls on the doctors, who continue to be an integral part of the treatment process. This integration cannot be ignored. Of fundamental importance is the patient's autonomy and his decisions in terms of limits of treatment. The AI decisions may not be in line with what the patient expresses.

4.2 Model-Driven Methods

In recent years, the frontiers of biological and medical research have become increasingly dependent on sophisticated modelling, analysis and computational techniques. Operations research and computer science have emerged as vital tools for studying complex biological systems. The interaction between operations research and biomedicine has been fruitful and synergistic; in turn, biomedical research has generated new mathematical developments in operations research and its interfaces with computer science [14]. Various models and approaches have been developed in this field.

For instance, target controlled infusion (TCI) is a clinically-available and widely-used computer-controlled method of drug administration, adjusting the drug titration towards user selected plasma- or effect-site concentrations, calculated according to pharmacokinetic-pharmacodynamic (PK-PD) models. This model uses the characteristics of a single patient (age, sex, etc.) to simulate a three compartments model to predict and control the concentration of anesthetic drugs in the blood. The model is currently used in many operating rooms [21].

The forecast of blood glucose (BG) levels with data collected is often useful for glycemic management. The BG dynamics are non-linear, complex and non-stationary, which can be represented by non-linear models. These forecasts have potential use for glycemic management as part of control systems [22].

A common feature is the creation of mathematical models capable of simulating a biological system, allowing the study of its behaviour as initial conditions vary. These applications can be very useful for gaining a deep understanding of biological systems and exploring the potential effects of changes on the systems themselves, without actually changing the system. A goal for scientific research is to make possible testing the effects of a treatment on a patient, without actually carrying it out, evaluating its risks and benefits and recalibrating therapies. The mathematical model, and therefore the simulation of the physiological processes involved, could be the solution to these problems. By modelling a process, function or action of a drug, a basis is obtained on which the doctor can experiment with different solutions. The "what if" question, that every doctor asks himself when starting a treatment, could be answered by simply introducing into the model the perturbation induced by the therapy, whatever it may be. In this sense, operations research algorithms offer a valid tool for building these models.

As already mentioned, when a pathophysiological model is defined, not all the relevant parameters are always known beforehand, but it is necessary to include as many of them as possible to make the model adherent enough to the actual patient. Data science helps by providing techniques and tools to search large databases for all the variables related to a phenomenon that one wants to simulate. In my vision, in the near future health decision science will use large databases to provide increasingly precise variables and data to mathematical models that will simulate biological phenomena.

Besides simulation, operations research also provides sophisticated mathematical optimization algorithms to answer "what's best" questions. From this point of view, a relevant change of paradigm characterizes the application of operations research to health care problems, compared to traditional applications to industrial production, logistics and other fields that have been typical for operations research studies all along the past century: this change of paradigm can be summarized in the word "robustness". While typical objective functions were concerned with minimization of cost and time of operations and maximization of throughput, productivity and revenues, in healthcare a "good" decision is characterized by its robustness much more than its efficiency. Since the system under study (the human body of the patient) is extremely, complex and not completely known in all details, uncertainty is unavoidable; at the same time, medical decisions have a profound impact on life and well-being of people. Therefore, in healthcare a good decision is first of all a robust one. By robust, we mean that small variations, errors or approximations in input data must produce small variations in the outcome more in terms of feasibility (success of therapy or treatment) than optimality. This justifies the development of specialized areas in mathematical optimization such as stochastic programming and robust optimization, that are conceived for this precise purpose, i.e. to deal with uncertainty in input data.

5 Conclusions

The traditional approach to decision-making in medicine, based on statistics, is no longer sufficient: we need more personalized methods, based on models that are built from the specific patient's data. This is why decision science not only represents a new revolution in medicine, but also a necessity to meet future challenges. Medical doctors are now faced with a choice: either to lead this revolution, while maintaining a leading role, or to be overwhelmed by it. The key for a correct management of this transition is the creation of interdisciplinary teams and institutions, populated by medical doctors as well as by mathematicians and computer scientists, who should work in close collaboration to define a common language leading to innovative, effective and robust solutions in the treatment of patients. In addition to creating and navigating large databases, a future challenge is the creation of increasingly accurate mathematical models that will give doctors the possibility to answer "what if" and "what's best" questions.

Acknowledgments. Many thanks to Prof. Giovanni Righini for fruitful collaboration and a valuable exchange of ideas.

Disclosure of Interests. No conflict of interests.

References

1. Sackett, D.L., Rosenberg, W.M., Gray, J.A., Haynes, R.B., Richardson, W.S.: Evidence based medicine: what it is and what it isn't. BMJ **312**(7023), 71–72 (1996)
2. Cosmacini, G.: L'arte lunga, storia della medicina dall'antichità a oggi, Laterza (2006)

3. Ippocrate. Opere, traduzione e commento di Giuliana Lanata, Torino, Boringhieri (1961)
4. Laurent, M.-P.-H.: Traité du calcul des probabilités (1873)
5. Tenny, S., Varacallo, M.: Evidence Based Medicine. StatPearls Publishing (2022)
6. https://orahs.di.unito.it/
7. https://hdl.handle.net/11579/142058
8. Winters, B.: Johns Hopkins patient safety team finds diagnostic "fatal flaws" in the ICU may account for as many annual deaths as breast cancer. Johns Hopkins University School of Medicine (2012)
9. Goldberger, A.L., et al.: Chaos and fractals in human physiology. Sci. Am. **262**(2), 42–49 (1990)
10. Baish, J.W., Jain, R.K.: Cancer, angiogenesis and fractals. Nat. Med. **4**(9), 984 (1998)
11. Demling, L.: Chaos research, fractals, fuzzy logic. From stereotypes to reality-consequences for medicine. Fortschr. Med. **110**(6), 77–79 (1992)
12. Mustafa, A.M., Day, S., Higginson, J., Sharp, I.: Determinants of lost theatre capacity. Br. J. Oral Maxillofac. Surg. **58**(9), 1139–1144 (2020)
13. DeGrave, A.J., Janizek, J.D., Lee, S.: AI for radiographic COVID-19 detection selects shortcuts over signal. Nat. Mach. Intell. **3**, 610–661 (2021)
14. Guo, J.: Application of Operations Research to Healthcare. IntechOpen (2023). https://doi.org/10.5772/intechopen.109919
15. https://journals.plos.org/plosmedicine/
16. https://physionet.org
17. Johnson, A., et al.: "MIMIC-IV" (version 2.2). PhysioNet (2023). https://doi.org/10.13026/6mm1-ek67
18. Fleuren, L.M., Thoral, P., Shillan, D., Ercole, A., Elbers, P.W.G.: Machine learning in intensive care medicine: ready for take-off? Intensive Care Med. **46**, 1486–8 (2020)
19. Hyun, S., Kaewprag, P., Cooper, C., Hixon, B., Moffatt-Bruce, S.: Exploration of critical care data by using unsupervised machine learning. Comput. Methods Programs Biomed. **194**, 105507 (2020)
20. Nagendran, M., Chen, Y., Lovejoy, C.A., et al.: Artificial intelligence versus clinicians: systematic review of design, reporting standards, and claims of deep learning studies. BMJ **368**, m689 (2020)
21. Vandemoortele, O., et al.: General purpose pharmacokinetic-pharmacodynamic models for target-controlled infusion of anaesthetic drugs: a narrative review. J. Clin. Med. **11**(9), 2487 (2022)
22. Sirlanci, M., Levine, M.E., Low Wang, C.C., et al.: A simple modeling framework for prediction in the human glucose-insulin system. Chaos **33**, 073150 (2023)

AI-Enhanced Decision-Making in Childhood Obesity Interventions

Raquel Soriano-Gonzalez[1(✉)], Noelia Fuster[1], Elena Perez-Bernabeu[1], and Gonzalo Ros-Cervera[2]

[1] Research Centre on Production Management and Engineering (CIGIP), Universitat Politècnica de València, Ferrandiz y Carbonell, s/n, 03802 Alcoy, Spain
rsorgon@epsa.upv.es, {nfuscom,elenapb}@upv.es
[2] Department of Pediatrics, University Hospital Vinalopö, Elche, Spain

Abstract. Childhood obesity represents a global public health challenge. This study employs artificial intelligence (AI) to enhance decision-making in pediatric interventions against obesity. A predictive model was developed using data from children participating in an obesity intervention program at the Pediatric Service of the Elx-Crevillent Health Department to identify which type of patient achieves a successful outcome after a standard obesity intervention. The results demonstrate the potential of AI to improve personalized healthcare and optimize resources in combating childhood obesity, specifically in defining which patients may benefit from standard procedures and which patients may require additional resources to achieve improvements. This pilot research contributes to a deeper understanding of the factors related to childhood obesity and lays the groundwork for future AI-driven innovations in pediatric health.

Keywords: Childhood Obesity · Machine Learning · Decision Making

1 Introduction

Childhood obesity represents one of the most pressing challenges for public health today, with alarming rates continuing to rise worldwide. This health issue has reached worrying proportions in Spain [1], as in many other countries, with recent studies indicating a significant prevalence among children of various ages [2]. This condition entails immediate health risks for children and projects a considerable economic and healthcare burden for the future [3].

Currently, various studies are addressing the prediction of childhood obesity using innovative approaches and longitudinal data. Gupta et al. [4] propose a method based on the analysis of electronic health records (EHR) using recurrent neural networks (RNN) with long short-term memory (LSTM) cells, allowing for early detection of children at risk of obesity. Gou et al. [5] evaluate the accuracy of existing prediction models, emphasizing the importance of lifestyle factors in childhood obesity. Butler et al. [6] focus on developing a model to predict

A. A. Juan et al. (Eds.): DSA ISC 2024, LNCS 14778, pp. 138–150, 2025.
https://doi.org/10.1007/978-3-031-78238-1_13

obesity in New Zealand children. Fu et al. [7] employ machine learning techniques to identify risk factors in premature newborns. However, unlike these studies focusing on predicting childhood obesity based on medical history, our approach aims to predict the success of obesity interventions in pediatric patients.

This study addresses various aspects of childhood obesity management to enhance the effectiveness of interventions and decision-making in the pediatric field [8]. To achieve this, a predictive model develops to anticipate success in standard childhood obesity intervention programs by identifying the factors influencing treatment outcomes. Additionally, an examination was conducted on the factors associated with childhood obesity and their impact on intervention effectiveness. This approach aims to contribute to more precise and personalized care for children affected by obesity by identifying which patient profile may succeed with standard treatment and which profile may require a more personalized and rigorous approach to achieve satisfactory outcomes. A comprehensive evaluation of diverse artificial intelligence techniques, with a predominant focus on machine learning algorithms, was undertaken to determine their efficacy in improving decision-making in pediatric care. Lastly, an investigation is carried out into how consensus models can integrate predictions generated by multiple machine learning algorithms, aiming to provide more reliable estimates of treatment outcomes, thereby contributing to more precise and personalized healthcare for children affected by obesity.

The article follows a structured format: Sect. 1, the introduction provides the study's context and establishes its objectives. Section 2 focuses on data acquisition, detailing the data preparation and study design techniques. Section 3 describes the methodology, including the classification method employed and the use of the Polisabio Advanced Classifier Optimizer. The study findings are presented in Sect. 4, where the performance of the classification model is evaluated, and consensus models are discussed. Finally, in Sect. 5, the conclusions are summarized, highlighting the implications of the results and the potential of artificial intelligence in childhood obesity treatment.

2 Data Processing

This article details the tasks carried out in this preparatory action, as well as the design of the retrospective observational and multicenter study that was conducted. The study population consisted of 96 children aged between 5 and 17 years who participated in an obesity intervention program at the Pediatric Service of the Elx-Crevillent Health Department.

Hospital professionals created a data collection book, combining a literature review and retrospective study to develop predictive models in childhood obesity. Strict ethical measures were applied, preserving patient confidentiality, obtaining informed consent, and following established ethical guidelines.

In Fig. 1, the 44 variables included in the study are grouped according to their typology. Additionally, the data type of each variable is detailed. In order to simplify data management, categorical variables belonging to sociodemographic variables were encoded as numeric through a data mapping process [9]. However, the nature of these data was considered when creating the models explained in Sect. 3. It's noteworthy that the dataset is complete, without any missing data.

All patients receive a standard treatment for combating obesity. This treatment, offered by the Health Department, comprises improvements in diet, increased hours of exercise outside of school, and better rest time. This treatment is not specific to individual patients, as all patients receive the same information and guidelines to follow.

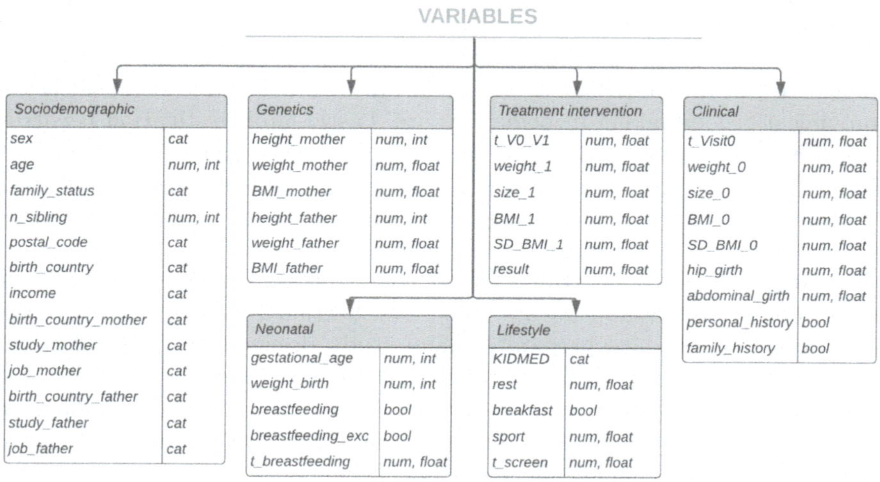

Fig. 1. Groups of variables defined in the project

3 Methodology

This section introduces a classification model to forecast treatment success in standard childhood obesity intervention programs. Using a uniform treatment approach, the model seeks to determine whether a patient achieves a positive outcome.

Patient classification variables are defined during the initial patient visit and encompass sociodemographic, genetic, neonatal, lifestyle, and initial clinical data. Leveraging these variables, patients are categorized into two groups: those with success in standard treatment outcomes and those without success.

The model is trained through historical data analysis, learning from previous outcomes. Once the training process is completed, the model can make predictions about the outcomes of new patients, whose data are stored in a designated

Excel file for prediction. This anticipatory capability facilitates informed clinical decision-making, as it can identify which type of patient requires more specific or differentiated treatments to achieve favorable results.

This approach anticipates new patients' treatment outcomes and provides opportunities for healthcare personalization and optimization. By basing predictions on predefined variables, the model contributes to more precise and efficient healthcare delivery aligned with each patient's specific characteristics.

3.1 Classificatory Variable

The variable used to predict the treatment success in new patients is based on the treatment outcome. Given the continuous growth of children, Body Mass Index (BMI) is not a directly applicable indicator. Thus, sex-differentiated tables considering age are employed to evaluate BMI in this demographic group. Specifically, the BMI (BMI-SDS) standard deviation is utilized. This metric quantifies the disparity between the observed BMI of the patient and the expected BMI for someone of the same age and sex [10]. Calculating this discrepancy requires information on sex, age, weight, and height. Utilizing resources such as the Spanish Society of Pediatric Gastroenterology, Hepatology, and Nutrition website ensures that comparisons are made regarding specific norms for the pediatric population in Spain based on WHO guidelines from 2006/2007 [11].

Our study measured BMI-SDS at the initial visit (BMI-SDS_0) and at the visit following treatment (BMI-SDS_1). Treatment outcome is the difference between BMI-SDS at the visit after starting treatment and BMI-SDS at the initial visit (Eq. 1). A negative result indicates an improvement in BMI standard deviation.

$$Result = (BMI - SDS\_1) - (BMI - SDS\_0) \tag{1}$$

Following the analysis conducted in the study [12] on the effects of reducing the BMI-SDS on cardiovascular risk factors, it was decided to establish a threshold of −0.15 in the outcome variable to define treatment success. This threshold was adjusted explicitly for the 6-month treatment period based on the findings provided by the aforementioned research. All results below this value indicate a more significant decrease in the BMI-SDS, representing treatment success, while values above mean treatment failure.

The variables that most significantly influence the outcome variable are: BMI-SDS_0 (Standard Deviation of the Body Mass Index), t_V0_V1 (time between doctor appointments) and BMI_mother (Body Mass Index of the patient's mother).

3.2 Polisabio Advanced Classifier Optimizer

The applied classification model is distinguished by its comprehensiveness and is highlighted as a robust tool, even within the context of a relatively small dataset comprising only 96 observations. Despite its limited size, this model has been strategically enhanced to improve its performance and enable the machine learning algorithms employed to yield highly accurate results. Figure 2 presents a flowchart depicting a summary of the classification method process.

The main distinguishing feature in data treatment is the contribution of the "Polisabio Advanced Classifier Optimizer" classification model, which addresses the challenge of working with small datasets by applying data oversampling and balancing techniques, using Python as a programming language due to its great popularity and efficiency for data science [13]. In this section, the model and the proposed characteristics are described.

The model requires two input datasets: historical data and prediction data. In this study, historical data refers to the database of pediatric obesity treatment patients. For this study, only the variables obtained before starting treatment (initial variables) and the treatment outcome (Section before) are necessary. In the prediction data, we have the initial variables of a new set of patients for whom the treatment outcome is unknown. This model aims to predict the standard treatment outcome for new patients.

Firstly, it must be checked that the columns of both data input sets are the same because the prediction data set must have the same variables as the data used to train the model, historical data. If this is not the case, the model would return an error, and the input would have to be corrected.

Based on the continuous variable "Result" and the defined threshold for classifying patients, we derive the variable "Outcome," which is the target variable for prediction. In this variable, patients is assigned a value of 0 (True) if they achieve success and a value of 1 (False) if they do not. So, if a patient is labeled as True, then it means that the regular treatment is working, otherwise, if labeled as False, then the patient should follow other type of treatment.

The next step is to separate the target variable "y" (Outcome) from the features "X," and the historical data set is divided into a training set (X_train, y_train) and a test set (X_test, y_test). One-Hot encoding is used for binary encoding of categorical variables, making it easier to include them in subsequent Machine Learning models [14].

If there is data class imbalance in the target classes, stratified sampling of X_train is used to ensure that all classes of the target variable y_train are equally represented [15].

Additionally, two oversampling strategies, including RandomOverSampler [16] and SMOTE [17], are employed to generate duplicate and synthetic data, respectively, aiming to address the limitation in the database size and thus generate more individuals.

The model uses the following predictive models: LogReg [18], decision trees [19], SVM [20], Naive Bayes [21], Random Forest [22], Ada Boost [23], and XGBoost [24]. The best samplers and oversampling ratios are selected from these models, generating y_pool, representing the best model found, in this case, it was XGB using SMOTE as a sampler and using oversampling with ratio = 3.

Algorithm 1 outlines the main components of the fine tuning of the machine learning (ML) models, which uses different criteria to refine the model and obtain the best predictions. The procedure receives two inputs, the base ML

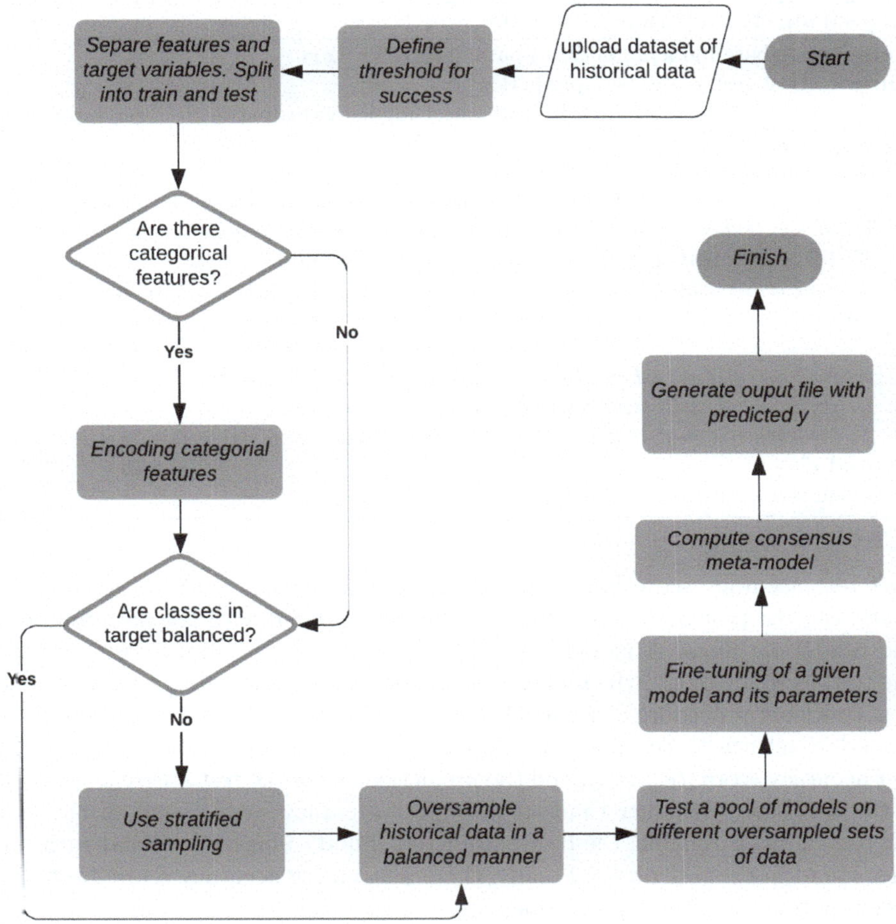

Fig. 2. Flowchart of Polisabio Advanced Classifier Optimizer

model (*b_model*), and a selection of parameters to test (*p_grid*). First, using standard cross-validation techniques, the bestTrainScore procedure computes the model with the best training score. The procedure returns three values, the model training accuracy score (*train_s_t*), the test accuracy score (*test_s_t*), and the predictions from the test set (*test_pred_t*). These values are assigned to be the best training accuracy score (*best_train_s_t*), best test accuracy score (*best_test_s_t*), and best predictions (*best_test_pred_t*) found so far. Next, the bestValScore procedure computes the model with the best validation score. The procedure splits the training set into two sets, reduced training and validation sets, and employs cross-validation techniques with the validation set to fine tune the model. The procedure returns three values, the model training accuracy score (*train_s_v*), the test accuracy score (*test_s_v*), and the predictions from the test set (*test_pred_v*). Suppose the validation test score is higher than the

Algorithm 1. Fine Tune ML Model

Require: A base model b_model, a parameter grid p_grid
Ensure: Best test score, best predictions

▷ Find the best model according to the best train score
$train\_s\_t, test\_s\_t, test\_pred\_t \leftarrow$ bestTrainScore(b_model, p_grid)
$best\_train\_s, best\_test\_s, best\_test\_pred \leftarrow train\_s\_t, test\_s\_t, test\_pred\_t$
▷ Find the best model according to the best validation score
$train\_s\_v, test\_s\_v, test\_pred\_v \leftarrow$ bestValScore(b_model, p_grid)
if $(test\_s\_v > best\_test\_s) \wedge (train\_s\_v \geq test\_s\_v)$ **then**
    $best\_train\_s, best\_test\_s, best\_test\_pred \leftarrow train\_s\_v, test\_s\_v, test\_pred\_v$
end if
▷ Find the best model according to the best test score
$train\_s\_e, test\_s\_e, test\_pred\_e \leftarrow$ bestTestScore(b_model, p_grid)
if $(test\_s\_e > best\_test\_s) \wedge (train\_s\_e \geq test\_s\_e)$ **then**
    $best\_train\_s, best\_test\_s, best\_test\_pred \leftarrow train\_s\_e, test\_s\_e, test\_pred\_e$
end if
return $best\_test\_s, best\_test\_pred$

best test accuracy score found so far, and the validation model is reliable. In that case, the best parameters found so far are updated with the ones found in the validation phase. A model is reliable if the training accuracy score exceeds the test accuracy score, which means the model is not underfiting. Moreover, the bestTestScore procedure computes the model with the test validation score. The procedure returns three values, the model training accuracy score ($train\_s\_e$), the test accuracy score ($test\_s\_e$), and the predictions from the test set ($test\_pred\_e$). If the test score is higher than the best test accuracy score found so far, and the test model is reliable, the best parameters found so far are updated with the ones found in the test phase. Finally, the best training accuracy score, best test accuracy score, and best predictions are returned from the procedure.

A classification report is obtained for the seven models, including precision in training and test sets. Additionally, a confusion matrix of the best result achieved in each model is presented, aiding in evaluating each model's performance by indicating the number of correct and incorrect predictions. This systematic approach aims to maximize the robustness and effectiveness of the classifier ensemble within the context of small and unbalanced datasets.

Algorithm 2. Consensus Meta-Models

Require: Predictions of different models y_pred
    $y\_predicted \leftarrow sort(y\_pred, by\ accuracyScore)$
    $y\_consensus\_all \leftarrow majorityVote(y\_pred)$
    $y\_pred\_top \leftarrow getFirst(y\_pred, 3)$
    $y\_consensus\_top \leftarrow majorityVote(y\_pred\_top)$

In Algorithm 2, the "Consensus" meta-model is defined to obtain the best result from the models' predictions. For this purpose, the predictions of the seven models are combined with y_pool, and the Consensus Models are applied. The Consensus All Model considers all predictions and selects the most frequent category among the seven models to provide an overall view. On the other hand, the Consensus Top Model relies solely on the three best models, selecting the most frequent category among their predictions and highlighting the influence of the most robust classifiers. The model's output displays the predictions and the obtained consensuses, offering a comprehensive perspective and highlighting the contribution of the most robust models.

4 Results

As discussed in the previous section, a classification report has been obtained for each of the seven models. As discussed in the last section, a classification report has been obtained for each of the seven models. This includes the accuracy of the training set and the test set. In addition, there is also a confusion matrix, which helps to evaluate the performance of the models by indicating the number of correct and wrong predictions [25]. This systematic approach seeks to maximize the robustness and efficiency of the classifier ensemble in the context of small and unbalanced data [26]. Table 1, presents the results of the Accuracy Test, which indicates the proportion of correct predictions made by the model on the test dataset, and Recall 0, which means the proportion of class 0 instances correctly identified by the model relative to all class 0 instances present in the test data, for each of the models used in the "Polisabio Advanced Classifier Optimizer", as well as the results of the consensus models.

Table 1. Results of predictive models

Model	Accuracy Test	Recall 0
LogReg	0.67	0.80
Decision Trees	0.87	0.60
SVM	0.60	0.60
Naive Bayes	0.67	0.60
Random Forest	0.87	0.80
XG-Boost	0.80	0.80
y_pool	0.87	1.00
Consensus All	0.87	1.00
Consensus Top	0.93	0.80

The results reveal that the best overall accuracy and recall of class 0 are achieved with the consensus models, underscoring the effectiveness of this algorithm in obtaining the most fitting and robust model. This is the distinctive feature of the "Polisabio Advanced Classifier Optimizer" through consensus meta-models. Figure 3 shows that the output file adds the seven proposed predictions of the models and, in addition, calculates two consensus predictions in Fig. 4.

	y_test_svm	y_test_logreg	y_test_naiveB	y_test_xgb	y_test_pool	y_test_trees	y_test_rforest
0	1	1	0	0	0	0	0
1	0	0	0	0	0	0	0
2	0	0	1	0	0	1	1
3	0	0	0	0	0	1	1
4	1	1	1	1	1	1	1
5	1	0	0	1	1	1	1
6	1	0	1	0	0	1	0
7	1	1	0	1	1	1	1
8	1	0	1	1	1	1	1
9	0	1	1	1	1	1	1

Fig. 3. Output: output of predictions of each model

On the one hand, it calculates "Consensus All", which considers the outcome predicted by each of the seven models and selects the most frequent category among all predictions, providing a comprehensive view based on the contribution of all models. The "Consensus Top" is based only on the results of the top three models and selects the most frequent category among the predictions of these top three models, in this case, random forest, decision tree, and the best model found in the pool (XGB). This approach considers only the top models' contributions, highlighting the influence of the most robust classifiers.

y_consensus_all	y_consensus_top	y_test
0	0	0
0	0	0
0	1	1
0	1	1
1	1	1
1	1	1
0	0	0
1	1	1
1	1	1
1	1	1

Fig. 4. Output: output of predictions of each consensus model

Its comprehensiveness characterizes the developed classification model and is a robust tool, even in a relatively small dataset with only 96 observations [27]. This model has been strategically enhanced to improve performance and enable the Machine Learning algorithms to provide highly accurate results.

In the same way that a detailed evaluation was performed on each classification model, one was generated for each consensus model. This evaluation comprises the accuracy, the confusion matrix, and the classification report. Figure 5 shows the report for the "Consensus All" and "Consensus TOP" model.

The confusion matrix in the Consensus All Model shows that, of the 15 test cases, the classifier correctly predicted 5 cases as negative (true negatives) and eight as positive (true positives). There were two cases where the model incorrectly predicted the negative label (false negatives), resulting in a perfect accuracy for the positive class (1.00) and an accuracy of 0.71 for the negative class.

The recall, also known as sensitivity, for the negative class was perfect (1.00), indicating that all true negative cases were correctly identified. However, the recall for the positive class was 0.80.

The model's overall accuracy was 0.87, meaning that 87% of the total predictions were correct. The average metrics, macro and weighted, for accuracy, recall, and F1 score were consistently high, with values between 0.86 and 0.90. These results indicate a balanced classifier performance in both classes.

On the other hand, the confusion matrix on the right-hand side of Fig. 5, illustrates the confusion matrix for the "Consensus Top" classifier, which achieved a higher test accuracy of 0.93.

The Consensus Top Model accurately predicted 4 cases as class 0 and 10 cases as class 1, with only 1 false positive and no false negatives. In terms of metrics (both individual per class and average metrics), the results are slightly higher than in Consensus All.

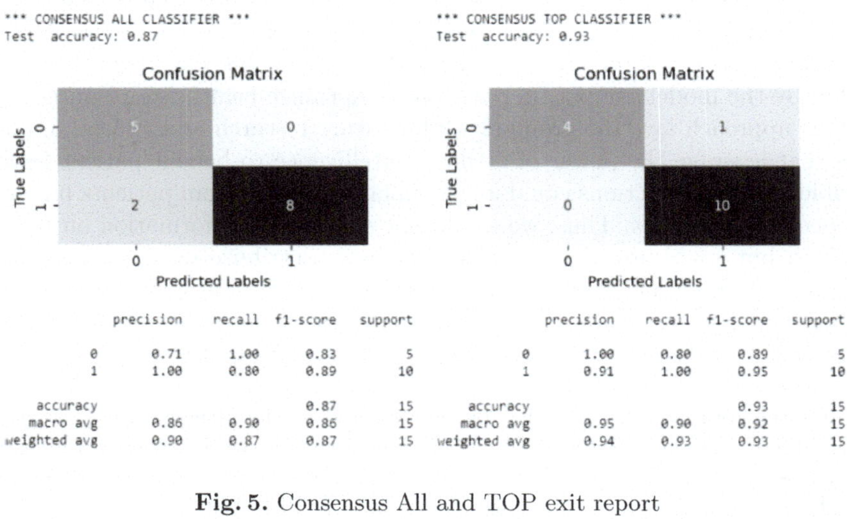

Fig. 5. Consensus All and TOP exit report

The comparison between the "Consensus All" and "Consensus Top" classifiers underlines the relevance of careful model selection in the development of classification systems [28]. Although both methods proved effective, the "Consensus Top" stood out for its superior accuracy, recall, and F1-score performance. This suggests that focusing on the predictions of the most robust models can improve classification system performance.

5 Conclusions

The development of a classification model to predict treatment success in pediatric patients with obesity using a standard treatment represents a significant advancement in personalized healthcare. This approach, grounded in various variables including sociodemographic, familial, habits, and clinical data, demonstrates the potential of artificial intelligence in informed clinical decision-making. Using historical data to train the model and subsequently apply it to new patients establishes a predictive framework that guides treatment selection and optimizes medical resources, considering that a standard treatment does not consume high resources, while a more rigorous and exhaustive one does.

Introducing the "Polisabio Advanced Classifier Optimizer" effectively addresses challenges associated with small and unbalanced datasets. Through techniques such as data oversampling and balancing, this model enhances predictive capability and improves prediction accuracy. Including multiple machine learning algorithms and selecting the best models based on specific criteria further strengthen the system's robustness.

Consensus models, "Consensus All" and "Consensus Top," represent a solid approach to integrating predictions from multiple models and improving the reliability of estimates. A comprehensive and accurate view of treatment outcomes is obtained by considering all models' contributions and focusing on the most robust models.

The results of this study underscore the promise of artificial intelligence in medicine, especially in areas where precision and personalization are crucial, such as pediatric obesity interventions. The high accuracy and effectiveness demonstrated by the model suggest its potential to optimize healthcare resources.

This approach lays the groundwork for future research and clinical applications that leverage the power of artificial intelligence to benefit patient health. Possible research directions could include analyzing data from patients undergoing specific treatments. Thus, we would not only obtain information on patient evolution but could also define which variables contribute most to each treatment and which patient profile is most suitable for each treatment. Furthermore, conducting long-term studies with treatment duration between one or two years would also provide valuable information on the evolution of treatment success.

Acknowledgments. This work has been supported by the Investigo Program of the Generalitat Valenciana (INVEST/2023/304) and the UNISALUT program of Fundaciö per al Foment de la Investigaciö Sanitária i Biomèdica de la Comunitat Valenciana (FISABIO) and Universitat Politècnica de València (POLISABIO22_AP06).

References

1. Bravo-Saquicela, D.M.: Has the prevalence of childhood obesity in Spain plateaued? A systematic review and meta-analysis. Int. J. Environ. Res. Public Health **19**(9), 5240 (2022)
2. Government of Spain. https://www.comisionadopobrezainfantil.gob.es/es/plan-estrategico-nacional-para-la-reduccion-de-la-obesidad-infantil. Accessed 6 Mar 2024
3. Pelone, F., et al.: Economic impact of childhood obesity on health systems: a systematic review. Obes. Rev. **13**(5), 431–440 (2012)
4. Gupta, M., Phan, T.L.T., Bunnell, H.T., Beheshti, R.: Obesity prediction with EHR data: a deep learning approach with interpretable elements. ACM Trans. Comput. Healthc. (HEALTH) **3**(3), 1–19 (2022)
5. Gou, H., Song, H., Tian, Z., Liu, Y.: Prediction models for children/adolescents with obesity/overweight: a systematic review and meta-analysis. Prev. Med. 107823 (2023)
6. Butler, É.M., et al.: A prediction model for childhood obesity in New Zealand. Sci. Rep. **11**(1), 6380 (2021)
7. Yuanqing, F., et al.: Integration of an interpretable machine learning algorithm to identify early life risk factors of childhood obesity among preterm infants: a prospective birth cohort. BMC Med. **18**(1), 1–10 (2020)
8. Kwan, J.L., et al.: Computerised clinical decision support systems and absolute improvements in care: meta-analysis of controlled clinical trials. Bmj **370** (2020)
9. Paassen, B., McBroom, J., Jeffries, B., Koprinska, I., Yacef, K., et al.: Mapping python programs to vectors using recursive neural encodings. J. Educ. Data Min. **13**(3), 1–35 (2021)
10. Guidelines for the prevention of overweight and obesity in childhood and adolescence. https://www.unicef.org/media/96096/file/Overweight-Guidance-2020-ES.pdf. Accessed 6 Mar 2024
11. Spanish Society of Pediatric Gastroenterology, Hepatology and Nutrition (2020). https://www.seghnp.org/. Accessed 6 Mar 2024
12. Reinehr, T., Lass, N., Toschke, C., Rothermel, J., Lanzinger, S., Holl, R.W.: Which amount of BMI-SDS reduction is necessary to improve cardiovascular risk factors in overweight children? J. Clin. Endocrinol. Metab. **101**(8), 3171–3179 (2016)
13. Raschka, S., Patterson, J., Nolet, C.: Machine learning in python: main developments and technology trends in data science, machine learning, and artificial intelligence. Information **11**(4), 193 (2020)
14 Seger, C.: An investigation of categorical variable encoding techniques in machine learning: binary versus one-hot and feature hashing. (2018:596):34 (2018)
15. Sadaiyandi, J., Arumugam, P., Sangaiah, A.K., Zhang, C.: Stratified sampling-based deep learning approach to increase prediction accuracy of unbalanced dataset. Electronics **12**, 4423 (2023)
16. Hayaty, M., Muthmainah, S., Ghufran, S.M.: Random and synthetic over-sampling approach to resolve data imbalance in classification. Int. J. Artif. Intell. Res. **4**(2), 86–94 (2020)
17. Douzas, G., Bacao, F., Last, F.: Improving imbalanced learning through a heuristic oversampling method based on k-means and smote. Inf. Sci. **465**, 1–20 (2018)
18. Miller, M.E., Hui, S.L., Tierney, W.M.: Validation techniques for logistic regression models. Stat. Med. **10**(8), 1213–1226 (1991)

19. Song, Y.-Y., Ying, L.U.: Decision tree methods: applications for classification and prediction. Shanghai Arch. Psychiatry **27**(2), 130 (2015)
20. Wang, W., Men, C., Weizhen, L.: Online prediction model based on support vector machine. Neurocomputing **71**(4–6), 550–558 (2008)
21. Kharya, S., Soni, S.: Weighted naive bayes classifier: a predictive model for breast cancer detection. Int. J. Comput. Appl. **133**(9), 32–37 (2016)
22. Speiser, J.L., Miller, M.E., Tooze, J., Ip, E.: A comparison of random forest variable selection methods for classification prediction modeling. Expert Syst. Appl. **134**, 93–101 (2019)
23. Bahad, P., Saxena, P.: Study of adaboost and gradient boosting algorithms for predictive analytics. In: International Conference on Intelligent Computing and Smart Communication 2019: Proceedings of ICSC 2019, pp. 235–244. Springer (2020)
24. Li, S., Zhang, X.: Research on orthopedic auxiliary classification and prediction model based on XGBoost algorithm. Neural Comput. Appl. **32**, 1971–1979 (2020)
25. Mullick, S.S., Datta, S., Dhekane, S.G., Das, S.: Appropriateness of performance indices for imbalanced data classification: an analysis. Pattern Recogn. **102**, 107197 (2020)
26. Luque, A., Carrasco, A., Martín, A., de Las Heras, A.: The impact of class imbalance in classification performance metrics based on the binary confusion matrix. Pattern Recogn. **91**, 216–231 (2019)
27. Xu, P., Ji, X., Li, M., et al.: Small data machine learning in materials science. npj Comput. Mater. **9**, 42 (2023)
28. Jain, D., Singh, V.: Feature selection and classification systems for chronic disease prediction: a review. Egypt. Inform. J. **19**(3), 179–189 (2018)

On the Evaluation of the Degree of Disability of the Elderly: A Case Study

Carlo Filippo Bonetta[1], Luca Alfredo Grossi[2], and Giovanni Righini[3](✉)

[1] University of Verona, Unità Operativa Complessa "Forensic Medicine", Verona, Italy
carlo.filippob@yahoo.it

[2] Azienda Socio-Sanitaria Territoriale di Crema, Unità Operativa Complessa "Primary Care", Crema, Italy
luca.grossi@asst-crema.it

[3] Department of Computer Science, University of Milan, Milan, Italy
giovanni.righini@unimi.it

Abstract. In multidimensional geriatric assessment, elderly patients are evaluated from different points of view to decide whether to accept or reject their requests for special benefits. The decision is taken in several stages and it does not currently rely on any decision support system. The need to make it more equitable and consistent gave rise to this case study, based on data relating to a retrospective cohort of patients over 67 years of age, who underwent a geriatric examination and geriatric multi-dimensional assessment. To avoid selection and evaluation bias, it was deemed appropriate to consider only patients evaluated with the same methodology and measurement scales. In a preliminary step some inconsistencies were identified in the available data-set. Then, a subset of patients in need for a re-evaluation was defined by a suitable combination of different methods: namely, threshold values related to single indicators, dominance analysis and search for inconsistent pairs. This subset of patients was re-evaluated and a tentative classification model was developed. Therefore, the outcome of the case study was a critical review of the data, the model and the decision-making process.

Keywords: Decision science · Classification · Integer linear programming

1 Introduction

Motivation. The national health care system in Italy provides special assistance to the elderly according to their needs: examples of facilitations are discounts on pharmaceuticals for patients affected by chronical diseases, dedicated parking lots for people with reduced mobility and a special allowance for paying an accompanying person for the most severe cases. The growing number of aged citizens combined with tighter and tighter budget restrictions put an increasing pressure on the concession of such facilities. This makes the decision problem

A. A. Juan et al. (Eds.): DSA ISC 2024, LNCS 14778, pp. 151–162, 2025.
https://doi.org/10.1007/978-3-031-78238-1_14

critical and raises the need for a decision support system to improve the level of fairness and reliability.

The decision process is articulated in several phases. When a patient issues a request, he undergoes a multi-dimensional geriatric visit. The visit consists of evaluating the disability level of the patient from two main points of view: mobility and mental conditions. For this purpose some standard tests are used; on the basis of the scores, the geriatrician draws up a yes/no proposal. Then, the patient is visited again by a commission of five physicians who decide, with a majority vote, whether to confirm the geriatrician's proposal or not. Finally, the request for allowance is sent to Istituto Nazionale di Previdenza Sociale (INPS), the Italian institute for social security, for an official and final confirmation.

This study was carried on in the city of Crema (Italy), where a single geriatrician is in charge for all multi-dimensional evaluations, which favors consistency in the proposals over time. Furthermore, it was observed that almost all final recommendations from the commission had been accepted by INPS (with a single exception in 144 cases). This suggested that a very good selection criterion had been found also at the commission level, in spite of its heterogeneity since its five members are not always the same.

Therefore, the request arose to capture these classification criteria in a mathematical model, in order to make them reproducible and to ensure that these same criteria could be consistently used in the future. However, inserting a model-based tool in a delicate decision process, dealing with assistance to the elderly, should be done with care. For this reason, the study turned out to be an interesting example of a decision science approach in health care. Details are reported in [1]. As for any good application of decision science, the main outcome was not just a solution but an increase in knowledge about the decision problem.

Main Contributions. The evaluation of disability in the elderly has been addressed in the medical literature, mainly for the purpose of validating multi-dimensional scales and indicators [3]. However, the decision problem concerning the use of such indicators to classify patients and to accept or refuse their requests for assistance allowance or other benefits has been shown to be highly critical and prone to a high degree of arbitrariness and biases.

The topic covered by this study is less studied in legal medicine than forensic pathology: only sporadic articles appear in the literature. Moreover, the related literature is country-specific, because every country has its own evaluation methods and legislation.

At the best of our knowledge, the classification problem considered in this paper has not been addressed so far in the same way we propose. In particular, the scientific literature on classification problems and methods is huge, both from the medical and the computer science side, but the common assumption is that data must be taken for granted and suitable classifiers must be sought. On the contrary, in this case study, we simultaneously subjected the available data and the classification methods to criticism.

The main outcome was threefold, concerning the data, the model and the process. A critical analysis of the data revealed some missing data and some inconsistencies, but the main criticism was put on the degree of arbitrariness in evaluating the patients' scores not from the viewpoint of the geriatrician but rather from the viewpoint of how the test outcomes are graded. A critical analysis of the model allowed to propose several alternative solutions, mainly based on the convexity of the set of points corresponding to patients whose application was rejected. A critical analysis of the decision process revealed that INPS decisions were based on partial data, selected by the commission to support its own requests. Therefore the almost 100% success rate in this step turned out to be less significant than expected.

As a final result, a tentative classifier was also computed, but it should be regarded as a temporary intermediate step in a longer and still ongoing continuous improvement process.

Paper Outline. In Sect. 2 we describe the tests that are used to score patients. In Sect. 3 we illustrate a preliminary analysis whose main outcome was to reveal some inconsistencies in the available data and the decision process, allowing to define some long-term and short-term research directions. In Sect. 4 we describe how some patients were selected for being re-classified, the outcome of their re-classification and its consequences on the separability of the two classes of patients. Some conclusions are outlined in Sect. 5.

2 The Decision Problem

The multi-dimensional evaluation of geriatric patients (see [8]) was based on four tests: (i) Instrumental Activities of Daily Living (IADL) [6]. This test was done in two different ways for men and women. Women were evaluated on a $[0, 8]$ scale by eight tests with binary outcome, while men were evaluated on a $[0, 14]$ scale by five tests whose integer scores range in $[0, 2]$ to $[0, 4]$. The difference in the two tests is meant to take into account different expected habits of patients and thus different significance of the questions (see also [10] and [9]). Furthermore, we observed that, for unknown reasons, this choice had not been strictly followed for all patients in the available data-set. To make the scores comparable, we normalized them in a $[0, 1]$ range. (ii) Barthel index [7]. This index is an integer ranging in $[0, 100]$. It is based on the evaluation of 11 basic activities, like eating, walking, dressing/undressing and using the toilet. (iii) Tinetti index [11]. This index is the sum of two indices: the former one, named Balance, ranges in $[0, 16]$ and it evaluates the patient's risk of fall when standing and sitting; the latter one, named Gait, ranges in $[0, 12]$ and it evaluates the patient's pace and stability while walking. (iv) Mini-Mental State Examination (MMSE) [4]. This neuropsychological test evaluates the patient's intellectual efficiency disorders and it detects the possible presence of cognitive deterioration. It is based on 19 questions and the outcome is an integer score in $[0, 30]$. Besides these four tests, patients' age was also taken into account.

On the basis of geriatricians' experience, each index is known to have some drawback: age by itself is rarely significant; IADL tends to be low for all patients who enter the system and are evaluated; Barthel index is self-assessed on the basis of what the patient declares to be able to do; Tinetti index weighs similar or strongly correlated abilities multiple times; MMSE typically yields non-robust outcomes, depending on many factors that are not under the geriatrician's control.

For each patient the proposal of the geriatrician, the request of the commission and the final decision of INPS are binary. At each step, patients are classified as G, meaning "grave" (serious impairment), and S, meaning "severo" (severe impairment). S-patients receive special facilitations or allowances, while G-patients do not.

In our case study the classification was the same in all three steps and this is the reason why a mathematical model was sought to capture the underlying criterion.

3 A Preliminary Analysis

Our preliminary analysis concerned the data, the model and the process. The corresponding goals were (a) to check the consistency of the provided data-set (Subsect. 3.1), to check for the existence of a model allowing for automatic classification (Subsect. 3.2) and to analyze the steps of the decision process (Subsect. 3.3). The outcome was the identification of some critical issues from all three viewpoints (Subsect. 3.4), setting the agenda for the next steps to be taken (described in Sect. 4).

3.1 Analysis of the Data-Set

For the purpose of the analysis a data-set of 144 patients was available. For eight patients the MMSE score was not available, because their request had been rejected on the basis of the other tests. It was apparent that they did not have intellectual disorders and thus the MMSE test was unnecessary. In other words, these patients were far from the border between S-patients and G-patients. Therefore, these eight patients could be simply neglected, reducing the data-set to 136 patients.

Only 14 requests out of 136 had not been confirmed by INPS. Hence we temporarily eliminated these 14 requests from the data-set and we concentrated on extracting the classification criterion suitable for the remaining 122 patients.

3.2 Search for a Classifier

Mapping the patients in a four-dimensional Euclidean space according to their four test scores did not allow finding any reasonable way to partition them in two sets with linear or non-linear separators. The main reason for this was the existence of several inconsistent pairs. By inconsistent pair we mean a pair of patients, say A and B, such that all scores of A are greater than or equal to the corresponding scores of B but A is classified as an S-patient, while B is classified as a G-patient.

All inconsistent pairs were detected with the help of a linear programming model solved by a free solver (glpsol). In a second round, the same search was done in a five-dimensional space, taking into account also the age of patients. Disregarding the age, the result consisted of sixteen inconsistent pairs. Taking the age into account, the number of inconsistent pairs was reduced to four. Curiously, in all these four (s, g) pairs, the G-patient is a man and the S-patient is a woman. This suggests the possible existence of a classification bias in favor of female patients.

3.3 The Decision Process

During this analysis, some characteristics of the decision process were also investigated. In particular, from the examination of the commission reports it appeared that sometimes the commission only reported some of the available scores, omitting others. This takes away meaning from the fact that INPS confirmed all the commission judgments: INPS was not provided with the necessary data to be able to possibly correct such judgments.

Another important aspect of the decision process is that all patients are examined again by the commission after having been graded through the four tests by the geriatrician. The second visit by the commission can follow the first one by several days or even some weeks. This is an additional reason for possible disagreement between the geriatrician recommendation and the commission judgment, the main reason being the variable composition of the commission over time.

3.4 Critical Issues and Research Directions

On the basis of the preliminary analysis, rather than trying to capture the classification criteria used so far in a mathematical model, it seemed appropriate to build a decision support system that allow the excessive arbitrariness of the commission judgments to be reduced, to prevent it from arriving at contradictory classifications. In turn, this implied the need not to consider all the available commission judgments as valid, but to identify some of them as possibly "incorrect" or "exceptional".

Another consequent critical issue concerns the reliability of the scores. The assumption that the numerical scores, as they are currently evaluated, are actually significant must not be taken for granted. Since the scores are based on the translation of empirical observations of non-numerical quantities into numbers, they include an intrinsic amount of arbitrariness, both in the definition of the scores attributed to each individual question and in the choice to sum them up to produce the final score. In a few words, "numerical" is not a synonymous for "objective". This remains true independently of the geriatrician, although he is always the same and he is perfectly consistent with himself over time.

At the end of this preliminary analysis, we could distinguish some short-term and long-term research goals.

1. Correcting the classification. Without any modification to the input scores and independently of the mathematical model of the classifier, it is necessary to eliminate all incompatible pairs, by suitable modifications of the final decision reported in the data-set. Three techniques were considered for this purpose.
   a. Detect the minimum number of re-classifications needed to make the data-set consistent (criterion 1).
   b. Identify a subset of critical (or Pareto-optimal) G-patients and S-patients, whose classification determines that of others (criterion 2).
   c. Re-classify the already found incompatible pairs (criterion 3).
2. Building a classifier, based on the new classification results and the given input scores. This should reproduce the typical procedure followed by the physicians, not only the final result. A rather important requisite for the classifier is its simplicity, leading to a better interpretability of its results.
3. Revising the grading process, so that the final result allows to partition the patients into well-separated S and G classes.

Research line 3 was not compatible with the time restrictions of the project and it was left as an open research direction for future investigation. We concentrated on the short-term objectives corresponding to research lines 1 (Subsects. 4.1 and 4.2) and 2 (Subsect. 4.3), as described in the remainder.

4 Re-classification

Once the need to rediscuss some of the classifications was made clear, we proceeded to identify the critical cases to be re-examined (Subsect. 4.1). In doing so, we also tried to better understand the classification problem and the criteria used by physicians (Subsect. 4.2), instead of only relying on the available data-set. The partially re-classified data-set was analyzed again and a tentative automatic classifier based on thresholds and two separating hyperplanes was defined (Subsect. 4.3).

Convexity and Subsets of S-Patients. A fundamental property that arose from our preliminary analysis concerns the convexity of the set of G-patients. For any two G-patients, any convex combination of their scores (in any n-dimensional Euclidean space) should correspond to a G-patient. The reason is that we can safely assume that a healthy condition corresponds to the satisfaction of a set of constraints, while a pathology corresponds to the violation of at least one of them. Assuming for simplicity that such constraints are represented by linear inequalities, the healthy condition would correspond to a polyhedron (which is convex) in the d-dimensional space of the d indicators used. Even assuming non-linear functions to separate the two regions, it is safe to assume that the difference between healthy and pathological conditions can be well represented by suitable thresholds on a distance function computed from an ideal point according to some p-norm in a d-dimensional metric space. In this case the region corresponding to healthy conditions would be a ball (i.e. a convex set). For instance, using a p-norm with $p = 2$, the region would be a sphere, or an ellipsoid if the scores are weighted.

As a consequence of the convexity of the G-patients set, the S-patients set is likely to be non-convex. Actually, it is expected to consist of two distinct (but possibly overlapping) subsets, named ND, meaning "Non Deambulante" (non-ambulatory), and AC, meaning "Assistenza Continua" (continuous assistance). The former one is likely to correspond to low Tinetti index values, while the latter one is likely to correspond to low MMSE values. However, this was not verified in the data-set. For instance, we found S-patients of type ND with (normalized) Tinetti index equal to 57% as well as S-patients of type AC with (normalized) MMSE index equal to 97%.

4.1 Identification of Patients to be Re-classified

To enforce the convexity property of the G-patients set and to correct for some inconsistent classifications some patients should have been identified as good candidates for re-classification. For this purpose we used the three criteria listed above and illustrated in the remainder: namely linear separability, dominance, inconsistent pairs.

Criterion 1: Linear Separation. We analyzed the convex hull of the given set of G-patients in the data-set. Observing the minimum score recorded for each index of G-patients the following threshold values were obtained: IADL = 12.5%, Barthel = 47/100, Tinetti Balance = 6/16, Tinetti Gait = 2/12, MMSE = 15.4/30. There were sixty-one S-patients falling below at least one of these thresholds, while the remaining twenty-three S-patients had all scores above the thresholds.

We also observed that no S-patient fell below the threshold for IADL only, nor for Tinetti Gait only. One could conclude that these two indices are less relevant than the others to distinguish S-patients from G-patients.

Moreover, several S-patients that are under-threshold for just one index, show a score of that index only slightly below the threshold. Owing to the subjective nature of the scoring system, their difference from the threshold value is hardly significant. Hence, to make the classification more robust, the set of patients classified as S because they fell under threshold had to be reduced, by suitably decreasing the threshold values.

Then, owing to the convexity property of the G-patients set, we tried to define an additional separating hyperplane in five dimensions (four indices and age) to separate the set of G-patients from the set of above-threshold S-patients, minimizing the number of misclassifications. This resembles the classical classification method based on linear Support Vector Machines [2]. The underline observation is that if the G-subset and the S-subset cannot be separated by a system of linear inequalities, then they cannot be separated by any system of non-linear inequalities preserving the convexity property of the G-subset. Therefore, a consistent data-set must allow for a set of separating hyperplanes.

The search for a single separating hyperplane was done with an ILP model and it was found that at least ten classifications should be reversed from G to S or vice versa to allow for linear separability of the two sets. Moreover, the resulting separating hyperplane was such to assign 63% weight to Tinetti Gait, 28% to Barthel and 9% to IADL, neglecting Tinetti Balance and MMSE, which does not look reasonable. For this purpose, we did not limit ourselves to this criterion to identify critical patients. The ILP model is discrete and its optimal solution is not necessarily unique. We observed the set of outliers corresponding to one of its optimal solutions and we defined them as *critical patients.*

Criterion 2: Dominance. We also analyzed the G-patients set to detect dominating G-patients, that is the vertices of the convex hull of the G set. If these patients are classified as G, they imply the classification of all the other G-patients. We found they were five, by means of another ILP model. Symmetrically, we identified ten non-dominated S-patients, i.e. S-patients whose classification is not implied by the classification of other S-patients. These fifteen patients were also declared critical.

Criterion 3: Incompatible Pairs. Finally, the sixteen incompatible pairs were examined: they were formed by nine G-patients and nine S-patients. These eighteen patients were declared critical.

4.2 Re-classification

By the three above criteria, a set of twenty-one critical patients was eventually produced and the two physicians co-authors of this paper took the task of re-classifying them in a consistent way, by considering all of them represented in a single table (see Table 1).

In doing this, they were also asked to explicitly describe the classification criterion they were following. The outcome was the following: (i) IADL was assigned little significance; (ii) both physicians considered age, Tinetti and MMSE as main indicators; one of them also considered Barthel; (iii) both gave importance to the presence of low values for some indicator, independently of others having high values (this is consistent with the convexity assumption for the G-patients set); (iv) both applied an intuitive threshold system, although using slightly different values for the thresholds: Barthel = 50, Tinetti = 10, MMSE = 10 in one case; Tinetti = 14, MMSE = 20 in the other; (v) both tend to classify a patient as S when one of his scores is largely under threshold or at least two scores are slightly under threshold; (vi) age is used as a multiplicative factor, rather than an index by itself; however, there is a threshold also for the age: patients above 90 years are considered "grandi anziani" (great elders) and this can be enough to classify them as S-patients.

Table 1. The critical patients, their scores and the outcome of their re-classification

ID	sex	Age	IADL/100	Barthel/100	Tinetti/28	MMSE/30	Initial	Revised
8	F	80	25	65	12.6	17.4	G	S/G
9	F	86	25	69	14.28	20.4	G	G
14	F	95	25	76	15.96	21.0	S	S
17	F	95	25	68	14.28	15.9	G	S
27	M	87	36	53	7.56	26.7	G	S
33	F	85	50	59	7.56	28.8	S	S
42	M	91	21	59	9.8	20.1	G	S
50	F	90	50	67	10.64	23.1	S	S
72	F	83	13	56	8.68	21.0	G	S
82	M	88	14	58	8.68	28.5	S	S
83	M	92	43	56	10.92	26.1	S	S
86	F	89	13	67	14.84	15.3	G	G
93	F	76	38	82	21.84	12.6	S	S
94	F	75	25	47	11.2	22.8	G	S
102	F	82	38	55	9.8	28.8	G	S
116	F	86	25	72	13.16	19.5	S	G
118	F	84	13	73	10.64	15.3	G	S
131	M	92	29	55	12.88	27.0	S	S
138	F	92	38	65	7.56	29.1	S	S
142	F	86	25	66	22.96	17.1	S	G
144	F	87	63	56	7.56	27.3	S	S

Analysis of the Re-classified Data-Set. The classification of about half of the critical patients was changed. After the re-classification no incompatible pairs remained. The number of G-patients was 44 and the number of S-patients was 92.

Linear separation was still impossible owing to a single patient. Patient 50 was still classified as S, but she was not linearly separable from the set of the G patients, because she has relatively high scores in all tests, in spite of being ninety years old. Therefore, she was temporarily re-classified as G.

The search for dominating and non-dominated patients in sets G and S respectively provided very different (and better) results. The vertices of the convex hull of the G set were found to be fifteen and the non-dominated S-patients were found to be twenty-seven. This result should be regarded positively, because the presence of misclassified patients tends to reduce the number of borderline patients in both partitions.

From the analysis of the minimum values of the four indices and the maximum age value of G-patients, the following thresholds were observed: Age = 95; IADL

= 13%; Barthel = 55/100; Tinetti = 8/28; MMSE = 16/30. Among the 27 non-dominated S-patients, 12 were under-threshold for one index, none for two indices and 3 for three indices. The remaining 12 non-dominated S-patients were not under-threshold for any index.

As before, some critical analysis of the thresholds was needed. As expected, IADL looked redundant: only two critical S-patients were under-threshold for IADL and both of them were under threshold also for two other indices. Tinetti index looked also redundant: no critical S-patient was under threshold for it; this was caused by a G-patient with a very low Tinetti score. In general, these threshold values are affected by several questionable classifications, also in relation to the distinction between ND and AC patients. Infact, it was not possible to linearly separate critical G-patients from critical S-patients, not even using two different hyperplanes for ND and AC. Therefore, we re-classified a few patients from AC to ND.

4.3 A Tentative Classifier

After this, it was finally possible to find two linear separators between critical AC patients and G-patients and critical ND patients and G-patients. For this purpose, an index to express the effect of age was introduced as the ratio between 80 years and the patient's age. The ND separator weighs the indices as follows: Age: 12%, IADL: 18%, Barthel: 22%, Tinetti: 48%, MMSE: negligible. The AC separator weighs the indices as follows: Age: negligible, IADL: 22%, Barthel: 73%, Tinetti: negligible, MMSE: 5%.

One can observe that IADL turns out to be more significant than age and Barthel index is significant in both cases. As expected, Tinetti index has a large weight to identify ND patients, but the same does not hold for MMSE and AC patients, for which Barthel is more significant than MMSE.

It should be remarked, however, that the separation of ND patients from G-patients remains questionable, because their weighted values are quite close to one another: the minimum weighted score of G-patients is 0.486 while the maximum weighted score of ND patients is 0.477. This result can be explained because only few patients have been re-classified, but the data-set still contains some questionable classifications as, for example, patient 25 who is classified as G in spite of a Tinetti score equal to 8/28.

The analysis of the re-classified data-set shows remarkable improvements with respect to the previous results on the initial data-set. However, it also suggests the need for a further re-classification step, guided by pair-wise or subset-wise comparisons between critical patients identified at the borders of the current classification.

5 Conclusions

Three main conclusions of this study concern the data, the model and the process.

Data. An open long-term research direction is a critical re-examination of how grades are assigned in the tests: the use of a double scale in the IADL test, the high correlation between some sub-grades in the Tinetti test, the lack of robustness of the MMSE test, the role of age as an index by itself or as a weight factor of the other scores are just some examples of issues that deserve further analysis.

Model. One of the main goals of the future developments of a classifier is the reproduction of criteria as similar as possible to those applied by physicians. All observations directly extracted from experts knowledge, and not only from the data-set, can be usefully combined with the available data to produce a decision support system able to suggest explainable and consistent classifications. A new model may include the use of more refined and sophisticated functions as in logistic regression [5].

Decision Process. An outcome of this project has been to trigger a feedback between the model and its corresponding results, i.e. the classification. In one direction one wants to optimally adjust thresholds, score weights and separating functions to reproduce a desired outcome, once a reliable classification of patients is given. In the other direction, given a reliable classifier, one wants to apply it, to suggest fair and consistent classifications or to detect and correct inconsistent or questionable classifications. The study of the interplay between the two problems, where the role of input and output are swapped, should be considered as a typical outcome of tackling a complex decision problem with a decision science approach, with apparent and peculiar differences from mathematical optimization as well as from machine learning approaches. In mathematical optimization, the input (scores and classifications) is given and an optimal model is computed from it, with a one-way process. In machine learning, the classification is given for granted and it is used to train an automatic classifier, not necessarily based on an explicit and interpretable mathematical model. In our approach neither the scores nor the model nor the classification can be assumed to be perfect. Rather, the interplay between them is investigated in a never-ending loop. This iterative process, in which neither the data nor the model are fixed, but rather both are subject to modifications, allows to gradually refine and improve the classification system, to make it more consistent, reliable and explainable over time. Mathematical optimization models and algorithms (integer linear programming) have not been used to compute solutions but to guide the iterative improvement process which is still ongoing and is likely to be developed during the next months or years.

As already mentioned in the introduction, assessment criteria and legislation are country-specific; this hampers transferring the results of specific studies from a country to others. However, although this study was limited to a single case (Azienda Socio-Sanitaria Territoriale, Crema, Italy), we hope to shed some light on generally applicable methods, that may possibly find application in a broader context.

Acknowledgments. The authors are grateful to three anonymous reviewers whose observations and suggestions have been helpful to improve both the content and the presentation.

References

1. Bonetta, C.F.: La valutazione multi-dimensionale (VMD) geriatrica nel riconoscimento dell'indennitá di accompagnamento presso l'A.S.S.T. Ospedale Maggiore di Crema, medical specialization thesis, University of Verona (2023)
2. Cortes, C., Vapnik, V.: Support-vector networks. Mach. Learn. **20**(3), 273–297 (1995)
3. De Vreese, L.P., Bruno, P., Giuliani, G., Scalori, G., Fabbo, A.: Medico-legal evaluation of dementia: validation of a multidimensional scale. G. Gerontol. **62**, 161–171 (2014)
4. Folstein, M.F., Folstein, S.E., McHugh, P.R.: Mini-mental state. A practical method for grading the cognitive state of patients for the clinician. J. Psychiatr. Res. **12**, 189–198 (1975)
5. Hilbe, J.M.: Logistic Regression Models. Chapman and Hall/CRC Press (2009)
6. Katz, S., Ford, A.B., Moskowitz, R.W.: Studies of illness in the aged. The index of ADI: a standardized measure of biological and psychological function. J. Am. Med. Assoc. **185**, 914–919 (1963)
7. Mahoney, F.I., Barthel, D.W.: Functional evaluation: the Barthel index. Maryland State Med. J. **14**, 61–65 (1965)
8. Rengo, G., Brandi, V., Bencivenga, L., Onder, G.: La valutazione multidimensionale geriatrica. In: Antonelli Incalzi, R., Cesari, M., Leosco, D., Pedone, C., Ungar, A., Volpato, S. (eds.) Manuale di geriatria, edra (2019)
9. Schön, P., Parker, M.G., Kåreholt, I., Thorslund, M.: Gender differences in associations between ADL and other health indicators in 1992 and 2002. Aging Clin. Exp. Res. **23**(2), 91–98 (2011)
10. Sheehan, C.M., Tucker-Drob, E.M.: Gendered expectations distort male-female differences in instrumental activities of daily living in later adulthood. J. Gerontol. B **74**(4), 715–723 (2019)
11. Tinetti, M.E.: Performance-oriented assessment of mobility problems in elderly patients. J. Am. Geriatr. Soc. **34**, 119–126 (1986)

Science

Innovative Approaches to Analyze Truck Dwell Times in Seaport Environments

Airam Expósito-Márquez, Israel López-Plata, and Christopher Expósito-Izquierdo(✉)

Universidad de La Laguna, C/ Padre Herrera s/n, 38200 San Cristóbal de La Laguna, Santa Cruz de Tenerife, Spain
cexposit@ull.edu.es
http://www.ull.es

Abstract. This work presents a set of intelligent techniques aimed at analyzing and solving various issues regarding truck arrivals and stays at seaports. These techniques aim to optimize the truck arrival process, reduce dwell times, and alleviate congestion situations. The proposed techniques offer a comprehensive analysis of truck status at seaports, including the use of machine learning techniques to predict future situations, data visualization tools for a concise overview of current status, and simulation models to analyze behavior in hypothetical scenarios. All proposed intelligent techniques are integrated into a decision-support system, providing port management with detailed planning of truck arrival statuses. This enables the planning and optimization of port spaces to reduce truck waiting times and entry door congestion.

Keywords: Trucks · Entry gates · Machine learning · Data visualization · Simulation model

1 Introduction

Maritime international trade is an activity with a significant impact on global commerce. The volume of goods transported by this means has grown considerably in recent years, although it stabilized after the growth caused by COVID-19. For instance, according to the UNCTAD 2023 report [14], nearly 2 billion tons of containers were loaded in various ports in 2022. Historically, scientific literature has predominantly focused on improving sea-side operations, while land-side operations have received substantially less attention in terms of research. However, the cost of land-side activities can exceed that of sea-side activities, with land-side operations representing between 40% and 80% of the total cost [1].

The proper management of entry gates is one of the key focal points in optimizing landside operations, given the critical impact they have on both the terminal's performance and the city in which it is situated. For this reason, digital systems are increasingly utilized to enhance these operations, as supported by [2], such as Truck Access Registration Systems. These systems utilize cameras and

A. A. Juan et al. (Eds.): DSA ISC 2024, LNCS 14778, pp. 165–170, 2025.
https://doi.org/10.1007/978-3-031-78238-1_15

sensors to gather data concerning truck access at ports, enabling managers to pinpoint potential bottlenecks at entry gates. Another alternative system is the Gate Appointment Systems, which permit transporters to schedule their preferred time window for port access. These systems aim to diminish queues at entry gates and are extensively examined by terminal managers, as indicated by [3].

This paper introduces a set of intelligent techniques designed to optimize truck arrival operations at seaports. These techniques enable the analysis of the current status of trucks on the landside of the terminal and facilitate the prediction of future situations. They are integrated into a decision-support system that assists terminal managers in making critical decisions.

The remainder of this paper is organized as follows. Section 2 analyzes the existing literature on the application of intelligent techniques in the landside of the terminal. Section 3 provides a brief description of the decision-support system which utilizes these techniques to assist the terminal manager in making appropriate decisions. Section 4 describes the intelligent techniques developed, and lastly, Sect. 5 presents the main conclusions and highlights some promising lines of work to be pursued in the future.

2 Literature Review

The congestion at port entry gates is a well-researched issue in scientific literature due to its significant impact on terminal performance. Poor optimization of these gates results in long queues of trucks and high waiting times for transporters, leading to corresponding economic and environmental impacts. Several authors have aimed to enhance gate management efficiency. For instance, [4] proposes a model employing multi-service queues to analyze congestion and subsequently quantify waiting times for all trucks. Furthermore, [5] specifies the maximum number of trucks a terminal can accommodate based on its available resources.

One possible solution to avoid congestion at entry gates is to create auxiliary structures like the *Chassis Exchange Terminal (CET)* [6], a dedicated section within the terminal for container interchange between trucks. This helps in reducing such operations within the terminal itself. It enables trucks to carry out their operations during low-traffic hours, thereby alleviating congestion during peak hours.

The creation of structures like CET is not possible for all situations; for this reason, it's important to explore alternative solutions such as optimizing the flow of trucks arriving at the terminal. In [7], different truck arrival planning rules are analyzed with the main goal of minimizing delays in their operations. With a higher level of detail, [8] applies different predictive strategies to calculate the congestion level of the entry gates.

Another solution is to create time windows to divide the flow of truck arrivals according to the arrival of vessels [9]. This solution implies the creation of a predictive model to generate the time windows, as well as an optimization process to adjust these windows. A set of genetic algorithms and Simulated Annealing were proposed. Lastly, [10] explains a collaborative planning system between trucks and the terminal.

3 Decision-Support System

The intelligent techniques proposed in this paper are integrated into a decision-support system designed to present important information to the terminal manager as understandably and simply as possible. With this software, the manager can establish guidelines to optimize the stay and transit of trucks in the terminal. Furthermore, it provides a general overview of the port's status, which helps the manager make the best possible decisions.

The information in the decision-support system is transformed from the raw data provided by various datasets such as sensors, databases, or files through four steps. The first step involves obtaining the data from the different available sources. Once the data is obtained, it undergoes transformation using an *ETL* process (extraction, transformation, and load) and is saved in a data warehouse. The third step involves applying intelligent techniques to the data stored in the system to obtain new information. Finally, this information is presented to the users through dashboards (Fig. 1(a)) or other visualization elements such as maps (Fig. 1(b)).

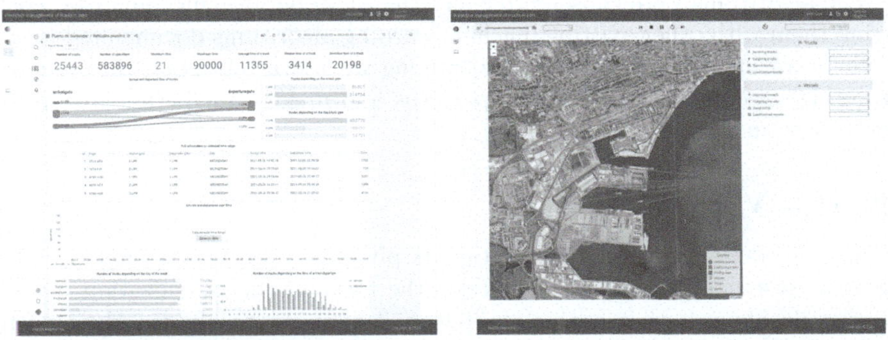

(a) Trucks dashboard (b) Simulation model

Fig. 1. Decision-support system

4 Intelligent Techniques

This section explains the set of intelligent techniques created to help terminal managers optimize truck operations using the decision-support system explained in Sect. 3. The main aim of these techniques is to extract useful knowledge from available data through various algorithms and simulation models.

4.1 Machine Learning Algorithms

The machine learning algorithms seek to predict new information about future situations. They use statistical models to analyze data about vehicle access to the port and make predictions about when new vehicles will arrive at the port in the future and how long they will stay inside the terminal. This implies identifying patterns and trends that can be used to make accurate predictions.

The process followed by these algorithms uses historical information to train the statistical model, and once the training is finished, it executes the trained model to make predictions. The data used to make these predictions include the time of day, type of cargo, vehicle size, among others. The machine learning algorithms integrated into the system are as follows:

- *C4.5* [11]. An algorithm that explores a decision tree using attributes with different characteristics. Additionally, it can utilize missing values on attributes to enhance the exploration of the decision tree.
- *XGBoost* [12]. Also known as "Extreme Gradient Boosting," this algorithm employs supervised learning for classification and regression tasks. It exhibits strong performance, particularly with large datasets.
- *Random forest* [13]. The algorithm described is based on combining multiple decision trees to make decisions, whether for classification or regression purposes. Its primary advantage lies in the flexibility it offers for use.

The algorithms were chosen after an exhaustive performance analysis, comparing them with other existing machine learning algorithms documented in scientific literature. Following extensive training with large datasets, they demonstrate high accuracy in recognizing patterns and making predictions regarding truck stays and arrivals.

4.2 Data Visualization

A correctly designed data visualization is pivotal in optimizing any complex logistical system. In this particular case, the focus of the visualization was on creating dashboards that enable the terminal manager to observe the current status of the entry gates and identify problematic situations such as gate congestions, delays, and the flow of traffic inside the terminal.

In particular, three dashboards were developed for this paper. The first dashboard provides real-time information about the trucks inside the terminal. The second one is focused on prediction; it utilizes the data returned by machine learning algorithms to predict the duration of trucks' stays at the terminal. Lastly, the third dashboard displays the relationship between the arrivals of trucks and vessels. This final dashboard enables the terminal manager to optimize the flow of goods upon arrival at the seaport. Figure 1(a) shows an example of the first dashboard, showing information like the number of trucks that pass through the gates, the flow of these trucks inside the terminal, and more.

4.3 Simulation Model

Terminal managers find simulation models invaluable as they provide a means to visualize terminal operations in hypothetical scenarios. In this case, a discrete-event simulation model was designed to simulate the arrival of trucks and vessels at the port, facilitating the analysis of their interrelation. When a vessel arrives, it triggers the assignment of trucks to handle its transshipment operations.

The simulation model incorporates various parameters to accurately depict terminal operations. These parameters include:

- *Arrival rates.* These define the frequency at which trucks and vessels arrive at the terminal.
- *Transshipment operations.* Parameters related to the handling of cargo, such as loading and unloading times, capacity of trucks and vessels, etc.
- *Port space.* Parameters describing the available space within the port for parking trucks, docking vessels, and conducting transshipment activities.
- *Route networks.* Information about the road network for trucks and waterways for vessels, including distances and travel times.
- *Simulation time.* The duration for which the simulation runs, allowing managers to observe operations over a specific time frame.
- *Performance metrics.* Criteria used to evaluate terminal performance, such as throughput, turnaround time, and resource utilization.

By adjusting these parameters, terminal managers can explore different scenarios, assess the impact of changes or improvements, and make informed decisions to optimize terminal efficiency and performance.

Figure 1(b) demonstrates an instance of the simulation model output, showcasing routes on a map alongside relevant data presented in textual format.

5 Conclusions and Further Research

This paper introduces a set of intelligent techniques aimed at helping terminal managers make decisions regarding the arrival and stays of trucks at the seaport. All these techniques are integrated into a decision-support system that presents the information provided by the techniques through a user-friendly interface. The first group of techniques comprises machine learning algorithms that enable the prediction of new data for future truck arrivals at the terminal. The machine learning techniques used include *C4.5*, *XGBoost*, and *Random Forest.* All these algorithms perform predictions with a high precision ratio. The second technique consists of correct data visualization using dashboards. These dashboards represent key information about truck stays and allow the manager to have a precise overview of the system. The last technique is a discrete-event simulation model, whose main goal is to demonstrate the behavior of the terminal in hypothetical situations, such as the arrival of different vessels at the port.

For future lines of work, it would be interesting to use these techniques in other issues that arise in a terminal, such as vessel arrivals or the management of internal transport vehicles. On the other hand, machine learning could be improved by adding new variables to the prediction, such as port capacity or the size of entry gates. Lastly, it is important to test the created decision-support system in real environments and make necessary improvements based on the feedback obtained.

Acknowledgements. This work has been partially funded by the University of La Laguna through the projects headed by young researchers of its own Research Plan 2023 (accounting code 2023/2305) and the Grants for Research Projects 2023, CajaCanarias Foundation and "La Caixa" Foundation, (project 2023DIG35).

References

1. Noteboom, T.E., Rodrigue, J.-P.: Port regionalization: towards a new phase in port development. Marit. Policy Manag. 297–313 (2007). https://doi.org/10.1080/03088830500139885
2. Inkinen, T., Helminen, R., Saarikoski, J.: Technological trajectories and scenarios in seaport digitalization. Res. Transp. Bus. Manag. (2021)
3. Minh, C.C., Nguyen, V.N.: Optimising truck arrival management and number of service gates at container terminals. Marit. Bus. Rev. **8**(1), 18–31 (2023)
4. Guan, C., Liu, R.: Container terminal gate appointment system optimization. Marit. Econ. Logist. 378–398 (2009). https://doi.org/10.1057/mel.2009.13
5. Huynh, N., Walton, M.: Robust scheduling of truck arrivals at marine container terminals. J. Transp. Eng. (2008). https://doi.org/10.1061/(ASCE)0733-947X(2008)134:8(347)
6. Dekker, R., Heide, S., Asperen, E., Ypsilantis, P.: A chassis exchange terminal to reduce truck congestion at container terminals. Flex. Serv. Manuf. J. 528–542 (2013). https://doi.org/10.1007/s10696-012-9146-3
7. Huynh, N.: Reducing truck turn times at marine terminals with appointment scheduling. Transp. Res. Rec. 47–57 (2009). https://doi.org/10.3141/2100-06
8. Sharif, O., Huynh, N., Vidal, J.M.: Application of El Farol model for managing marine terminal gate congestion. Res. Transp. Econ. 81–89 (2011). https://doi.org/10.1016/j.retrec.2011.06.004
9. Chen, G., Govindan, K., Yang, Z.: Managing truck arrivals with time windows to alleviate gate congestion at container terminals. Int. J. Prod. Econ. 179–188 (2013). https://doi.org/10.1016/j.ijpe.2012.03.033
10. Phan, M., Kim, K.H.: Collaborative truck scheduling and appointments for trucking companies and container terminals. Transp. Res. Part B: Methodol. 37–50 (2016). https://doi.org/10.1016/j.trb.2016.01.006
11. Quinlan, J.R.: C4.5 Programs for Machine Learning. Morgan Kauffman (1993)
12. Chen, T., Guestrin, C.: XGBoost: a scalable tree boosting system. In: Proceedings of the ACM SIGKDD International Conference on Knowledge Discovery and Data Mining, pp. 785–794 (2016)
13. Breiman, L.: Random forests. Mach. Learn. 5–32 (2001)
14. UNCTAD. Review of Maritime Transport 2023. https://unctad.org/publication/review-maritime-transport-2023

Emotion Recognition from Text and Audio Dataset Using Cumulative Attribute Graph Neural Networks (CA-GNN)

Hussein Farooq Tayeb Alsaadawi(✉) and Resul Das

Technology Faculty, Department of Software Engineering, Firat University Technology, 23119 Elazig, Turkey
husseintaeyb@gmail.com, rdas@firat.edu.tr

Abstract. Emotions play a critical role in human understanding and interpersonal communication. Deciphering emotions from text and audio sources presents significant challenges in Affective Computing and Human-Computer Interaction. The challenge intensifies for those with hearing or speech impairments, necessitating accessible technologies. Our study proposes a new approach, employing advanced AI to tackle text and audio data emotion recognition complexities. This approach directly caters to the unique needs of deaf and speech-impaired communities, striving to enhance the accuracy of emotion recognition and understanding. Our study innovates by integrating text and audio features via the CMU-Multimodal Opinion Sentiment and Emotion Intensity CMU-MOSI dataset, employing t-distributed Stochastic Neighbor Embedding (t-SNE) visualizations for comprehensive feature representation. This fusion approach aims to boost emotion detection accuracy by harnessing more information from each modality. Using CA-GNNs for classification, our method leverages the combined feature set for accurate emotion identification, thereby enhancing model effectiveness.

Keywords: Bimodal emotion recognition · t-distributed Stochastic Neighbour Embedding (t-SNE) · Emotion recognition · cumulative attribute-weighted graph neural network (CA-GNNs)

1 Introduction

Envision a future where deaf and Hard-of-Hearing (HoH) individuals effortlessly overcome communication barriers. In our digital age, technological integration is crucial, especially for persons with disabilities. Approximately 20% of the global population experiences deafness and muteness, according to the World Health Organization (WHO) [23]. Traditional communication methods like sign language require mutual proficiency, often a rarity. Our study presents a groundbreaking AI solution designed to understand how emotion transforms how people with hearing and speech impairments communicate. This system is multilingual

A. A. Juan et al. (Eds.): DSA ISC 2024, LNCS 14778, pp. 171–185, 2025.
https://doi.org/10.1007/978-3-031-78238-1_16

and utilizes sophisticated machine-learning methods to enhance communication skills greatly.

Organizations such as the World Association of Sign Language Interpreters and the World Federation of the Deaf, as highlighted in the United Nations' 2019 release [17], have strived to deploy human interpreters. The growing dependence on technology for translation and comprehension underscores the importance of accessible solutions. These tools should focus on the literal translation and capture the emotional context of words for clearer understanding. Our study introduces the Bimodal-Cumulative Attribute Graph Neural Network (Bi-CAGNN) algorithm, adept at interpreting and responding to both vocal and textual cues. This surpasses current technologies like O'Connor's gesture-sensing glove and Islam et al.'s sensor-equipped glove for the Bengali alphabet [1]. Our approach integrates wearable technology with advanced machine learning, as seen in the Automated Sign Language Interpreter (SLI) [19], providing a comprehensive solution for the HoH community. This article explores the Bi-CAGNN development and its potential as a transformative tool in enhancing communication accessibility for the deaf and HoH.

The Bi-CAGNN algorithm utilizes a neural network framework to process textual and vocal inputs, enabling the detection of a wide emotional spectrum. This is vital for the HoH community, offering insights into the subtleties of spoken language that extend beyond text. Additionally, the algorithm conducts an in-depth analysis of vocal characteristics such as tone, pitch, and rhythm, coupled with textual analysis. This dual approach provides a comprehensive understanding of communication, empowering HoH individuals with a deeper grasp of the speaker's emotions, going beyond simple text analysis capabilities. This adaptability is instrumental in making it a dynamic tool in the field of ER technology. Furthermore, it aims to diminish communication barriers for deaf individuals by facilitating a nuanced understanding of emotional nuances in conversations. This fosters greater inclusivity in social interactions and bridges a vital gap in text-only communication methods. Emotion recognition and audio-to-text technologies enhance communication for the hard of hearing by transcribing speech and capturing emotions, making conversations more accessible. Tools like Otter.ai and Ava provide live transcription, especially useful when lip-reading is obstructed, by offering features such as shared notes and instant captioning, thereby facilitating better engagement in various scenarios [8].

1.1 Study Contributions and Novelties

Our research investigates the limitations inherent in current text-based communication platforms designed for individuals with hearing impairments. While these tools proficiently support functional dialogue, they notably fall short in conveying emotional nuance and depth. We emphasize the untapped potential of affective computing and AI to improve emotion recognition and expression in digital communications, particularly noting the current methods' limited emotional range. By introducing a novel approach that combines text and audio analysis using the Bi-CAGNN model and analyzing the CMU-MOSI dataset, we offer a promising solution to enhance emotion interpretation for individuals with hearing impairments. This research moves towards bridging the gap in accurately conveying and interpreting emotional cues in text-based communication, aiming to develop more effective assistive technologies for the deaf and hard of hearing.

The study is meticulously organized to guide the reader through its extensive findings, starting with an introductory section. It then delves into a thorough literature review in Sect. 2, outlines methodologies and proposed approaches in Sect. 3, presents results in Sect. 4, and engages in a detailed discussion in Sect. 5. The paper concludes with Sect. 6.

2 Literature Review

In 2009, significant strides were made with Daassi-Gnaba and Jaime Lopez Krahe's creation of an artificial talking head to aid lip readers, combining speech and emotion recognition (ER) technologies [5]. The same year, Millett P. evaluated speech-to-text captioning's accuracy in educational settings for the deaf or HoH students, emphasizing the ongoing challenges and advancements [14]. Progressing to 2020, M. Shezi and A. Ade-Ibijola introduced Deaf Chat, a mobile app enhancing speech-to-text conversion for multiple speakers using advanced technologies [21]. By 2021, Alam F. et al. developed an integrated sign language translation system, marking a significant leap in communication tools for the deaf and HoH [1]. The year 2022 saw the introduction of the Deep Emotional Arousal Network (DEAN) for multimodal sentiment analysis and ER by Zhang et al. [24], alongside Arun Prasath G and Annapurna Panaiyappan's voice recognition system for the HoH community [7].

In 2024, Alsaadawi and Das demonstrated the critical role of GNNs in affective computing, enhancing emotion detection capabilities [3]. Additionally, the integration of Natural Language Processing (NLP) with ER technologies significantly advanced communication aids for the deaf or HoH, as highlighted in research utilizing models like BERT and RestNet50v2 [11,16]. Recently, the CA-WGNN model emerged as a breakthrough in emotion recognition, demonstrating superior performance in classifying emotions from multimodal data [2]. Additionally, S. Anithaelavarasi evaluated the effectiveness of various methodologies in recognizing MNIST Sign Language, stressing the significance of language and sentiment understanding for the HoH population [18]. In multimodal affective computing, A. Shahid et al. showed how t-SNE is instrumental in ER, particularly in evaluating emotional audio in synthesized speech [20].

3 Methodology and Implementation

In this section, we elaborate on our methodology, which is anchored in the principles of affective computing and focuses on a multimodal analysis framework. Utilizing the CMU-MOSI Dataset, comprising audio and textual data, Our methodology synthesizes insights drawn from seminal works in the field to refine computational models for emotion recognition.

Our analysis integrates extracted features into a single array for further examination or use in machine learning models. This integration is reflected in our classification algorithms, demonstrating a comprehensive approach to handling multimodal data. Figure 1 depicts the structure of the algorithm. Drawing on principles from [2], our algorithm is designed to convert text and audio into formats that are more accessible to the hard of hearing, using emotional cues to improve understanding. This approach underscores the importance of a multifaceted methodology in affective computing, especially for increasing accessibility and comprehension for those with hearing impairments.

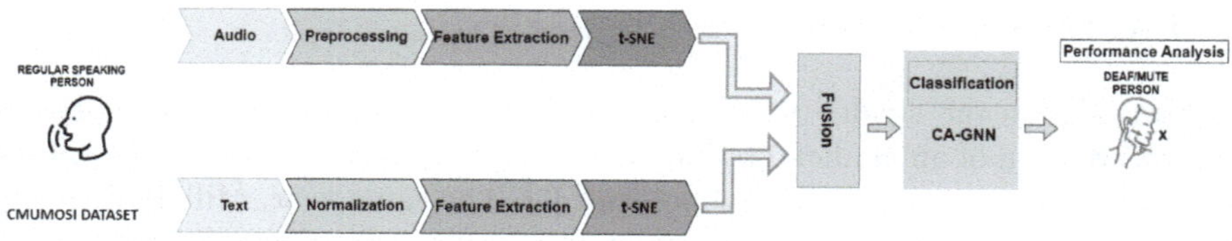

Fig. 1. HoH: Analysis of Bi-CAGNN Algorithm.

3.1 Dataset

The CMU-MOSI (Multimodal Corpus of Sentiment Intensity) dataset [10], forms a comprehensive resource for research in multimodal sentiment analysis and emotion recognition. The CMU-MOSI dataset comprises 2,199 opinion video clips, each presenting an individual's viewpoint. Sentiment labels in this dataset range from −3 (strongly negative) to 3 (strongly positive), accompanied by extensive annotations. These include subjectivity (objective vs. subjective opinions), sentiment intensity, visual features (annotated per frame and opinion), and detailed millisecond-level audio features. These pre-processed audio files are pivotal for extracting acoustic features, serving as inputs for deep-learning algorithms to classify emotions effectively. Expanding the research scope, the dataset encompasses over 23,500 sentence utterances from more than 1,000 YouTube speakers, featuring video snippets from YouTube movie reviews annotated for sentiments and a range of emotions like happiness, sadness, anger, fear, disgust, and surprise. This dataset is particularly noteworthy for its combination of text, audio, and visual data, making it a rich resource for sentiment analysis, emotion recognition, and multimodal machine learning tasks. Recent studies [10] highlight the significance of these datasets, particularly the CMU-MOSI, in advancing the Emotion Recognition (ER) field. The CMU-MOSI datasets, which incorporate both visual and auditory data, are crucial for their rich emotional content, making them essential for researchers in multimodal affective computing, natural language processing, and machine learning. These datasets are key for developing sophisticated emotion recognition models that interpret real-world human communication.

3.2 Emotion Classification Using BERT

After data preprocessing, the study leverages the BERT (Bidirectional Encoder Representations from Transformers) model, a preeminent model in the field of natural language processing, for emotion classification. The model, adept at understanding the nuances of language, categorizes each textual document into distinct emotions. This process showcases the application of pre-trained models in emotion analysis and underscores the model's adaptability across diverse linguistic datasets. BERT (Bidirectional Encoder Representations from Transformers) represents a seminal advancement in the domain of natural language processing (NLP), as introduced by [6]. This model significantly enhances machine understanding of human language by leveraging bidirectional training, wherein it simultaneously evaluates the context of each word in a sentence from both sides. Predicated on the Transformer architecture, which eschews traditional sequential processing methods for a mechanism focused on global dependencies between words, BERT is distinguished by its ability to grasp the nuanced meanings of words within their specific contexts. The training methodology for BERT encompasses two stages: pre-training and fine-tuning. In the pre-training phase, the model is exposed to a vast corpus of text and learns language patterns through two unsupervised tasks: Masked Language Model (MLM) and Next Sentence Prediction (NSP). These tasks aid the model in understanding language structure and flow. After pre-training, BERT can be fine-tuned with just an additional output layer, enabling its application to a wide array of NLP tasks, such as question answering and sentiment analysis, with minimal modifications. BERT's introduction has precipitated a paradigm shift in NLP, facilitating the development of models that achieve unprecedented accuracy in language understanding tasks. Its methodology of contextual analysis, pre-training on extensive text data, and adaptability to various tasks have set new benchmarks in the field. This breakthrough has paved the way for the emergence of numerous BERT-inspired models and underscored the model's versatility across different linguistic challenges, affirming its significant impact on advancing machine comprehension of human language [6].

3.3 Visualization and Dimensionality Reduction: Creating a Valence-Arousal Graph

This outlines the process of categorizing emotions, retrieving emotional scores, employing analysis techniques, and graphically representing emotional spectrums. The core of this computational analysis is establishing an *emotion_mapping* dictionary. This dictionary assigns discrete emotions, such as "anger", "happiness", and "sadness", to specific valence and arousal scores. Valence scores represent the emotion's positivity or negativity, and arousal scores indicate the emotion's intensity. These scores are based on a widely accepted understanding of emotional experiences, offering a standard approach to quantifying emotions in textual data. Following the mapping of emotions, the process involves scanning a text corpus to extract emotions from each document. Using the *emotion_ mapping*, it assigns the relevant valence and arousal scores. These scores are collected into two arrays: *valence_ scores* and *arousal_ scores*, converting qualitative emotional expressions into quantitative data for analysis. This methodology enhances understanding of emotional content in text datasets through visualization and detailed analysis, improving digital communication for applications in sentiment analysis, psychology, and marketing. It makes the complex emotional content within large datasets more accessible, giving us a deeper understanding of human emotions. Figure 2 illustrates the distribution of valence and arousal scores as derived from our analysis.

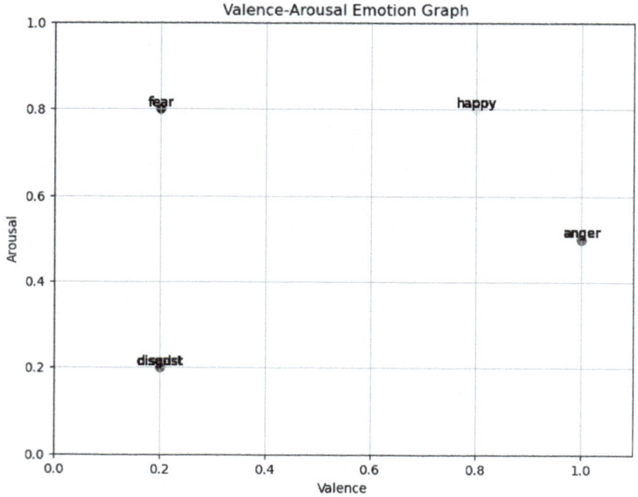

Fig. 2. Text-Valence-Arousal Emotion Graph.

3.4 Dimensionality Reduction in Text Data Using T-SNE

The procedure for reducing the dimensionality of text data commences by encoding the textual information numerically through the Term Frequency-Inverse Document Frequency (TF-IDF) method. This step prepares the data for the subsequent reduction in dimensionality. The t-distributed Stochastic Neighbor Embedding (t-SNE) technique is applied. Renowned for its capability to illustrate high-dimensional data in a comprehensible manner, t-SNE effectively models the similarities between data points as joint probabilities and strives to minimize the Kullback-Leibler divergence between the embeddings in the original high-dimensional space and the reduced-dimensional space. Despite its sensitivity to initial conditions and the non-convex nature of its cost function, which may lead to variable results, t-SNE unveils intricate structures within the data.

Such revelations are particularly beneficial for the domain of affective computing, which seeks to decode human emotions through diverse channels. Specifically, t-SNE facilitates the discernment of intricate relationships among various emotional states, thereby augmenting our understanding of emotional expressions. Figure 3 displays the dimensionality reduction of text data, illustrating the distribution and relationships within the dataset in a 2D space. The t-SNE method effectively preserves the data's local structures while simplifying its representation. In the context of affective computing, the efficacy of t-SNE can be attributed to several factors:

- **Perplexity:** This critical parameter influences the accuracy of the visualization by indicating the anticipated density of data around a point, thereby affecting the balance between local and global aspects of the data structure.
- **Data Quality:** The fidelity of the initial feature extraction be it from text, audio, or visual data-significantly impacts the resulting visualization. The presence of noise and irrelevant features may obscure the genuine categories of emotions.
- **Emotional Variability:** t-SNE's proficiency in clustering similar emotions while distinctly separating dissimilar ones underscores the diversity within emotional expressions, necessitating a refined interpretation of these variations.
- **Visualization and Hyperparameters:** Additional visualization techniques, such as the valence-arousal graph, and meticulous adjustment of t-SNE hyperparameters, including the learning rate and number of iterations, are crucial for achieving optimal visual representations.
- **Comparative Analysis:** While t-SNE facilitates the comparative study of emotional responses elicited by different stimuli, its stochastic nature can pose challenges to straightforward interpretation.

Through its ability to unveil latent structures within data, t-SNE significantly contributes to our comprehension of complex datasets, thereby propelling further analytical and modeling endeavors in emotion recognition, as it identifies semantic correlations within neural. word embeddings [12].

3.5 Data Fusion and Feature Integration

Multimodal data fusion and analysis The cornerstone of this research lies in its multimodal approach, integrating audio data with textual analysis. Audio files are transcribed, and the transcriptions mirror the textual data processing steps. Using both textual and auditory data in parallel highlights the multifaceted nature of emotion recognition. Feature-level fusion, a pivotal technique in data integration, plays a significant role in multimodal affective computing as shown in Fig. 4. This method combines text, audio, and visual data into a unified feature set, starting with feature extraction using t-SNE for dimensionality reduction and visualization. The process involves validating and concatenating these features to form a comprehensive 'fused features' array. Feature-level fusion enhances machine learning efficacy by merging unique insights from

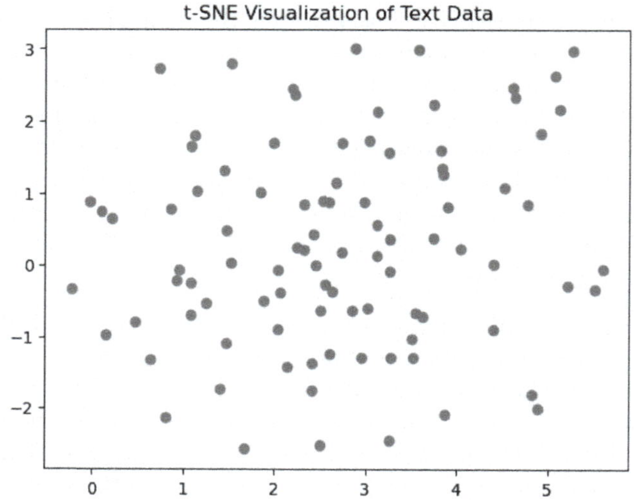

Fig. 3. t-SNE Visualization of Text Data.

multiple modalities, significantly improving predictive and classification model performance. It is crucial for effectively advancing multimodal data analysis and integrating diverse data sources.

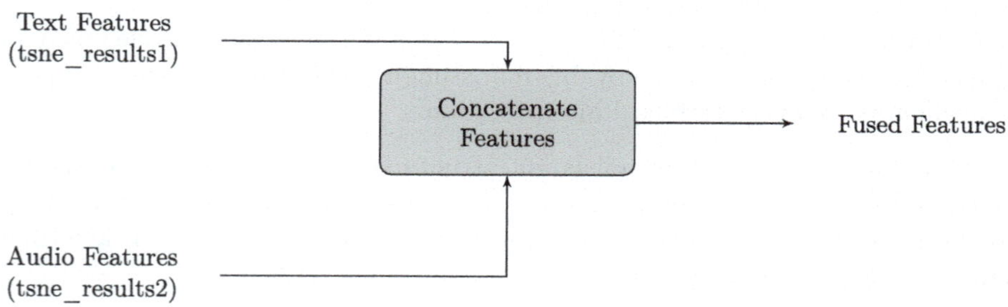

Fig. 4. Block diagram illustrating the feature-level fusion process.

3.6 Bi-CAGNN Approach for Emotion Classification

CA-GNN is a powerful machine learning model that integrates multimodal data with emotion identification. First, a neural network is built for binary classification, then it is trained and evaluated. Cumulative Attribute Graph Neural Networks (CA-GNNs) detailed in Fig. 5, aggregate attributes from a node's direct neighbors and cumulatively from nodes at greater distances at each layer. This aggregation can be weighted based on distance, node importance, or other graph metrics, allowing for a nuanced integration of attributes representing an advanced framework within the domain of graph neural networks (GNNs), specifically designed to enhance the learning of node representations by incorporating local and non-local structural and attribute information. This academic discussion delves into the core concepts, methodologies, and potential applications of CA-GNNs, offering a comprehensive understanding of their significance in machine learning and network analysis.

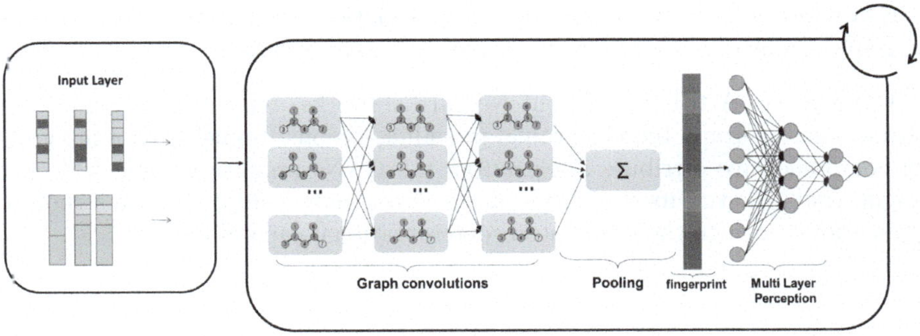

Fig. 5. The CAGNN architecture [2].

Graph Neural Networks (GNNs) are a significant method for analyzing graph-structured data, useful in fields like social network analysis, bioinformatics, and recommendation systems. Traditional GNN models, like Graph Convolutional Networks (GCNs) and Graph Attention Networks (GATs), mainly gather features from direct neighbors to update node representations. However, these models often fall short in utilizing the full hierarchical structure of graphs, especially in recognizing long-range dependencies and incorporating global information.

Concepts. CA-GNNs extend traditional GNN architectures by incorporating mechanisms to aggregate and integrate attributes from direct neighbors and nodes further away in the graph, effectively capturing cumulative attribute information. This approach enables the model to leverage a more comprehensive set of features, encompassing the immediate locality and broader structural contexts. The key components of CA-GNNs include:

- **Attribute Aggregation:** CA-GNNs aggregate attributes from both direct neighbors and distant nodes at each layer, with weightings based on distance or node significance, enabling nuanced attribute integration.
- **Feature Transformation:** CA-GNNs apply learnable transformations to node features to refine the aggregated data for better prediction accuracy.
- **Hierarchical Representation Learning:** By layering, CA-GNNs develop hierarchical representations that capture broader graph dynamics, essential for analyzing complex structures.

Applications: CA-GNNs find applications across a wide array of domains where graph-structured data is prevalent:

- **Social Network Analysis**: Understanding community structures, influence propagation, and user behavior prediction.
- **Bioinformatics**: Protein function prediction, gene expression analysis, and biomolecular interaction modeling.
- **Recommendation Systems**: Leveraging user-item interaction graphs to improve recommendation accuracy and personalization.
- **Knowledge Graphs**: Entity resolution, relation prediction, and knowledge graph completion for enhancing semantic web applications.

Cumulative Attribute Graph Neural Networks represent a significant advancement in graph-based learning, providing a robust framework for integrating comprehensive attribute information. CA-GNNs promise to unlock deeper insights and achieve superior performance across many complex graph structure applications through their enhanced representational capabilities.

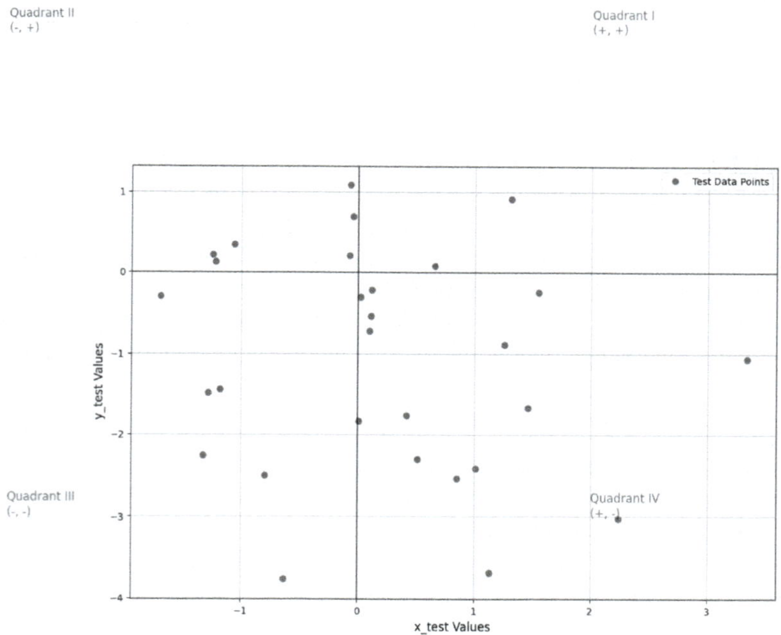

Fig. 6. X, Y Plot of Test Data.

The scatter plot depicted in Fig. 6 elucidates the relationship between two variables, x_{test} and y_{test}. Each red point on the plot signifies a pair of corresponding values from these arrays. The horizontal axis represents the x_{test} values, while the vertical axis corresponds to the y_{test} values. The data distribution, as shown, does not exhibit a distinct linear pattern, indicating an absence of a straightforward linear relationship between the variables. This is further evidenced by the spread of data points across all four quadrants of the plot. Specifically, Quadrant I (top right) contains points where both x_{test} and y_{test} are positive, Quadrant II (top left) with negative x_{test} but positive y_{test}, Quadrant III (bottom left) where both are negative, and Quadrant IV (bottom right) with positive x_{test} and negative y_{test}. This variability indicates a diverse dataset and suggests that a simple linear model may not be sufficient for accurate prediction. Additionally, the size of the training data, denoted as $x_{\text{train}}.shape$ and $y_{\text{train}}.shape$, is reported as (114,), indicating that each training dataset comprises 114 samples. This size is a crucial aspect in determining the learning capability of the model, where generally, a larger dataset is preferable for enhanced model performance.

4 Experimental Results

This study introduces the Bi-CAGNN algorithm, an advanced ER system enhancing text and audio analysis for the HoH community. It accurately classifies nuanced emotions and can be customized to meet specific research and application needs. The table delineates the model's performance in classifying data into Negative, Neutral, and Positive categories. The model achieves an overall accuracy of 82%, indicating a high proficiency level in correctly identifying the class labels of unseen data. This accuracy is substantiated by the precision and recall metrics, which also stand at 82%, illustrating the model's balanced ability to minimize false positives and false negatives across classes.

Table 1. Model Performance Metrics

Metric Name	Value (%)
Accuracy	0.82
Precision	0.8224
Recall	0.82
F1-Score	0.8202

As seen Table 1, the F1 score, a harmonic mean of precision and recall, mirrors these metrics closely with a value of 0.8202, underscoring the model's consistent performance across both dimensions of precision and recall. This balance is crucial in applications where both the reduction of false positives and the capture of true positives are equally important. A closer examination of the classification report reveals nuanced performance differences across the categories. The model exhibits the highest precision (0.86) in identifying Negative instances, albeit with a slightly lower recall (0.78), suggesting a conservative approach in classifying instances as Negative. Conversely, while having the lowest precision (0.76), the Positive class enjoys the highest recall (0.83), indicating the model's propensity to more liberally classify instances as Positive, albeit at the cost of a higher false positive rate. As seen Table 2, the Neutral category demonstrates the most balanced performance with closely matched precision (0.84) and recall (0.85) metrics, leading to the highest F1 score (0.85) among the three classes. This suggests that the model is particularly adept at identifying genuinely neutral instances, with a good balance between avoiding false positives and capturing true positives. The model may need to be tuned to improve precision and recall, and perhaps more sophisticated methods should be used.

Table 2. Detailed Classification Report

Sentiment	Precision	Recall	F1 Score
Negative	0.86	0.78	0.82
Neutral	0.84	0.85	0.85
Positive	0.76	0.83	0.79
Average/Total	**0.82**	**0.82**	**0.82**

5 Discussion

The Bi-CAGNN algorithm enhances emotional recognition (ER) by integrating audio and text analysis, significantly aiding the hard-of-hearing (HoH) community. It innovatively captures emotions during audio-to-text conversion, improving traditional interpretation methods by offering a nuanced textual representation of spoken emotions. Performance assessments reveal the algorithm's proficiency in sentiment classification, achieving an 82% accuracy. However, its tendency to over-identify positive sentiments suggests the need for refinement to balance precision and recall across different emotional categories, particularly by mitigating false positives in positive sentiment detection without compromising its overall recall capabilities. This research employs multimodal affective computing to aid individuals with hearing impairments, leveraging the CMU-MOSI Dataset to develop a model that accurately classifies emotional states from text and audio. By utilizing advanced feature extraction, dimensionality reduction through t-SNE, and the Bi-CAGNN model, this study not only provides enhanced emotional comprehension for those with hearing loss but also opens new avenues in ER technology by suggesting potential adaptations for diverse languages and integrating visual data to enrich emotional recognition. Moreover, translating audio to text is outlined as a pivotal advancement in communication technologies, greatly benefiting accessibility, comprehension, and content analysis. It facilitates multitasking and extends content reach globally through easier translation into multiple languages. These capabilities are vital in educational, professional, and legal contexts, providing indispensable benefits across various scenarios. Future research could explore its extension to various linguistic contexts and possibly integrate visual data to enhance its multimodal capabilities further. Using affective computing, assistive technologies could be revolutionized, reducing misunderstandings and improving communication for the hearing impaired (Table 3).

Table 3. Consolidated Emotion Recognition Performance on CMU-MOSI Dataset for Multimodal framework

References	Authors	Year	Approach	Accurate(%)
[4]	Chudasama et al.	2022	M2FNET	66.52
[9]	Han et al.	2021	BBFN	86.20
[13]	Mai et al.	2022	MIB	84.40
[15]	Rahman et al.	2020	MAG-BERT	72.16
[22]	Si et al.	2024	SPACE+WavLM	20.90
This study	Alsaadawi et al.	2024	Bi-CAGNNs	82.00

6 Conclusion

This study represents a significant leap in multimodal affective computing, highlighting the field's dynamic evolution. The core of the research focused on applying the Bi-CAGNN algorithm for effective emotion recognition from both vocal and textual inputs, aiming to benefit the deaf community significantly. Utilizing the CMU-MOSI Dataset, the study analyzed the algorithm's performance across various metrics, including accuracy, precision, recall, and F1 score, achieving commendable outcomes. However, the research also illuminated challenges, such as class imbalances, pointing toward the necessity for continued model refinement. An integral aspect of the study was the utilization of the Valence-Arousal Emotion Graph, providing a sophisticated visual representation of emotional states. This innovative approach not only deepens our understanding of the complex interplay between valence and arousal in emotional responses but also underscores the Bi-CAGNN's potential to revolutionize digital communication through precise emotion classification. The research delineates future pathways for enhancing the model, incorporating visual data, and broadening its application spectrum, thereby setting a new course for affective computing research and its societal implications.

References

1. Alam, R., Munir, M., Ishrak, S., Hussain, S., Shalahuddin, M., Islam, M.N.: A machine learning based sign language interpretation system for communication with deaf-mute people. Association for Computing Machinery, New York (2021). https://doi.org/10.1145/3471391.3471422

2. Alsaadawi, H.F.T., Das, R.: TER-CA-WGNN: trimodel emotion recognition using cumulative attribute-weighted graph neural network. Appl. Sci. **14**(6), 2252 (2024). https://doi.org/10.3390/app14062252, https://www.mdpi.com/2076-3417/14/6/2252
3. Alsaadawi, H.F.T., Das, R.: Multimodal emotion recognition using bi-LG-GCN for meld dataset. Balkan J. Electr. Comput. Eng. **12**(1), 36–46 (2024). https://doi.org/10.17694/bajece.1372107
4. Chudasama, V., Kar, P., Gudmalwar, A., Shah, N., Wasnik, P., Onoe, N.: M2FNet: multi-modal fusion network for emotion recognition in conversation. In: Proceedings of the IEEE/CVF Conference on Computer Vision and Pattern Recognition, pp. 4652–4661 (2022)
5. Daassi, H., Krahe, J.: Universal combined system: speech recognition, emotion recognition and talking head for deaf and hard of hearing people (2009). https://www.researchgate.net/publication/237605884, https://doi.org/10.3233/978-1-60750-042-1-503
6. Devlin, J., Chang, M.W., Lee, K., Toutanova, K.: BERT: pre-training of deep bidirectional transformers for language understanding. arXiv:1810.04805 [cs] (2019). https://doi.org/10.48550/arXiv.1810.04805
7. Prasath, A.: Design of an integrated learning approach to assist real-time deaf application using voice recognition system. Comput. Electr. Eng. **102**, 108145 (2022). https://doi.org/10.1016/j.compeleceng.2022.108145, https://www.sciencedirect.com/science/article/pii/S0045790622003950
8. Goldstein, L.A.: Speech-to-text apps for the deaf community. Hearing Like Me (2020). https://www.hearinglikeme.com/speech-to-text-apps-for-the-deaf-community/
9. Han, W., Chen, H., Gelbukh, A., Zadeh, A., Morency, L.P., Poria, S.: Bi-bimodal modality fusion for correlation-controlled multimodal sentiment analysis. In: Proceedings of the 2021 International Conference on Multimodal Interaction, pp. 6–15 (2021). https://arxiv.org/abs/2107.13669
10. Huddar, M.G., Sannakki, S.S., Rajpurohit, V.S.: Multi-level context extraction and attention-based contextual inter-modal fusion for multimodal sentiment analysis and emotion classification. Int. J. Multimed. Inf. Retrieval **9**(2), 103–112 (2020). https://doi.org/10.1007/s13735-019-00185-8
11. Khan, K., Aslam, D.N., Abid, K., Munir, S.: Robot assist sign language recognition for hearing impaired persons using deep learning. VAWKUM Trans. Comput. Sci. **11**(1), 245–267 (2023). https://doi.org/10.21015/vtcs.v11i1.1491, https://www.vfast.org/journals/index.php/VTCS/article/view/1491
12. Liu, S., et al.: Visual exploration of semantic relationships in neural word embeddings. IEEE Trans. Vis. Comput. Graph. 1 (2017). https://doi.org/10.1109/TVCG.2017.2745141
13. Mai, S., Zeng, Y., Hu, H.: Multimodal information bottleneck: learning minimal sufficient unimodal and multimodal representations. IEEE Trans. Multimed. **25**, 4121-4134 (2023). https://doi.org/10.1109/tmm.2022.3171679, http://dx.doi.org/10.1109/TMM.2022.3171679
14. Millett, P.: Accuracy of speech-to-text captioning for students who are deaf or hard of hearing. J. Educ. Pediatric (Re)Habilitative Audiol. **25**, 1–13 (2021). https://www.edaud.org/journal/2021/1-article-21.pdf
15. Rahman, W., et al.: Integrating multimodal information in large pretrained transformers. In: Proceedings of the Conference on Association for Computational Linguistics. Meeting, vol. 2020, p. 2359. NIH Public Access (2020). https://arxiv.org/pdf/1908.05787.pdf

16. Rehman, U., Sohail, A.: Exploring the role of verbal and non-verbal communication skills of nursing in patients healing. Research Square (2023). https://doi.org/10.21203/rs.3.rs-3376906/v1
17. Release, U.N.P.: International day of sign languages (2019). https://www.un.org/en/observances/sign-languages-day
18. Elavarasi, S.A., Jayanthi, J.: Role of machine learning and deep learning in assisting the special children's learning process. J. Algebraic Stat. **13**(2), 2327–2334 (2022). https://publishoa.com/index.php/journal/article/view/430
19. San-Segundo, R., Pérez, A., Ortiz, D., D'Haro, L., Torres, M., Casacuberta, F.: Evaluation of alternatives on speech to sign language translation. In: Proceedings of the Annual Conference of the International Speech Communication Association, Interspeech, vol. 1, p. 2532. International Speech Communication Association (2007). https://doi.org/10.21437/Interspeech.2007-672
20. Shahid, A., Latif, S., Qadir, J.: Generative emotional AI for speech emotion recognition: the case for synthetic emotional speech augmentation. Appl. Acoust. (2023). https://doi.org/10.48550/arXiv.2301.03751
21. Shezi, M., Ade-Ibijola, A.: Deaf chat: a speech-to-text communication aid for hearing deficiency. Adv. Sci. Technol. Eng. Syst. J. **5**, 826–833 (2020). https://doi.org/10.25046/aj0505100 https://doi.org/10.25046/aj0505100 https://doi.org/10.25046/aj0505100
22. Si, S., et al.: SpokenWOZ: a large-scale speech-text benchmark for spoken task-oriented dialogue agents (2024). https://arxiv.org/pdf/2305.13040.pdf
23. Vaidya, O., Gandhe, S., Sharma, A., Bhate, A., Bhosale, V., Mahale, R.: Design and development of hand gesture based communication device for deaf and mute people. In: 2020 IEEE Bombay Section Signature Conference (IBSSC), pp. 102–106 (2020). https://doi.org/10.1109/IBSSC51096.2020.9332208, https://ieeexplore.ieee.org/document/9332208
24. Zhang, F., Li, X.C., Lim, C.P., Hua, Q., Dong, C.R., Zhai, J.H.: Deep emotional arousal network for multimodal sentiment analysis and emotion recognition. Inf. Fusion **88**, 296–304 (2022). https://doi.org/10.1016/j.inffus.2022.07.006, https://linkinghub.elsevier.com/retrieve/pii/S1566253522000653

Estimating Solid Waste Generation in Tenerife by Machine Learning Models

Alejandro Corbea-Pérez, Julio Brito, and José Andrés Moreno-Pérez(✉)

Instituto Universitario de Desarrollo Regional, Universidad de La Laguna, 38206 Santa Cruz de Tenerife, Spain
{acorbeap,jbrito,jamoreno}@ull.edu.es

Abstract. Solid waste management is a major challenge in municipalities and cities due to population growth and consumption habits. Therefore, the need for efficient use of resources, based on a reliable estimate of the waste generated, is the reason why artificial intelligence systems are being increasingly used. This will result in appropriate planning, which is particularly important in island territories. In response to this problem, the purpose of this work is to implement and test models to predict the volumes of mixed waste and biodegradable waste in Tenerife, Spain, using linear regressions and different artificial intelligence methods.

Keywords: Machine learning · Waste Management · Prediction

1 Introduction

Solid waste management is one of the main economic, social, and environmental challenges in municipalities and cities due to population growth and consumer habits. Municipal solid waste are an essential factor in climate change studies, as it is one of the main contributors to urban emissions. The efficient use of resources for the collection of these waste, as well as the observation of optimal sanitary conditions, is also of vital importance. To achieve efficient management, it is essential to have a reliable estimate of the volumes of waste generated in each city, which is particularly difficult in places with high tourist flows, which implies a substantial increase in population. Generation estimation is an essential instrument for designing and planning waste management policies and measures. Information on demand allows planning and making medium-term decisions about equipment, infrastructure and management models, and garbage collection models.

Some of the most reliable solutions for this forecasting are based on Artificial Intelligence, for example, Artificial Neural Network (ANN) models, which can train data for classification and prediction [2]. Insufficient planning is known to be one of the factors leading to poor waste management [9]. Previous studies used time series to generate Municipal Solid Waste (MSW) models [2,8,12]. Machine Learning (ML) approaches have also been used to forecast waste volumes in both large [11] and small-scale areas [13].

A. A. Juan et al. (Eds.): DSA ISC 2024, LNCS 14778, pp. 186–193, 2025.
https://doi.org/10.1007/978-3-031-78238-1_17

Adequate planning is especially important in island territories, taking into account space limitations and the fact that they cannot always count on optimal infrastructures for the final processing of such waste. In response to this problem, we are involved in a project named "Smart Management of Urban Solid Waste in Tourist Destinations project". The objective of this work is to create models to predict the volumes of mixed municipal waste and biodegradable waste on Tenerife Island, Spain, using linear regressions and different artificial intelligence methods. The intended application of this work is the prediction of waste data in Tenerife to facilitate planning. Once the model is trained and validated, it is used to obtain predictions for the following years. Partial results are presented here, since new variables are currently being experimented with in the model.

In addition to demographic and economic variables, solid waste management planning must take into account specific aspects such as container systems, where economic or aesthetic criteria generally take precedence over environmental ones [16]. Data exploration at smaller scales, for example, at the container level, will allow us to obtain more reliable patterns. Finer granularity will lead to more effective resource planning, adapted to specific needs and challenges at the individual level. For this reason, one of the future directions of our research is to include information on containers. Although not currently available, local authorities have committed to making this information available in the short term.

The remainder of this document is organised as follows. In Sect. 2 we describe the methodological aspects of the prediction task. In Sect. 3 we present some preliminary results obtained using three ML methods. Finally, Sect. 4 includes the main conclusions and future work.

2 Methodology

2.1 Study Area

We used waste and population data from Tenerife Island between 2010 and 2022. However, data for 2020 were excluded to avoid biases due to the anomalous behaviour of all indicators due to the COVID-19 pandemic. Tenerife Island has some of the most important tourist destinations in the Canary Islands and Spain. In 2023, the Canary Islands received almost a million more tourists than in 2019, 65% of whom chose Tenerife. In addition to the complexity of dealing with such a high volume of tourist population, the population forecast for 2026 continues to indicate an increase. Therefore, it is essential to project the demand for the resources necessary for the management of solid waste. The study area is shown in Fig. 1, which is the metropolitan area of the island and its most important tourist destination. The metropolitan area is the joint area of the municipalities of Santa Cruz de Tenerife and San Cristóbal de La Laguna, and the cost area of the municipality of Adeje, named "Costa Adeje", is the most important tourist destination in Tenerife.

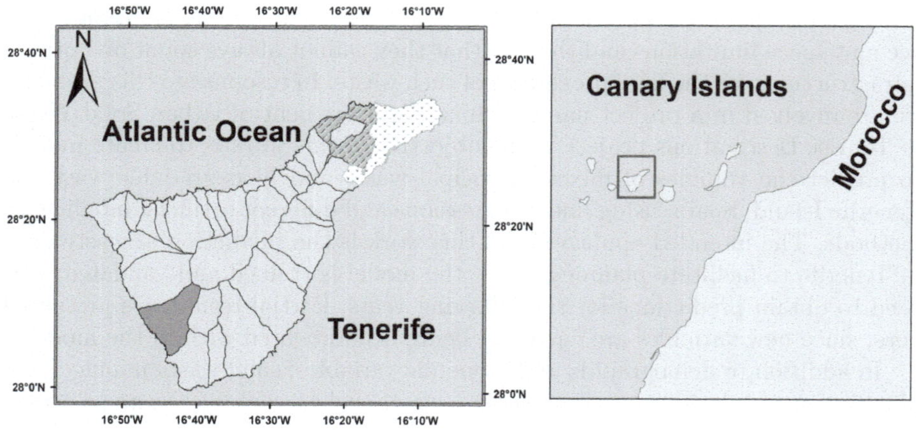

Fig. 1. Location of Tenerife (left) in the Canary Islands (right). Municipalities of Santa Cruz de Tenerife (white), San Cristóbal de La Laguna (yellow with dashed lines) and Adeje (orange). (Color figure online)

2.2 Data

From the monthly mixed municipal waste and biodegradable waste data series for the period 2010–2022, annual data per municipality were obtained. This parameter as a predictor variable allows us to reflect spatial variability throughout the island, given the uneven distribution of the population. Population and waste data were obtained from the Canary Institute of Statistics (ISTAC) (https://www.gobiernodecanarias.org/istac/). Population estimation data for the years 2024, 2025 and 2026 were calculated from Spanish Institute of Statistics (INE) (https://www.ine.es/) projections for Tenerife. All these data are open access. The predicted variables are the weight of mixed municipal waste and biodegradable waste. The year, municipality, and population are the predictor variables, often used in this type of study [4,5].

2.3 Prediction Models

Multiple Linear Regressions (MLR) allow the model of a target variable to be linearly based on a series of independent predictors [3]. In simple regression, the model uses a single predictor variable, while in multiple regression, several factors are included as predictors. Correlation and regression techniques are useful to predict waste volumes using demographic and socioeconomic factors [12]. The dynamic properties of waste generation require, in addition to MLR, new dynamic approaches to predict waste with reasonable accuracy [15].

Furthermore, we obtained two models, one based on Decision Trees (DT) and the other on Artificial Neural Networks (ANN), both of which were used in previous studies to estimate waste volumes. The DT model used the Classification and Regression Tree algorithm (CART). This is a nonparametric model that uses the partitioning of the data of the predictive variables into multiple homogeneous regions that allows the handling of categorical variables and therefore does not require data transformation [14].

Artificial neural networks, with high learning capacity [11], allow the construction of models with uncertain data sets. Their operation simulates the reactions of a biological nervous system. We used single hidden layer feedforward neural networks, as in previous studies [11]. This Neural Network architecture consists of three layers: an input layer, a hidden layer, and an output layer. Each layer consists of a number of nodes linked to each node in the subsequent layer by directed weighted edges [7]. Artificial Neural Networks are a robust method with a high level of fault tolerance, widely used in the prediction of waste generation, given their ability to capture complex relationships in multivariate systems and model complex non-linear behaviour [6]. However, the method tends to overfitting and does not allow for delimitation of which input factor has the highest weight in the analysis. The minimum leaf size and the number of neurons in the hidden layer determine the complexity of the Decision Tree and Neural Network models, respectively. For Decision Trees, we select a leaf size of 10 and 5, and in ANN the hidden layer has 5 and 10 neurons, for mixed municipal waste and biodegradable waste, respectively. The number of neurons in the hidden layer was determined by trial and error, as in previous studies [1].

3 Results

We used the following modules in Python: TensorFlow, Scikit-learn, and Pandas for the implementation of the prediction models.

Data from 2010 to 2019 were chosen to adjust the models, and data from 2021 and 2022 were chosen to validate the results. We used the Pandas Python module for preprocessing data. The LinearRegresion function of Scikit-learn module was used for MLR and DecisionTreeRegressor for CART DT. For the ANN the Keras library and TensorFlow module were used.

To test the accuracy of the models, they were evaluated on the basis of the following statistics: the Mean Absolute Error (MAE), the Root Mean Square Error (RMSE) and the coefficient of determination denoted by R^2 using the corresponding functions of Scikit-learn module.

Figure 2 shows the prediction obtained for mixed municipal waste with the regression model for three municipalities on the island, La Laguna and Santa Cruz, which make up the metropolitan area of Tenerife, and Adeje, the municipality with the highest number of tourists. This is a first approximation; we have worked with the population data without including the equivalent population, based on the number of tourists, which could explain the worse fit obtained in Adeje. Tables 1 and 2 show the MAE, RMSE and R^2 values for the three models obtained for municipal mixed waste and biodegradable waste, respectively. The model created with ANN turns out to be a better predictor, as it obtains slightly higher R^2 and lower RMSE than the one obtained with Decision Trees. These results are consistent with previous studies that indicate that ANN models are better predictors than MLR models [10] and Decision Trees [11].

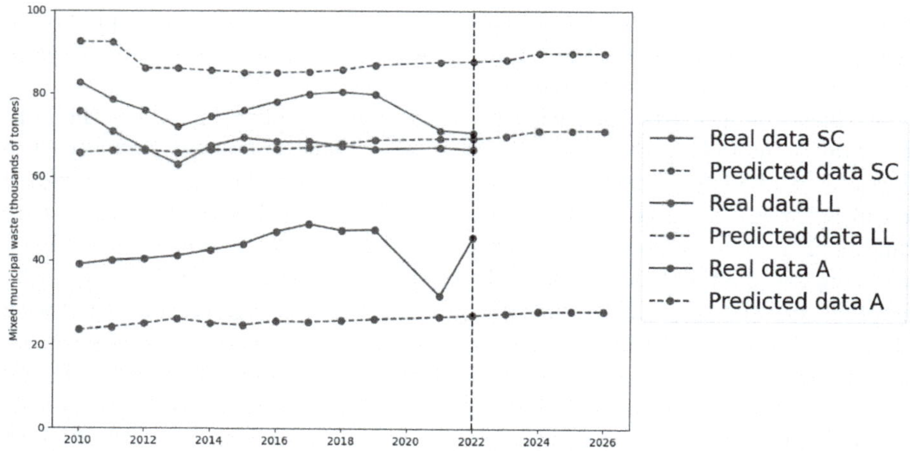

Fig. 2. Multiple Linear Regression (MLR) results for mixed municipal waste for Adeje (A), La Laguna (LL) and Santa Cruz (SC)

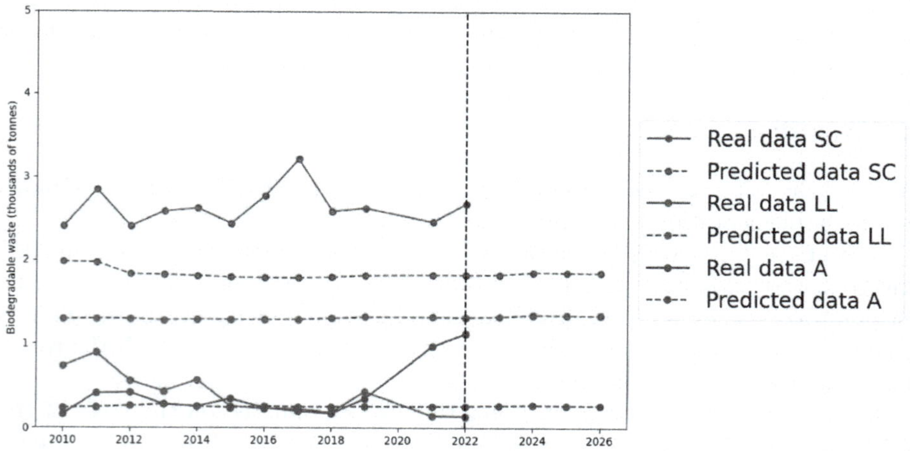

Fig. 3. Multiple Linear Regression (MLR) results for biodegradable waste for Adeje (A), La Laguna (LL) and Santa Cruz (SC)

Figure 3 shows the results obtained for biodegradable waste using MLR for the municipalities of La Laguna, Santa Cruz, and Adeje. A worse fit is observed than the fit for mixed municipal waste, which can also be seen in Table 1, which shows the MAE, RMSE and R^2 values for the three models obtained to forecast annual mixed municipal waste and to forecast annual biodegradable waste, respectively. Except for MLR for biodegradable waste, R^2 values greater than 0.8 were obtained with all models. In terms of error, the ANN and DT models obtained the lowest values. In the case of tourist municipalities, such as Adeje, a better estimate would be obtained that includes socioeconomic data and the equivalent population. Table 3 shows the prediction of mixed waste and biodegradable waste for the years 2024–2026. The results show that these models are suitable for forecasting the waste generated in Tenerife.

Table 1. Accuracy of Multiple Linear Regression (MLR), Decision Tree (DT) and Artificial Neural Network (ANN) models to forecast annual mixed municipal waste.

Mixed waste	R^2	RMSE	MAE
MLR predictor	0.89	6.01	4.42
DT predictor	0.90	5.52	3.10
ANN predictor	0.91	5.26	3.75

Table 2. Accuracy of Multiple Linear Regression (MLR), Decision Tree (DT) and Artificial Neural Network (ANN) models to forecast annual biodegradable waste levels.

Biodegradable waste	R^2	RMSE	MAE
MLR predictor	0.59	0.34	0.20
DT predictor	0.86	0.20	0.09
ANN predictor	0.85	0.21	0.12

Table 3. Estimates with the ANN model for mixed municipal waste and biodegradable waste (thousands of tonnes) for Tenerife.

Waste type	2024	2025	2026
Mixed municipal waste	537.90	556.54	574.67
Biodegradable waste	5.87	5.96	5.96

4 Conclusions and Future Works

In this study, developed within the framework of the "Smart Management of Urban Solid Waste in Tourist Destinations" project, we have implemented three methods to predict the generation of Municipal Solid Waste (MSW) from Multiple Linear Regression (MLR), Decision Trees (DT), and Artificial Neural Networks (ANN), for mixed municipal waste and biodegradable waste in Tenerife Island, including demographic data and municipalities as predictor variables, to reflect spatial variation. In mixed municipal waste, the differences between the three models are few. For biodegradable waste, the Multiple Linear Regression model works worse than the rest of the models. The best models were the Artificial Neural Networks model for mixed municipal waste and the Decision Trees model for biodegradable waste, respectively. However, the error values of the Artificial Neural Networks model for biodegradable waste are very close to those of the Decision Trees model. The results demonstrate that machine learning is a useful tool for planning waste management.

In future work, we will include data from 2020, which was excluded in this study due to its atypical behaviour caused by the COVID 19 pandemic, to see its influence on the models. In addition, we will add different socioeconomic factors as predictor variables and the equivalent population as a demographic variable, and we will analyse the behaviour of seasonality, all of which will result in greater precision of these models. We will also try to include seasonality in our models and predictors.

Acknowledgement. This study is funded by Fundación CajaCanarias and Fundación La Caixa (Project 2021ECO20).

References

1. Abbasi, M., El Hanandeh, A.: Forecasting municipal solid waste generation using artificial intelligence modelling approaches. Waste Manage. **56**, 13–22 (2016)
2. Abdallah, M., Abu Talib, M., Feroz, S., Nasir, Q., Abdalla, H., Mahfood, B.: Artificial intelligence applications in solid waste management: a systematic research review. Waste Manage. **109**, 231–246 (2020)
3. Abdi, H.: The method of least squares, pp. 1–7 (1974)
4. Abdoli, M.A., Falahnezhad, M., Behboudian, S.: Multivariate economic approach for solid waste generation modeling: impact of climate factor. Environ. Eng. Sci. **28**(9), 627–633 (2011)
5. Chen, M., Giannis, A., Wang, J.: Application of system dynamics model for municipal solid waste generation and landfill capacity evaluation in Singapore. Macrotheme Rev. **1**, 101–114 (2012)
6. Dreiseitl, S., Ohno-Machado, L.: Logistic regression and artificial neural network classification models: a methodology review. J. Biomed. Inform. **35**(5), 352–359 (2002)
7. Duda, R.O., Hart, P.E., Stork, D.G.: Pattern Classification. Wiley, Hoboken (2000)
8. Goel, S., Ranjan, V., Bardhan, B.: Forecasting solid waste generation rates. In: Sengupta, D., Agrahari, S. (eds.) Modelling Trends in Solid and Hazardous Waste Management, pp. 1–171. Springer (2017)
9. Hannan, M.A., Arebey, M., Begum, R.A., Mustafa, A., Basri, H.: An automated solid waste bin level detection system using Gabor wavelet filters and multi-layer perception. Resour. Conserv. Recycl. **72**, 33–42 (2013)
10. Jahandideh, S., et al.: The use of artificial neural networks and multiple linear regression to predict rate of medical waste generation. Waste Manage. **29**(11), 2874–2879 (2009)
11. Kannangara, M., Dua, R., Ahmadi, L., Bensebaa, F.: Modeling and prediction of regional municipal solid waste generation and diversion in Canada using machine learning approaches. Waste Manage. **74**, 3–15 (2018)
12. Kolekar, K., Hazra, T., Chakrabarty, S.: A review on prediction of municipal solid waste generation models. Procedia Environ. Sci. **35**, 238–244 (2016)
13. Kontokosta, C.E., Hong, B., Johnson, N.E., Starobin, D.: Using machine learning and small area estimation to predict building-level municipal solid waste generation in cities. Comput. Environ. Urban Syst. **70**, 151–162 (2018)

14. Murthy, S.K., Salzberg, S.: Decision tree induction: how effective is the greedy heuristic? In: KDD, pp. 222–227 (1995)
15. Perera, C., Fernando, M.: Comparison of multiple linear regression and artificial neural network models for the prediction of solid waste generation in Sri Lanka. Trans. Mach. Learn. Data Min. **13**(1), 3–25 (2020)
16. Rives, J., Rieradevall, J., Gabarrell, X.: LCA comparison of container systems in municipal solid waste management. Waste Manage. **30**(6), 949–957 (2010)

Generic Planning Algorithm Adaptable for Order Decision to Automated Clinical Analysis Equipment

Plácido Fernández-Cuevas(✉), Saturnino Vicente-Diaz, and Antón Civit-Balcels

Robotics and Technology of Computers Lab., Universidad de Sevilla, Sevilla, Spain
pfernandez9@us.es

Abstract. Many of the systems used in industry can be considered real-time systems, meaning that the validity of their outputs is related to the instant in time at which they are produced. Since the appearance of computers, real-time systems often make use of computer-assisted decision making algorithms. In this category, scheduling is one of the most extended computer-assisted decision making solutions to manage and organize the tasks they must execute and ensure real-time compliance, but it is not necessary in all cases. There are few examples of schedulers applied in biomedical robotics, usually designed for a specific machine and relying on pre-existing code libraries. We studied the different needs of a biomedical analysis robot, and we defined and modeled test protocols which mimic real processes carried out during real operation. Then, we propose the implementation of a simple scheduler for this robot and these test protocols which uses custom libraries designed to be easily ported to other similar machinery. Timing comparisons were made between an unplanned system and a planned system, concluding that the time can be reduced up to 80% by using our scheduler.

Keywords: Scheduler · Real-time Systems · Robot

1 Introduction

Bioanalysis [10] is a branch of health science, which is responsible for studying samples taken from an individual to know their health status. This study is usually carried out with the purpose of predicting or controlling the diseases of the individual who undergoes said analysis. This specialty began more than a hundred years ago by health professionals who carried out this type of work in clinical laboratories and has expanded worldwide. The specialist who carries out this type of analysis is called a bioanalyst. Bioanalysis is carried out using different techniques, which specialize in searching for certain biological entities. The most used techniques are the following:

Immunohistochemistry [7] is a chemical process whose objective is to detect, amplify and make visible a specific antigen within a sample. This technique thus

A. A. Juan et al. (Eds.): DSA ISC 2024, LNCS 14778, pp. 194–200, 2025.
https://doi.org/10.1007/978-3-031-78238-1_18

allows the location of a specific substance to be identified at the tissue or cellular level. It focuses on the use of antibodies that specifically bind to a substance to be identified.

Immunofluorescence [4] is based on immunolabeling by antibodies, which chemically bind to a fluorescent substance to highlight the presence of a certain molecule in a sample.

The polymerase chain reaction (PCR) [1] is a technique focused on molecular biology, which was developed in 1986 and has gained great popularity in recent years as a result of the world pandemic of the COVID-19 virus. The main objective of this technique is to obtain a large number of copies of a DNA fragment, starting from a minimum.

In situ hybridization (ISH) [5] is a technique used mainly for the localization and detection of specific DNA and RNA sequences in cells, chromosomes or preserved tissues. This is possible by forming a hybrid molecule between an endogenous RNA or DNA molecule of the cell and a complementary single-stranded RNA or DNA probe.

For a conventional clinical laboratory, these techniques must be performed by hand, which takes time and a high chance of human error. To solve this problem, automation is usually chosen using biomedical analysis robots. Although it may look and sound like something current, clinical laboratory automation has been in development for a long time. So much so, that the first fully automated clinical laboratory was created in 1984 by Dr. M. Sasaki at Kochi Medical School in Japan. [3] These biomedical analysis robots are designed and specialized for a single technique, which is translated into a series of movements that are executed periodically. This offers the possibility of great optimization at the technical level, but the inflexibility of a closed and cyclic system.

In this paper, a dynamic scheduler is implemented, which allows the application of a biomedical robot to be generalized to multiple techniques simultaneously. To achieve this, the scheduler offers a decision order consistent with the needs of each technique, complying with the requirements imposed. Unfortunately, there is no literature on a scheduler that offers this capability in other robots, so testing uses a comparison of the same robot applying a schedule and without applying it. For this project, a scheduler library has been developed, which is applied to a specific robot, but is extensible to other biomedical robots. Unfortunately, it is not possible to reveal information since confidentiality contracts have been signed with the company for the development of this project.

2 Scheduling for Decision Making

The scheduler of any system can be defined as a program that determines the order of the tasks using computer-assisted decision making algorithms [2]. The implementation of a scheduler is not general, since it is completely dependent on the system for which it is going to be used [8,9]. Every scheduler is based on the priorities of the tasks they execute.

Two large groups of schedulers can be distinguished according to how they interact with this priority:

- Dynamic scheduler: Modify task priorities at run time.
- Static scheduler: They define priorities at compile time. Priorities can change due to the tasks themselves or due to functions external to the scheduler.

On the other hand, schedulers can also be classified based on when they are executed and under what conditions, thus being able to differentiate two types:

- Collaborative scheduler: The scheduler switches to the highest priority task upon completion of the current task. Here it is the tasks that decide when to stop executing (yield).
- Preemptive scheduler: The scheduler always executes the task with the highest priority ready to be executed, expelling the lower priority tasks.

Scheduling periodic tasks is a problem that Liu and Layland [6] solved in 1973, which is based on a series of assumptions:

- All tasks are periodic and the deadline of each task coincides with its period.
- Tasks do not synchronize with each other or share resources.
- The priority is fixed, static and unique for each task
- Always execute the task with the highest priority

The drawback that arises when applying static planning to an analysis robot is that it is forced to carry out planning for each type of analysis and technique, so that the instrument is anchored to a single specialized function.

3 Implementation

In the case of this study, there is a robot that has a series of tools that are used in the laboratory such as a reagent dispensing probe, a series of pumps that are capable of dispensing some liquids used during the different techniques, a probe washing station that allows you to wash the probe to use different types of reagents without cross contamination, a compressor as a blower...

Starting from the capabilities of the robot and taking into account the steps followed in biochemical analysis techniques, we choose to represent each of the steps of the technique as a process, which has a priority. These processes are planned using the Scheduler to offer the possibility of performing different techniques without having to change robots, in addition to seeking improvements in the performance of the entire series. These processes will be formed by the user and will depend on the specific system to which the scheduler wants to be applied, since, depending on the robot, it may be of interest or may be capable of carrying out a series of processes that others cannot and vice versa. The steps of the techniques must be followed in a specific order, which requires linking a process with its previous process, so that given a technique formed by process A followed by process B, the execution of process B cannot begin until process A has not finished. The scheduler proposed for this purpose is a collaborative type scheduler and also has dynamic priorities. The fact that it is collaborative means that the processes at the end of their execution will be the ones that invoke the scheduler so that it chooses the next process that should be executed. On the other hand, the fact that it is dynamic allows us to define a series of priorities that allow processes to be managed in a more efficient and controlled way.

To control the execution of processes, different execution states are defined. These states are the following:

- NOT READY: The process is not ready for execution
- READY: The process is ready for execution
- EXEC: The process is running
- EXEC OK: The process has completed its execution

These minimum states are necessary so that the scheduler can carry out a correct planning of the processes, releasing those processes that can be executed when their precondition (process that precedes it) has been completed and executing those that are in the READY state at the time. Figure 1 shows the execution trace according to the states.

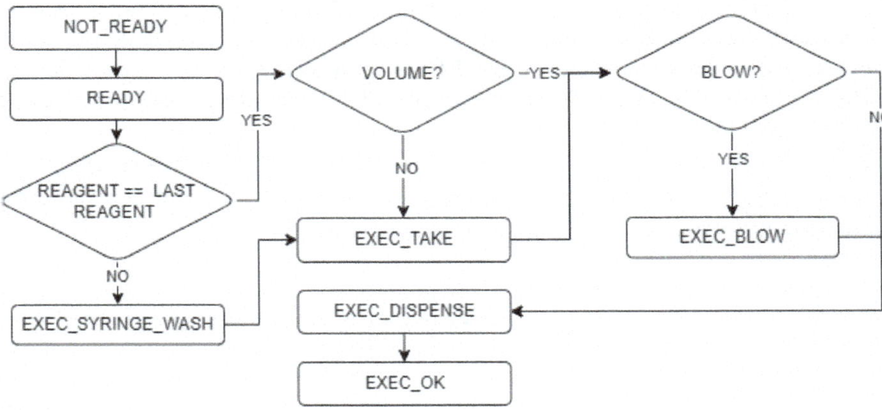

Fig. 1. Dispense process execution flow

Each specific process will have different execution states depending on the tools it must use to complete its task in addition to the order and number of actions it must perform. The techniques that have been used for this project make use of incubations. An incubation consists of waiting for a while until the next process is executed. Not only do you have to wait, but you also have to execute the next process right when the incubation time ends. Because of this, the incubation process is the only process that cannot terminate itself, but must be terminated by the next process to run. Slides are loaded into the robot that have pieces of tissue or biological material to which the different techniques must be applied. In this way, each technique can be applied to one or more slides within the robot and the scheduler will be in charge of guaranteeing efficient execution that respects all the limitations expressed when generating the processes.

4 Experimental Results

To check the correct operation of the system, a technique was developed that includes different processes normally used in the laboratory, such as dispensing, incubation and washing the slides. This technique was applied to two different slides, so the scheduler must be able to efficiently execute the two techniques taking advantage of the robot's downtime during incubations. Figure 2 shows the results of this test.

To understand test results, it is necessary to understand the scheme followed in a protocol that represents a medical technique. As mentioned previously, a protocol is a set of steps which we have called processes. These steps must be executed in a specific order on a sample to complete a technique. The proposed experiment is based on the execution of the same technique on two different samples. This means that the steps that make up the protocol must be executed on both samples, trying to follow a coherent and similar order of execution, so that both samples progress in a similar way over time. Incubate processes consist of leaving the sample for a period of time determined by the user without interacting with it. They need to be finished by a dispense, so the scheduler must wait until the incubation is close to its end time before the next dispense can be executed. The incubations can be used to execute other processes that depend on the robot arm in other different samples, thus simultaneizing the work as much as possible.

In the graph of the results, it can be seen that on the left are the names of the processes that are executed followed by a number that identifies the order within the protocol.

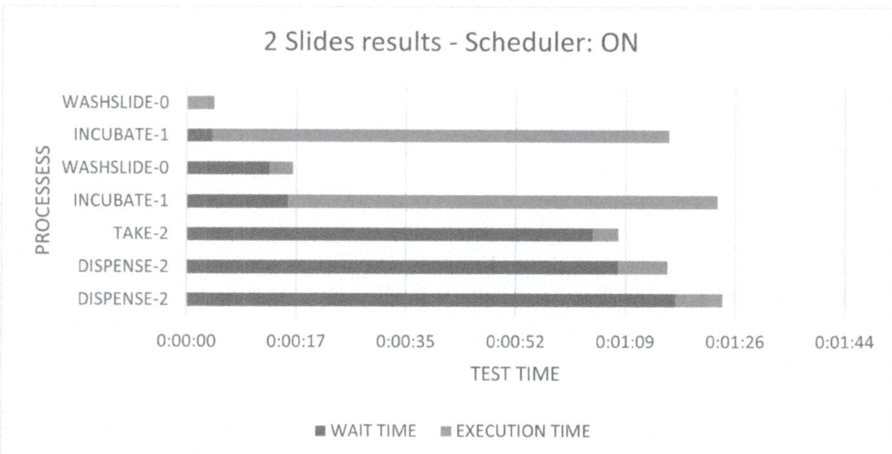

Fig. 2. Time results of the test with active scheduler

As can be seen in the results, between the first incubation and the next wash, the scheduler has determined that it is necessary to leave a time gap of a few seconds, which is reflected by the waiting time, until the next process is executed. Furthermore, the execution of the process that attends to the first incubation began in advance of the end of the incubation to finish the incubation exactly at the time specified by the user, thus managing to attend to the two incubations without interfering between them. If we modify the scheduler so that it completes the techniques that have been associated with it as a sequential execution, we obtain the execution trace what is represented in Fig. 3.

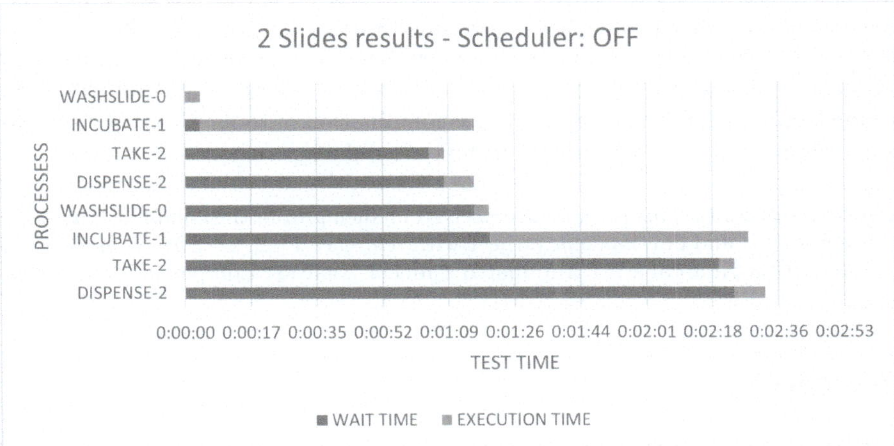

Fig. 3. Time results of the test with non active scheduler

One of the main differences between both results is that the second contains two Take processes, while the first only contains one. This is because when a planned system is used, it can perform actions in the future, taking enough reagent to execute two dispensations in a single Take process. On the other hand, when the system is not planned, decisions cannot be made in the future; it can only be limited to performing an action in the present, which is why it performs a Take process for each Dispense.

The completion time of the first case with the scheduler active is 0:01:24 min while the completion time of the second case in which the scheduler is disabled is 0:02:32 min, which is a difference of 0:01:08 min, practically twice as long as in the planned case. In the Eq. 1 is shown how the acceleration was obtained. In this equation, $t_{SchedulerOff}$ is the total completion time without using the scheduler in seconds, while $t_{SchedulerOn}$ is the total completion time with the scheduler active, also in seconds. The acceleration value obtained was 1.8095, which translates in 80.95% improvement.

$$Acceleration_{SchedulerOn->SchedulerOff} = \frac{t_{SchedulerOff}}{t_{SchedulerOn}} = \frac{152}{84} = 1.8095 \quad (1)$$

5 Conclusions

This project starts from scratch since a scheduler is something specific for each problem, that is, there is no universal library that can adapt to a specific system. Furthermore, no robot with similar characteristics has been found that includes a planner of this type.

After carefully observing the experimental results obtained, it is verified that the scheduler provides an improvement in performance compared to an unscheduled system. This improvement will depend on the number of Slides that you want to analyze, in addition to the techniques used. On the other hand, one of the main characteristics it has is the generalization of robotic equipment that incorporates it, since it eliminates the limitation of specializing equipment to a specific technique. With the appropriate combination of tools, this scheduler can execute practically any laboratory technique, reducing costs.

Acknowledgment. This project has received financing from the company Vitro S.A. as an innovation and development project with reference P022-21/E03 "Generic Planning Algorithm Adaptable to Automated Clinical Analysis Equipment (Vitroscheduler)".

References

1. Bartlett, J.M., Stirling, D.: A short history of the polymerase chain reaction. PCR Protocols 3–6 (2003)
2. Brehmer, B.: Dynamic decision making: human control of complex systems. Acta Physiol. (Oxf.) **81**(3), 211–241 (1992)
3. Felder, R.A.: The clinical chemist: masahide sasaki, MD, PhD (August 27, 1933-September 23, 2005). Clin. Chem. **52**(4), 791–792 (2006). https://doi.org/10.1373/clinchem.2006.067686
4. Im, K., Mareninov, S., Diaz, M.F.P., Yong, W.H.: An introduction to performing immunofluorescence staining. In: Biobanking: Methods and Protocols, pp. 299–311 (2019)
5. Jensen, E.: Technical review: in situ hybridization. Anat. Rec. **297**(8), 1349–1353 (2014)
6. Liu, C.L., Layland, J.W.: Scheduling algorithms for multiprogramming in a hard-real-time environment. J. ACM (JACM) **20**(1), 46–61 (1973)
7. Ramos-Vara, J.A.: Technical aspects of immunohistochemistry. Vet. Pathol. **42**(4), 405–426 (2005)
8. Ripps, D.L.: An Implementation Guide to Real-Time Programming. Prentice-Hall, Inc. (1989)
9. Sha, L., Goodenough, J.B.: Real-time scheduling theory and ADA. IEEE Comput. **23**(4), 53–62 (1990)
10. Venn, R.F.: Principles and Practice of Bioanalysis. CRC Press, Boca Raton (2008)

Conceptual Design of a Decision Support System for the Allocation of Electric Vehicle Charging Stations

José-Luis Guisado-Lizar[1,2(✉)], Antonio Navas-Orozco[2], David Ragel-Díaz-Jara[1], José Morera-Figueroa[2], Daniel Cascado-Caballero[1,2], María José Morón-Fernández[1,2], Gabriel Jiménez-Moreno[1,2], and Fernando Diaz-del-Rio[1,2]

[1] Research Institute of Computer Engineering (I3US), Universidad de Sevilla, Avenida Reina Mercedes s/n, 41012 Sevilla, Spain
jlguisado@us.es

[2] Department of Computer Architecture and Technology, Universidad de Sevilla, Avenida Reina Mercedes s/n, 41012 Sevilla, Spain
https://grupo.us.es/sanevec/en/

Abstract. We propose the conceptual design of a Decision Support System (DSS) for the optimal allocation of electric vehicle (EV) charging stations (CSs) within an urban area under criteria of social and environmental benefits. The proposed DSS employs simulation, optimization, and a machine learning metamodel to suggest decisions that optimize traffic patterns in the city, operation of the electric grid, air quality, and carbon footprint of the city by taking into account all the variables involved and their mutual influence.

Keywords: Decision Support Systems · Urban Electric Vehicle Infrastructure · Charging Stations · Simulation · Optimization · Artificial Intelligence · Model-driven DSS · Metamodel

1 Introduction

Decision Support Systems (DSSs) are a class of information systems designed to support decision-making activities [11]. Following Keen [4], DSSs are a balanced combination of 'decision', 'support' and 'systems', which means the use of computer technology to design and implement systems that can support decision makers to make better informed decisions. DSSs extend traditional techniques, such as simulation and multi-criteria decision making, by incorporating elements

This work is part of the projects SANEVEC TED2021-130825B-I00, and SAINEVRA (PID2023-151065OB-I00), funded by the Ministerio de Ciencia, Innovación y Universidades (MICIU), Agencia Estatal de Investigación (AEI) of Spain, MICIU/AEI/10.13039/501100011033, and by the European Union NextGenerationEU/PRTR.

A. A. Juan et al. (Eds.): DSA ISC 2024, LNCS 14778, pp. 201–207, 2025.
https://doi.org/10.1007/978-3-031-78238-1_19

crucial for making informed decisions, including expert judgments, dynamic visualizations, and extensive databases [6]. A comprehensive historical analysis of DSSs is provided by Power [8].

In this work, we propose the conceptual design of a DSS for the optimal allocation of electric vehicles (EVs) charging stations (CSs) within an urban area under criteria of social and environmental benefits, in the context of the SANEVEC research project [3]. It will integrate all the variables involved and their mutual influence: location of the charging stations, traffic patterns, characteristics of the electric grid, charging fees, air quality, and pollution. The DSS will employ simulation techniques using agent-based models and cellular automata, as well as artificial intelligence to optimize CS placement and minimize computing time.

The remainder of this work is organized as follows. Section 2 presents related work. Section 3 describes the methodology. Finally, Sect. 4 presents the conclusions and future research directions.

2 Background

DSSs can be classified in different categories according to the expanded framework of DSSs proposed by Power [7], which updates a previous taxonomy proposed by Alter [1]. Power's expanded framework systematically classifies DSSs into five distinct categories: data-driven, model-driven, knowledge-driven, document-driven, and communication-driven DSSs, each addressing different facets of decision support. Data-driven DSSs underscore the importance of structured data analysis. Model-driven DSSs leverage various mathematical and simulation models for problem-solving. Knowledge-driven DSSs, on the other hand, harness expert knowledge to provide specialized recommendations. Document-driven DSSs facilitate the manipulation and retrieval of unstructured data. Finally, communication-driven DSSs emphasize enhancing collaborative decision-making through advanced communication tools. This classification not only offers a systematic method for understanding the range of capabilities within DSSs, but it also underscores the importance of flexible frameworks to effectively grasp the evolving relationship between technology and the decision-making process.

The DSS system proposed in this work belongs to the model-driven DSSs category, which involves using algebraic, decision analytic, financial, simulation, and optimization models to provide decision support [9]. In recent works, there has been significant emphasis on integrating artificial intelligence (AI) into DSSs to enhance their decision-making capabilities. AI technologies, including machine learning, natural language processing, and expert systems, are increasingly being incorporated to improve the analysis, prediction, and optimization aspects of DSSs. This integration aims to make DSSs more adaptive, intelligent, and efficient in handling complex decision-making processes across various domains. Integration of these techniques is particularly interesting for model-driven DSSs that employ simulation. In that area, the seminal work of Li et al. [5] studied how to incorporate metamodels (mathematical approximations of simulation models, which can reduce the computational burden associated with running such simulation models) into DSSs to improve their efficiency and allow larger and more complex models to be effectively analyzed with decision support. Typical metamodeling techniques include artificial neural networks (ANN), multivariate adaptive regression splines, radial basis functions, and support vector regression.

Greenhouse gas emissions from road vehicles constitute an essential contribution to the worsening of the current climate crisis. In response to this issue, EVs have been proposed as an alternative. As the number of EVs increases, deploying a public charging infrastructure becomes necessary. As shown in [2], the distribution of CSs within a city directly affects traffic patterns, which significantly influence air pollution. Therefore, CSs allocation should not be based simply on past traffic data; their synergy must also be considered to guarantee the goodness of our solution. In the SANEVEC research project, traffic patterns, combining both internal combustion vehicles and EVs, and CS locations, as well as their relation to the electric grid and air pollution and quality within an urban area, are simulated using a hybrid microscopic model based on cellular automata (CA) and agent-based models (ABMs) [10]. CA are temporally and spatially discrete mathematical models that display local interactions and synchronous parallel evolution. In our case, the CA simulates evolving properties of a city that are related to traffic. ABMs are stochastic models based on autonomous individual agents that interact in a defined way. In our case, each car is an agent that is heading towards a destination, has properties such as its remaining battery level, and makes decisions like the CS it goes when its battery is running low. The recommended decision of the DSS we propose is the optimal (or near-optimal) set of CS locations and CS sizes according to the social and environmental benefits sought (minimizing traffic jams and therefore optimizing air quality and urban mobility, as well as finding a CS deployment that is consistent with the good operation of the electric grid of the city).

The proposed system provides a better tool for the allocation of CSs than other current methods because it takes into account the feedback and synergies between all the variables involved. In particular, the projected location of new stations will change traffic patterns and alter the output of all the indicated variables. Our system will improve urban mobility by recommending specific configurations for CS placements and charger quantities that will minimize urban traffic jams. This reduction of traffic congestion will reduce transport times in the city. As a result, air pollution, carbon footprint, and noise pollution of the city will be reduced because there will be less polluting fossil fuel vehicles involved in traffic jams. In addition, the optimization of EV charging operations in the city will improve user experience of EVs by reducing waiting times, thus helping to increase the proporcion of EVs versus fossil fuel vehicles. The combination of all these benefits will help to increase the quality of life of citizens.

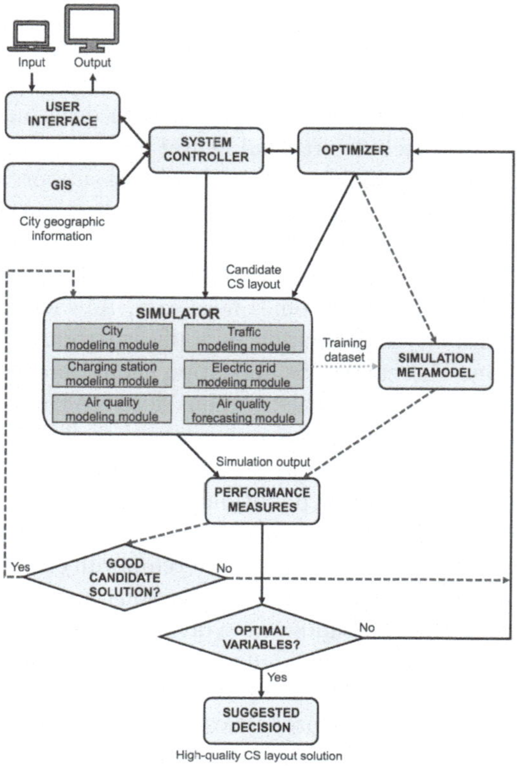

Fig. 1. Diagram of the proposed Decision Support System.

3 Methodology

Our DSS proposal incorporates three main components: simulation, optimization and a machine learning metamodel. The first one, simulation, is employed to investigate the effect of a candidate CS layout solution on the significant variables of the city, such as the efficiency of the CS layout to improve EV mobility, the traffic congestion, air pollution, carbon footprint of the city, electric grid operation, load balancing usage of the different CSs, and charging fees. Urban traffic is a case of a complex system consisting of many interacting components that give rise to emergent properties at a global level, which are difficult or impossible to predict in advance, except for simulation. In a city including EVs, urban traffic is also coupled with the operation of the public CSs as another component of the complex system. In addition, the operation of the CSs is also coupled with the operation of the city electric grid. Finally, urban traffic congestion is also linked to air pollution. Therefore, determining the optimal location of urban CSs is a complex problem governed by the mutual interrelationship between all these components, which must be simulated to predict their combined evolution. The second main component, optimization, is integral to our DSS, which aims to assist decision-makers in identifying the best (optimal or near-optimal) decision policy. Our proposal merges simulation models with optimization techniques to discover high-quality solutions, resulting in optimal outputs for the aforementioned city variables. The third component is a machine learning metamodel. The execution of a simulation of a large city including all those elements requires an extensive computing time. To optimize the CS layout, many simulation executions corresponding to different candidate solutions must be done, needing a considerable computing time. By incorporating this metamodel, our proposal aims to efficiently reduce the total runtime needed for the optimization.

The conceptual design of the proposed DSS is shown in Fig. 1. Our design is based on the generic simulation metamodel proposal in a DSS from [5], but with a significant difference. In their proposal, the rationale for incorporating the metamodel stems from the potential variability in the available computational resources, which may or may not allow the direct optimization of the simulation model. If the computational resources are not sufficient for direct optimization, the system utilizes the metamodel for optimization instead of direct simulation. On the other hand, in our present proposal, the execution of the simulations can have such a high computational cost that the aim of the metamodel is to make a preliminary evaluation of candidate solutions before executing them in the simulator. Thus, the metamodel acts as a filter that discards unpromising candidate solutions so that only good candidate solutions are effectively executed in the simulator. In this way, running the DSS for a very large, complex scenario, such as a complete large city, can be affordable.

The core components of our proposed DSS conceptual design are the simulator, the optimizer, and the simulation metamodel. There can be two modes of operation in the DSS. The standard one, represented in Fig. 1 by continuous blue arrows, involves only the usage of the simulator and the optimizer. The enhanced mode of operation also involves the previous evaluation of candidate solutions using the simulation metamodel. The simulator executes a simulation of the operation of the urban traffic in the city, including EVs and internal combustion engine vehicles. Vehicles travel to target destination points, but when the battery of an EV is low, it drives to a CS to recharge, and after that, it continues its journey. The operation of the CSs, its effect on the city electric grid, and the effects of the urban traffic on air pollution are also simulated. For each candidate CS layout solution, a simulation is run for several simulated hours. After that, performance measures are computed by calculating the values of the city variables (traffic congestion, air pollution and carbon footprint, electric grid operation, and load balance utilization between different charging stations) for the simulation output. The optimizer module applies an optimization technique to evaluate different candidate CS layout solutions. Since the problem is not simple, we propose the usage of metaheuristic optimization techniques, such as a genetic algorithm, or simulated annealing. The simulation metamodel can be an ANN, that will be trained by a training dataset initially produced by running a number of simulations for different candidate CS layout solutions and computing their performance measures. A small part of this dataset will be used to validate the output of the metamodel by comparing its output with the output of the simulator. There will be an initial running time cost for producing the training dataset and training the ANN. But after it has been successfully trained, the system can use the simulation metamodel (i.e. the ANN) for inference to make a first evaluation of candidate solutions, that will be order of magnitudes faster than the execution of a simulation in the simulator. Candidate solutions that are not promising according to the simulation metamodel will be discarded, and only promising solutions will be actually evaluated by simulation in the simulator.

4 Conclusions and Future Work

In this work, we present a novel conceptual design for a DSS dedicated to optimizing the distribution of EV charging infrastructure in urban areas. The system recommends specific configurations for charging station placements and charger quantities to achieve optimal outcomes. Utilizing a blend of simulation, optimization, and machine learning, the DSS aims to improve urban traffic dynamics, electric grid performance, air quality, and minimize the carbon emissions of the city, acknowledging the complexity of their interactions. As a future work, we plan to implement the DSS, and test it for a real city. Also, we plan to work on issues like fine tuning the training of the metamodel, and determining the minimum size of the training dataset.

References

1. Alter, S.: Decision Support Systems: Current Practice and Continuing Challenges. Addison-Wesley, Reading (1980)
2. García-Suárez, A., Guisado-Lizar, J.L., Diaz-del Rio, F., Jiménez-Morales, F.: A cellular automata agent-based hybrid simulation tool to analyze the deployment of electric vehicle charging stations. Sustainability **13**(10), 5421 (2021). https://doi.org/10.3390/su13105421
3. Guisado-Lizar, J.L., et al.: Designing EV charging stations deployment through holistic simulations: the SANEVEC project. In: EVI: Charging Ahead (EVI 2023), pp. 133–139. IEEE Xplore, Glasgow (2023). https://doi.org/10.1049/icp.2023.3140
4. Keen, P.G.: Decision support systems: the next decade. Decis. Support Syst. **3**(3), 253–265 (1987). https://doi.org/10.1016/0167-9236(87)90180-1
5. Li, Y.F., Ng, S.H., Xie, M., Goh, T.: A systematic comparison of metamodeling techniques for simulation optimization in decision support systems. Appl. Soft Comput. **10**(4), 1257–1273 (2010)
6. Parnell, G.S., Driscoll, P.J.: Decision Making in Systems Engineering and Management, 2nd edn., pp. 1–501. Wiley (2011). https://doi.org/10.1002/9780470926963.ch1
7. Power, D.J.: Supporting decision-makers: an expanded framework. Inf. Sci. **1**, 1901–1915 (2001)
8. Power, D.J.: Decision support systems: a historical overview. In: Burstein, F., Holsapple, C. (eds.) Handbook on Decision Support Systems 1: Basic Themes, pp. 121–140. Springer, Heidelberg (2008). https://doi.org/10.1007/978-3-540-48713-5_7

9. Power, D.J., Sharda, R.: Model-driven decision support systems: concepts and research directions. Decis. Support Syst. **43**(3), 1044–1061 (2007)
10. Ragel-Díaz-Jara, D., et al.: Integrating efficient routes with station monitoring for electric vehicles in urban environments: simulation and analysis. In: Guisado-Lizar, JL., Riscos-Núñez, A., Morón-Fernández, MJ., Wainer, G. (eds) SIMUtools 2023. LNICST, vol. 519, pp. 166-181. Springer, Cham (2024). https://doi.org/10.1007/978-3-031-57523-5_13, https://link.springer.com/book/9783031575228
11. Sprague Jr, R.H., Carlson, E.D.: Building Effective Decision Support Systems. Prentice Hall Professional Technical Reference (1982)

Virus Machines at Work: Computations of Workflow Patterns

Antonio Ramírez-de-Arellano[1,2](✉), Francis George C. Cabarle[1,2,3], David Orellana-Martín[1,2], Agustín Riscos-Núñez[1,2], and Mario J. Pérez-Jiménez[1,2]

[1] Research Group on Natural Computing, Department of Computer Science and Artificial Intelligence, Universidad de Sevilla, Avda. Reina Mercedes s/n, 41012 Seville, Spain
{aramirezdearellano,fcabarle,dorellana,ariscosn,marper}@us.es
[2] SCORE Lab, I3US, Universidad de Sevilla, 41012 Seville, Spain
[3] Department of Computer Science, University of the Philippines Diliman, 1101 Quezon City, Philippines
fccabarle@up.edu.ph

Abstract. In this work we use for the first time an unconventional and bio-inspired model of computation known as a *virus machine* (in short, VM) to model workflow patterns. VMs are novel computing paradigms inspired by biological viruses and their transmissions. VMs are capable of building sequential computing algorithms (they are Turing-universal). In this work, a young variant of VMs, called parallel VMs (in short, PVMs) is used for defining and analysing workflow patterns. More precisely, fundamental workflow patterns, such as sequences and various types of parallel splits and joins, are defined using PVMs. Besides the mathematical formulation of PVMs, their graphical representations have attractive aspects for workflow patterns. Such patterns can form the basis for modelling tasks, consumption of resources, to support for instance in decision making.

Keywords: unconventional computing · virus machines · workflow patterns · natural computing

1 Introduction

In many organisations such as scientific, government, or commercial ones, often there are several *works* which can be split into *resources* and *tasks*: resources can be people, funding, time, or equipment, while tasks can be procuring such resources, deciding which to combine or distribute, in which order or sequence, and so on. In this way the ability to model, monitor, and control the *flow of work*, that is, the *workflow*, is thus important across various organisations.

F.G.C. Cabarle is supported by the *QUAL21 008 USE* project (PAIDI 2020 and FEDER 2014–2020 funds). A. Ramírez-de-Arellano is supported by the Zhejiang Lab BioBit Program (Grant No. 2022BCF05).

A. A. Juan et al. (Eds.): DSA ISC 2024, LNCS 14778, pp. 208–219, 2025.
https://doi.org/10.1007/978-3-031-78238-1_20

One conventional and formal way to study workflows are *workflow nets,* which are a restricted type of Petri nets [11,12]. Petri nets, and by extension workflow nets, allow for theoretical and graphical analyses of workflows [1]. More information on Petri nets, their theory and applications can be found in monographs or surveys [9,11,17]. Specifically, Petri nets are used to model *workflow patterns*, that is, to abstract in a formal way many recurring patterns in various contexts of organisations [16]. Workflows and Petri nets are used for instance to support the making of decisions in organisations [18]. An exhaustive list and analyses of workflow patterns, Petri nets, and software systems, commercial products for them is in [8].

Besides the conventional ways of modelling, such as with Petri nets, the fields of *unconventional computing* and *natural computing* are fast becoming even richer sources of models, ideas, inspirations. Unconventional computing looks at algorithms, techniques, media that are out of the mainstream models or ways of computing [2]. Natural computing has had many successes in the past, taking inspirations from the way natural processes manipulate information: neural networks, DNA, quantum, and evolutionary computing based on the brain, bio-molecules, quantum physics, and evolution, respectively, and so on [4].

The goal of this work is to use for the first time an unconventional and bio-inspired model of computation known as a *virus machine* (in short, VM) to model workflow patterns. At present, VMs are studied mainly for their computing power, that is, what kinds of problems they can(not) solve. Since the introduction of VMs and subsequent works, it is known that VMs are *Turing complete*, that is, they are as powerful as Turing machines [3,7,15]. Put in another way, VMs are algorithms capable of general purpose computations. VMs are novel and alternative models for the formal and graphical representation, modeling, and analyses of workflows, coming from an unconventional and bio-inspired perspective. Previously, other unconventional and bio-inspired models for workflows include spiking neural P systems and variants [5,6,19].

Basic workflow patterns such as sequence, and various types of splits and joins, are modelled using VMs. A VM consists of a *host graph* which is a directed graph where nodes are *hosts* and edges between nodes are *channels.* Hosts contain multiple copies of one symbol v to represent the copies of a virus. The viruses are transferred between hosts using the *instruction graph*: a directed graph where nodes are *instructions* and edges between nodes control which instruction to activate next. Edges can be weighted, thus giving priorities to certain instructions. The instructions are used to control channels between hosts: channels are closed by default, so an instruction opens a channel to allow the transfer of viruses between hosts. In this way, hosts can represent resources owing to the number or magnitude of virus objects v, while the host graph together with the instruction graph controls the sequence and flow of such objects.

The present work is organized as follows: Sect. 2 gives the formal definition of VMs, specifically parallel VMs. Section 2 also includes a short example to give the reader a better idea of the syntax and semantics of VMs. Section 3 provides our results: parallel VMs are used to model the fundamental workflow patterns *sequence*, the AND-, OR-, and XOR- *splits* and *joins*. Lastly, Sect. 4 provides conclusions and directions for future work.

2 Parallel Virus Machines

Virus Machines are a biologically inspired computing paradigm which constructs non-deterministic computing models [7]. Lately, authors in [14] proposed the variant *Parallel Virus Machines* (PVMs, for short), as a first step towards parallelism through these devices. VMs initially are sequential; now PVMs more than instruction can be activated for working in a parallel manner. This novel parallelism behavior will allow us to define workflow patterns such as parallel splits. In this section, the PVMs is formally defined, concluding with an explicit example.

First, the **syntax** is defined as follows:

Definition 21. *A Parallel Virus Machine of degree* (p, q), *for* $p \geq 1$ *hosts and* $q \geq 1$ *instructions, is defined as the following tuple* $\Pi = (\Gamma, H, I, D_H, D_I, G_C, n_1, \ldots, n_p, I_0, h_{out})$, *where:*

- $\Gamma = \{v\}$ *is the singleton alphabet:*
- $H = \{h_1, \ldots, h_p\}$ *and* $I = \{i_1, \ldots, i_q\}$ *are ordered sets such that* $v \notin H \cup I$, $H \cap I = \emptyset$ *and* $h_{out} \notin I \cup \{v\}$*: either* $h_{out} \in H$ *or* $h_{out} \notin H$ *(in this case,* h_{out} *represents the environment and it is denoted by* h_0*);*
- $D_H = (H \cup \{h_{out}\}, E_H, w_H)$ *is a weighted directed graph (WDG for short), where* $E_H \subseteq H \times (H \cup \{h_{out}\})$ *such that* $(h, h) \notin E_H$*, for each* $h \in H$*, out-degree*$(h_{out}) = 0$*, and* w_H *is a mapping from* E_H *onto* $\mathbb{N} \setminus \{0\}$*;*
- $D_I = (I, E_I, w_I)$ *is a WDG, where* $E_I \subseteq I \times I$*,* w_I *is a mapping from* E_I *onto* $\mathbb{N} \setminus \{0\}$ *and the out-degree of each node is less than or equal to 2;*
- $G_C = (V_C, E_C)$ *is an undirected bipartite graph, where* $V_C = I \cup E_H$*, being* $\{I, E_H\}$ *the partition associated with it, in addition each edge connect an element from* I *with, at most, an arc from* E_H*;*
- $I_0 \subseteq I$ *and* $n_j \in \mathbb{N}$*, for each* j*,* $1 \leq j \leq p$*, where* $\mathbb{N}$ *is the set of natural numbers.*

Regarding the **semantics** associated, a *configuration* or *instantaneous description* at an instant $t \geq 0$ is defined as $\mathcal{C}_t = (a_{1,t}, \ldots, a_{p,t}, I_t, a_{0,t})$, where for each $j \in \{0, \ldots, p\}$, $a_{j,t}$ represents the amount of viruses at each host h_j, and $I_t \subseteq I \cup \{\#\}$ represents the set of instructions that will be activated in the next transition step. The *initial configuration* is $\mathcal{C}_0 = (n_1, \ldots, n_p, I_0, 0)$, that is, the output always starts at 0 viruses. On the other hand, a *halting configuration* will have the set of instructions $I_t = \{\#\}$.

Every channel $(h, h') \in E_H$ is closed unless at least one activated instruction opens it, in that case if the host h is not empty, then there is *virus transmission*, that is, a virus is consumed from the host h, replicated $w_H(h, h')$ times and sent to the host h', in case the host h is empty then there is no virus transmission. It is important to note that, if from a host h, more than one channel is opened simultaneously. Suppose that $k > 0$ channels are opened simultaneously, two cases are differentiated depending on the number of viruses contained at the host $n \geq 0$:

- If $n \geq k$, then k viruses are consumed from the host h and one virus travels in each opened channel.
- If $n < k$, then n viruses are consumed, and n channels are non-deterministically chosen, from those channels there is virus transmission and from the others there is none.

The set of instructions that will be activated in the following step depends on the instruction graph and if there was virus transmission of not. For each $i \in I_t$, the following instruction or symbol $\#$, i' is obtained as follows:

- If for the channel associated there have been virus transmission, then it follows the highest weight path in the instruction graph. Otherwise, it follows the least weight path.
- In case the highest or least weight path is not unique, it is chosen non-deterministically.
- If there is no possible path, then $i' = \#$.

Explicit Example

To clarify the behavior of these devices, let us show an explicit example, let Π_{ex} the PVM of degree $(3, 5)$ visually be described in Fig. 1. In this representation, the hosts are represented as squares, and the instructions nodes are represented as blue dots, the initial ones as orange dots. The edges of the channel-instruction graph are drawn as red dotted lines. For simplicity, the weights of the arcs with weight 1 are omitted, as well as the 0 number in empty hosts. In this example $h_0 = h_{out}$ and it is represented as the environment.

The computation will be briefly described. First, the initial configuration is $\mathcal{C}_0 = (1, 0, 0, \{i_1, i_2\}, 0)$, two channels are opened (h_1, h_2) and (h_1, h_3), as the number of viruses is 1, then it is nondeterministically chosen in both channels:

- If the channel (h_1, h_2) is chosen, then from i_1 we go to the unique possible path, that is instruction i_3, and from i_2 the least weight path is not unique, so it has to be chosen in a non-deterministic way:
  - If we choose i_4, the configuration is $C_1 = (0, 1, 0, \{i_3, i_4\}, 0)$, as instruction i_4 is attached to channel (h_3, h_0), host h_3 is empty, and instruction i_3 is not attached to any channel. There is no virus transmission; moreover, there is no possible path in the instruction graph, thus a halting configuration is reached $C_2 = (0, 1, 0, \{\#\}, 0)$.
  - If we choose i_5, the configuration is $C_2 = (0, 1, 0, \{i_3, i_5\}, 0)$, and again there is no virus transmission, so from the instruction i_5 we follow the least weight path, leading the configuration $C_3 = (0, 1, 0, \{\#, i_3\}, 0)$, atfer this, a halting configuration is reached $C_4 = (0, 1, 0, \{\#\}, 0)$.

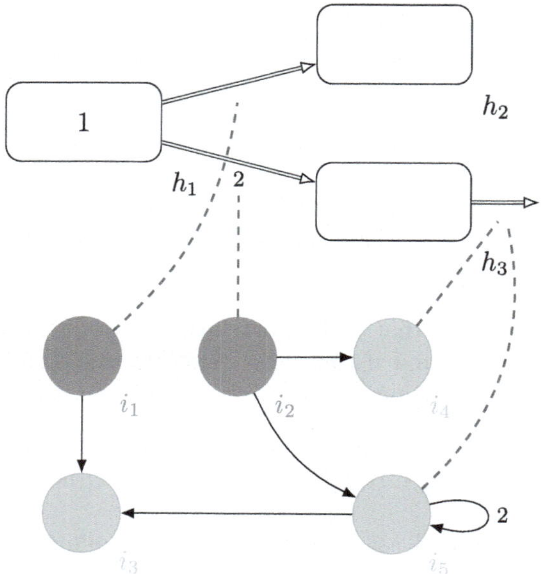

Fig. 1. Example of a parallel virus machine Π_{ex}

- If the channel (h_1, h_3) is chosen, one virus is replicated due to channel weight of 2, and sent to the host h_3. Again we have the same nondeterministic decision as the highest weight path is not unique:
  - If we choose i_4, the configuration is $C_1 = (0, 0, 2, \{i_3, i_4\}, 0)$, now host h_3 is not empty, thus there is virus transmission through channel (h_3, h_0) and one virus is sent to the environment (the output host), so the halting configuration is reached $\mathcal{C}_2 = (0, 0, 1, \{\#\}, 1)$.
  - if we choose i_5, the configuration is $C_1 = (0, 0, 2, \{i_3, i_4\}, 0)$, and the same channel as before is opened, but now the following instruction follows the highest weight path as it is not unique, that is again instruciton i_5: $\mathcal{C}_2 = (0, 0, 1, \{\#, i_5\}, 1)$, this process is repeated again $\mathcal{C}_3 = (0, 0, 0, \{\#, i_5\}, 2)$, and in the following step there is no virus transmission, thus the least weight path is taken $\mathcal{C}_4 = (0, 0, 0, \{\#, i_3\}, 2)$, after this, we have the halting configuration $\mathcal{C}_2 = (0, 0, 0, \{\#\}, 2)$.

3 Parallel Virus Machines for Workflow Patterns

In this section, the fundamental workflow patterns between the tasks from [8] are defined by PVMs. For convenience, explicit definitions of the tuples are replaced by their graphical representations as in Fig. 1.

First, an explicit definition of a task should be defined in the scope of virus machines. Let us fix that a PVM Π_k defines a given task labeled l_k, then we say that task l_k is *initialized* by the moment the instruction $i_{l_k}^{start}$ is activated and the host $h_{l_k}^{start}$ has only one virus, the task will be *finished* by the moment the host $h_{l_k}^{end}$ has one virus and after the instruction $i_{l_k}^{end}$ is activated. While the task is defined by Π_k, a virus will be present in the host $h_{l_k^{aux}}$. The structure of this can be visualized in Fig. 2.

It is important to clarify that all the patterns mentioned in this work are for single use only, that is, they cannot be used more than once.

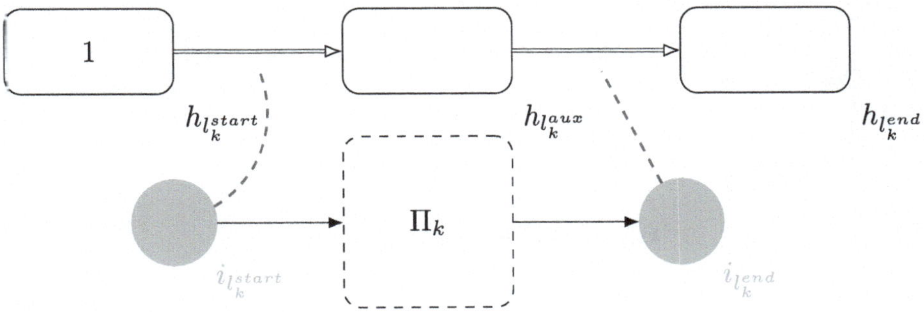

Fig. 2. A visual representation of a *task* l_k, using PVMs.

Pattern 1 (Sequence)

Description. An activity in a workflow process is enabled after the completion of another task in the same process. It can also be known as sequential routing.

The PVM that defines this pattern from task l_k to task $l_{k'}$ is shown in Fig. 3, it is interesting to note that we can include some delay $d \geq 0$ in the process. Here, the instruction $i_{SEQ_{k,k'}}$ is activated until the host $h_{l_k}^{end}$ is not empty, as if it follows the least weight path that takes us again to the same instruction. When the host is not empty, then one virus is transmitted to the initial host of the following task, now the delay of d steps is made by sequentially activating the instructions from i_1 to i_d. Thus the Pattern 1 has been properly defined.

Pattern 2 (Parallel Split)

Description. A point in process where a single control thread is split into multiple threads of the workflow that can be executed in parallel, thus allowing activities to be executed simultaneously or in any order. It can also be known as AND-split, or parallel routing.

The PVM that defines this pattern from task l_k to tasks $l_{k_1}, \ldots, l_{k_m}$ is shown in Fig. 4. Initially, the instruction $i_{split_{k_1,\ldots,k_m}-l_k}$ is activated until the task l_k is completed, at that instant the host $h_{l_k}^{end}$ has a virus, which is sent in the following step to $h_{split_{k_1,\ldots,k_m}-l_k}$ and replicated by m. Simultaneously, instructions $i_1, \ldots, i_m$ are activated, and once the m viruses reach the previous host, one virus is sent to each starting host $h_{l_{k_1}^{start}}, \ldots, h_{l_{k_m}^{start}}$ and the corresponding tasks are initialized.

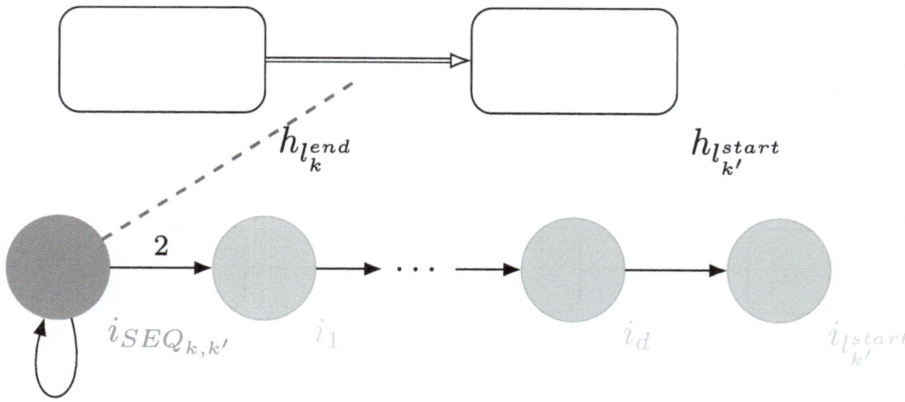

Fig. 3. PVM for sequential pattern.

Pattern 3 (Synchronization)

Description. A point in the workflow process where multiple parallel branches converge into one single thread of control, thus synchronizing multiple threads. It can also be known as AND-join, rendezvous, synchronizer.

The PVM that defines this pattern from tasks $l_{k_1}, \ldots, l_{k_m}$ to task l_k can be depicted in Fig. 5. In this module, the instructions from $i_{l_{k_1}}$ to $i_{l_{k_m}}$ are waiting until the respective task is completed, once completed, the corresponding host $h_{l_{k_1}}^{end}$ will contain one virus, and after that, the virus is sent to the host $join_{k_1,\ldots,k_m} - l_k$ and the next activated instruction follows the highest weight path, that is i_{HALT}. In the meantime, instruction i_1 is activated until one of the previous viruses is sent, after that it goes to instruction i_2 waiting for the second virus. Once all m tasks have been completed, instruction i_{m+1} will be activated, starting the task l_k.

Pattern 4 (Exclusive Choice)

Description. A point in the workflow process where, based on a decision or workflow control data, one of several branches is chosen. It can also be known as XOR-split, conditional routing, switch, or decision.

The PVM that defines this pattern to choose between tasks $l_{k_1}, \ldots, l_{k_m}$ can be depicted in Fig. 6. The decision of which task will be encoded in hosts $h_{l_{k_1}}, \ldots, h_{l_{k_m}}$ in the following way: for each $j \in \{1, \ldots, m\}$ the task l_{k_j} is chosen if and only if the host $h_{l_{k_j}}$ has one virus; otherwise, it will remain empty.

The idea of this module is that, for each $j \in \{1, \ldots, m\}$ the instructions i_{5j-4} and i_{5j-3} will be activated alternately; the first verifies if the task l_{k_j} is chosen, the other instruction verifies if any other task has been chosen, in that case that branch of instruction goes to i_{HALT}. If task l_{k_j} is chosen, then instructions i_{5j-2} and i_{5j-1} verify if any other task has also been chosen; in that case, they go to i_{HALT}. Otherwise, the instruction $i_{l_{k_j}}^{start}$ will be activated, and a virus is

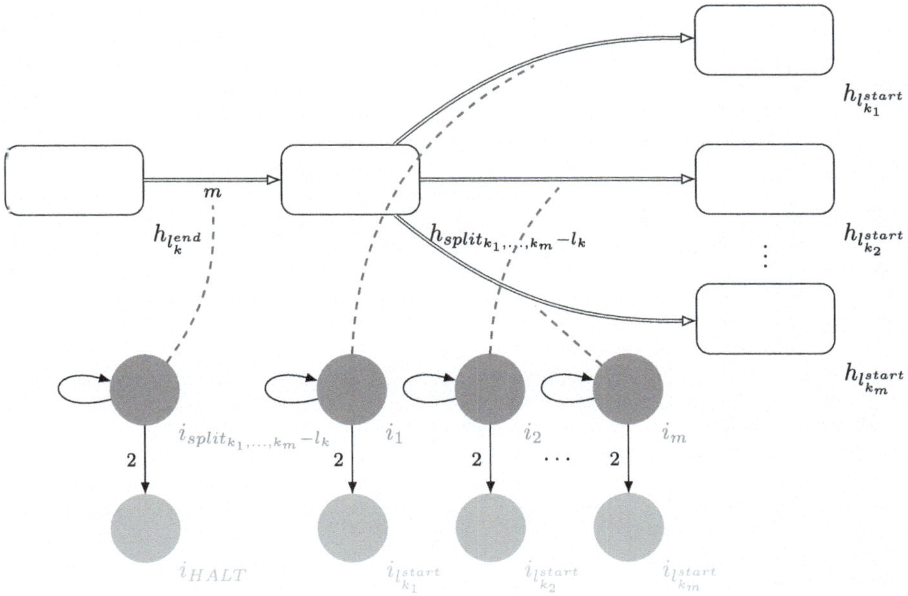

Fig. 4. PVM for the parallel split (AND-split) pattern.

transmitted to the host $h_{l_{k_j}}^{start}$. Thus, only one task is chosen if and only if one task has been previously chosen.

Pattern 5 (Simple Merge)

Description. A point in the workflow process where two or more alternative branches come together without synchronization. It is an assumption of this pattern that none of the alternative branches is ever executed in parallel. The difference with pattern 3 is that all the branches must be activated while in pattern 5 exactly one must be activated. It can also be known as XOR-join, asynchronous join, or merge.

The PVM that defines this pattern to merge tasks $l_{k_1}, \ldots, l_{k_m}$ in task l_k can be depicted in Fig. 7. For each $j \in \{k_1, \ldots, k_m\}$ when each task l_j is completed, the corresponding virus is transmitted to the host $h_{join_{k_1,\ldots,k_m}-l_k}$. If more than one is reached simultaneously, then it halts; otherwise, task l_k is initialized.

Pattern 6 (Multi-choice)

Description. A point in the workflow process where, based on a decision or workflow control data, a number of branches are chosen. It can also be known as conditional routing or OR-split.

The PVM that defines this pattern to choose among tasks $l_{k_1}, \ldots, l_{k_m}$ can be depicted in Fig. 8. The codification of the decision of which task is chosen is equal

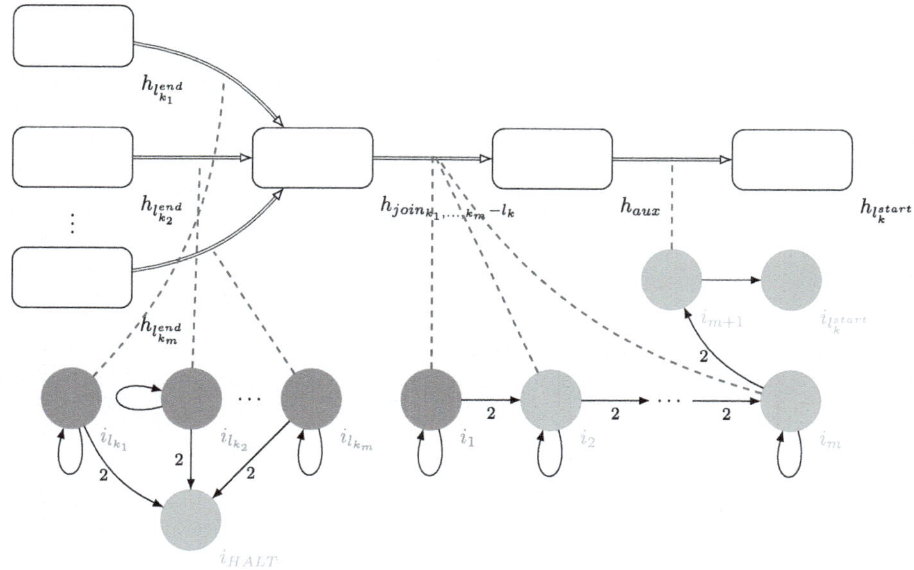

Fig. 5. PVM for the pattern synchronisation (AND-join).

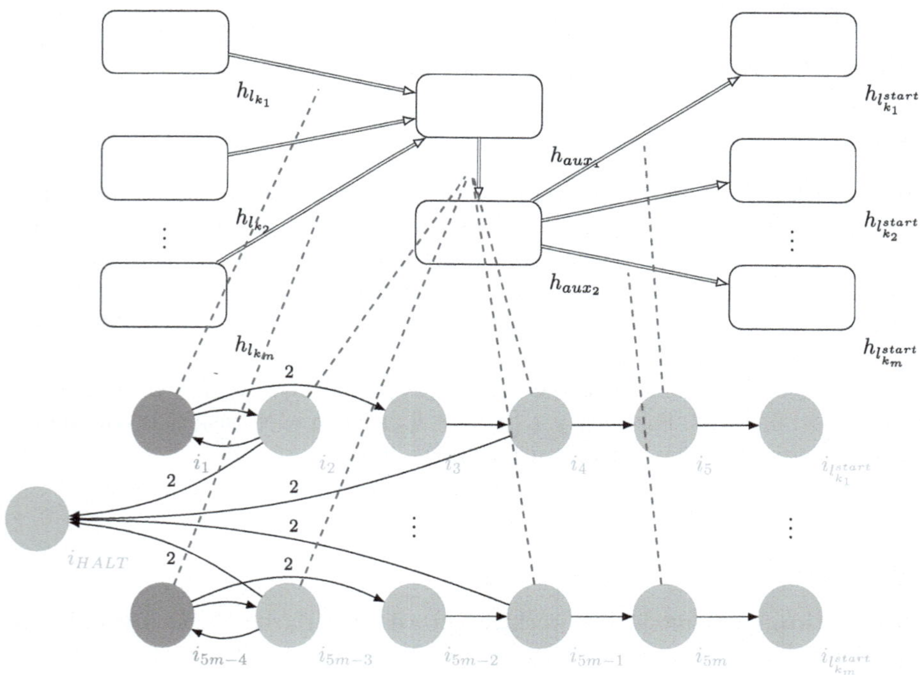

Fig. 6. PVM for the pattern exclusive choice (XOR-split).

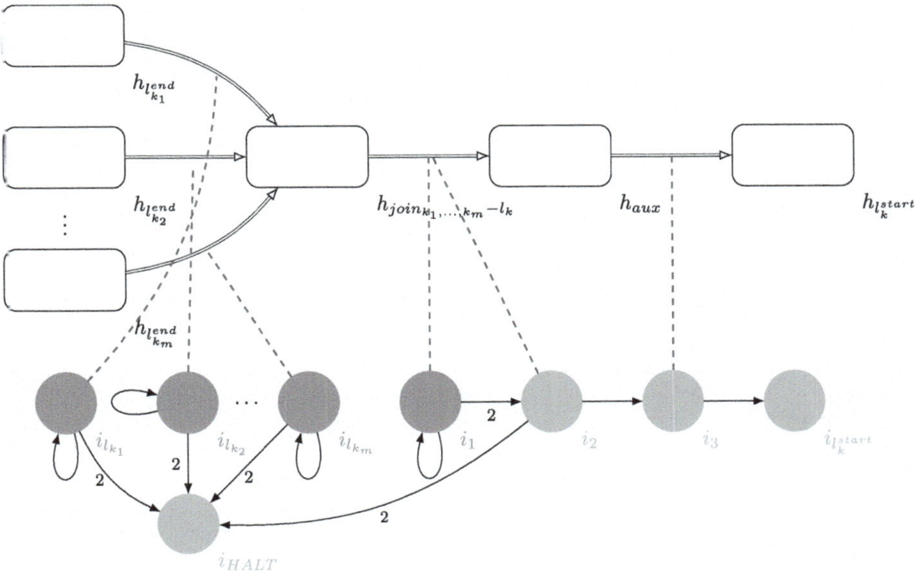

Fig. 7. PVM for the pattern simple merge (XOR-join)

to Pattern 4 (Exclusive Choice). Here, as it is not an exclusive choice, the unique problem is that it has to be single-use: for each $j \in \{1, \ldots, m\}$, instruction i_{3j+2} verify if any other task has been initialized.

4 Final Remarks

In the present work, we defined several and fundamental workflow patterns, using PVMs. Not only the formalism but also the graphical nature of PVMs are useful in the definitions. In the sequel works we aim to define more patterns such as those in [8]. Definitions of more complex patterns in workflows include loops or cycles, which is a limitation of the present work: the PVM modules used to define the patterns cannot be reused. Thus, designing modules that can be reused is part of our next works.

We also aim to consider the trade-offs of channel parallel VMs instead of instruction parallel VMs in this work, for defining workflow patterns. An expected effect of designing modules in the present work to be deterministic is their increased complexity, mainly in the sizes of the instruction graphs. In other works such as [5], the decision or choice is given by nondeterminism. In this way the complexity of the computing modules may decrease.

Recently, VMs were represented using vectors and matrices. The computations of VM are given by linear algebra operations [13]. Besides extending the representation to PVMs, it is interesting to investigate invariants, behavioral, and structural properties of VMs using such representations. Such investigations can benefit extensions of the present work, and vice-versa. A preliminary

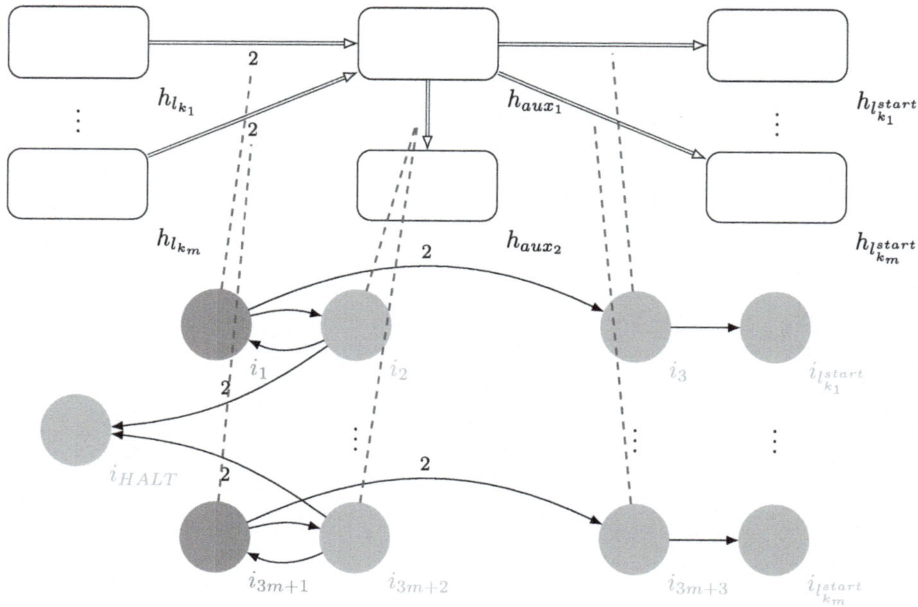

Fig. 8. PVM for the pattern OR-split.

simulator can be found in [10], a further work should be the implementation of these workflow patterns.

References

1. Van der Aalst, W.M.: The application of petri nets to workflow management. J. Circuits Syst. Comput. **8**(01), 21–66 (1998)
2. Adamatzky, A.: Handbook Of Unconventional Computing (In 2 Volumes). World Scientific (2021)
3. Ramírez-de Arellano, A., Orellana-Martín, D., Pérez-Jiménez, M.J.: Using virus machines to compute pairing functions. Int. J. Neural Syst. **33**(05), 2350023 (2023). https://doi.org/10.1142/s0129065723500235
4. Bäck, T., Kok, J.N., Rozenberg, G.: Handbook of Natural Computing. Springer, Heidelberg (2012)
5. Cabarle, F.G.C., Adorna, H.N.: On structures and behaviors of spiking neural p systems and petri nets. In: Membrane Computing: 13th International Conference, CMC 2012, Budapest, Hungary, 28–31 August 2012, Revised Selected Papers 13, pp. 145–160. Springer (2013)
6. Cabarle, F.G.C., Buño, K.C., Adorna, H.N.: Time after time: notes on delays in spiking neural p systems. In: Nishizaki, S., Numao, M., Caro, J., Suarez, M.T. (eds.) Theory and Practice of Computation, Tokyo, Japan, pp. 82–92. Springer (2013)
7. Chen, X., Pérez-Jiménez, M.J., Valencia-Cabrera, L., Wang, B., Zeng, X.: Computing with viruses. Theor. Comput. Sci. **623**, 146–159 (2016). https://doi.org/10.1016/j.tcs.2015.12.006

8. van Der Aalst, W.M., Ter Hofstede, A.H., Kiepuszewski, B., Barros, A.P.: Workflow patterns. Distrib. Parallel Databases **14**, 5–51 (2003)
9. Murata, T.: Petri nets: properties, analysis and applications. Proc. IEEE **77**(4), 541–580 (1989)
10. Orellana-Martín, D., Ramírez-de Arellano, A., Pérez-Jiménez, M.J.: Simulating and validating virus machines. PREPRINT (Version 1) Research Square (2023). https://doi.org/10.21203/rs.3.rs-2714903/v1
11. Peterson, J.L.: Petri nets. ACM Comput. Surv. (CSUR) **9**(3), 223–252 (1977)
12. Petri, C.A.: Communication with automata: Volume 1 supplement 1. DTIC Document (1966)
13. Ramírez-de-Arellano, A., Cabarle, F.G.C., Orellana-Martín, D., Pérez-Jiménez, M.J., Adorna, H.N.: Matrix representation of virus machines. In: 10th International Conference on the Interplay between Natural and Artificial Computation (IWINAC/ICINAC2024), Olhao, Portugal, 4–7 June 2024 (to appear)
14. Ramírez-de-Arellano, A., Orellana-Martín, D., Pérez-Jiménez, M.J.: Parallel virus machines, submitted to J. Membrane Comput
15. Ramírez-de-Arellano, A., Orellana-Martín, D., Pérez-Jiménez, M.J.: Generating, computing and recognizing with virus machines. Theor. Comput. Sci. **972**, 114077 (2023). https://doi.org/10.1016/j.tcs.2023.114077
16. Riehle, D., Züllighoven, H.: Understanding and using patterns in software development. Tapos **2**(1), 3–13 (1996)
17. Rozenberg, G., Reisig, W.: Lectures on petri nets I: Basic models. Lecture Notes in Computer Science, vol. 1491 (1998)
18. Rozinat, A., Wynn, M.T., van der Aalst, W.M., ter Hofstede, A.H., Fidge, C.J.: Workflow simulation for operational decision support. Data Knowl. Eng. **68**(9), 834–850 (2009)
19. Song, T., Zeng, X., Zheng, P., Jiang, M., Rodriguez-Paton, A.: A parallel workflow pattern modeling using spiking neural p systems with colored spikes. IEEE Trans. Nanobiosci. **17**(4), 474–484 (2018)

Integrating Simulation Budget Management into Drum-Buffer-Rope: A Study on Parametrization and Reducing Computational Effort

Balwin Bokor, Wolfgang Seiringer(✉), and Klaus Altendorfer

Department for Production and Operation Management, University of Applied Sciences Upper Austria, Wehrgrabenstraße 1, 4400 Steyr, Austria
{balwin.bokor,wolfgang.seiringer,klaus.altendorfer}@fh-steyr.at

Abstract. In manufacturing, a bottleneck workstation frequently emerges, complicating production planning and escalating costs. To address this, Drum-Buffer-Rope (DBR) is a widely recognized production planning and control method that focuses on centralizing the bottleneck workstation, thereby improving production system performance. Although DBR is primarily focused on creating a bottleneck schedule, the selection of planning parameters is crucial, as they significantly influence the scheduling process. Conducting a comprehensive full factorial enumeration to identify the ideal planning parameters requires substantial computational effort. Simulation Budget Management (SBM) offers an effective concept to reduce this effort by skipping less promising parameter combinations. This publication introduces a method for integrating SBM into multi-stage multi-item DBR planned and controlled production system with limited capacity, aimed at determining the optimal planning parameters. Furthermore, we conduct a simulation study to analyze the effects of different production system environments, i.e., varying levels of shop load and process uncertainty, on both the performance and parameterization of DBR and the efficacy of SBM. Our results show significant reduction in simulation budget for identifying optimal planning parameters compared to traditional full factorial enumeration.

Keywords: Production Planning · Simulation · Bottleneck · Drum-Buffer-Rope · Heuristics

1 Introduction

Production systems often face bottlenecks, described by Goldratt [1] in the Theory of Constraints as capacity constraint resources. Bottlenecks, where the slack between required and available capacity is minimal or nonexistent, play a pivotal role in determining a production system's overall throughput and significantly impact key performance indicators (KPIs) [2]. While bottlenecks in production systems can take various forms, including specific employees or machine tools [3], they are most commonly associated

A. A. Juan et al. (Eds.): DSA ISC 2024, LNCS 14778, pp. 220–234, 2025.
https://doi.org/10.1007/978-3-031-78238-1_21

with particular workstations [4]. To ensure favorable KPIs, it is crucial to identify bottlenecks [5] and strategically manage bottleneck capacity, focusing on aligning production with customer demand [6]. Hence the entire production system must be subordinated to the bottleneck workstation to ensure optimal performance. This principal is emphasized by the widely recognized production planning and control method Drum-Buffer-Rope (DBR). The selection of planning parameters significantly influences the bottleneck schedule and, consequently, affects KPIs and costs [7].

Given that real-world production systems must operate in stochastic environments with intricate interdependencies, simulation emerges as a particularly effective approach for determining these planning parameters [8]. However, to attain statistically reliable results in a stochastic environment, a substantial number of replications, involving simulations with equal parameterization but varying random number perturbations, is essential [9]. The full factorial enumeration approach, despite its precision, is time-consuming and may even be impractical as combinatorial complexity rises, particularly when it involves a complete enumeration of the entire feasible solution space [10]. A simple and fast concept to reduce simulation time while maintaining solution quality is achieved through Simulation Budget Management (SBM). This concept minimizes computational effort by skipping non-promising replications, as demonstrated by Seiringer et al. [11] in the context of a material requirements planning (MRP) production system. Extending its integration into other production planning and control methods will validate the broad applicability of SBM. Therefore, this publication integrates the SBM concept into a DBR planned and controlled multi-item multi-stage make-to-order flow shop production system with limited capacity. A real-world scenario similar to our observed setup includes automotive assembly line and electronics manufacturing assembly lines, both organized as flow shops with critical bottlenecks at complex stations like engine installation and smartphone assembly.

The primary objective is to reduce computational efforts when determining optimal planning parameters for DBR through a simulation-based approach. From a scientific standpoint, the study explores whether the SBM concept yields the same parameterization as a full factorial enumeration. Additionally, it explores the impact of different levels of shop load and process uncertainty on the parameterization of DBR and the performance of the SBM concept. From an economic perspective, decision-makers are presented with a practical method to reduce computational complexity in simulation studies and insights into how shop load and process uncertainty affect optimal planning parameters of DBR and costs. Hence, the following research questions are addressed:

RQ1: How do different production system environments, i.e., changes in shop load and process uncertainty, influence the performance and optimal planning parameters of DBR?

RQ2: Do different production system environments, i.e., shop load and process uncertainty, influence the performance of SBM?

RQ3: What is the risk of overlooking the optimal solution when employing SBM concepts?

RQ4: What potential savings in simulation budget (effort) can be realized through the implementation of the SBM concept at DBR, and what constitutes the optimal SBM parameterization?

The rest of the article is organized as follows: Sect. 2 provides a brief introduction to the production planning and control method DBR and introduces the concept of SBM for achieving optimal solutions within limited simulation resources, along with a discussion of related strategies. In Sect. 3, we explore the production system, describe the abstracted simulation model, and offer insights into the numerical study. Section 4 presents the outcomes of our simulation study, along with a comprehensive examination and discussion of the findings. Finally, in Sect. 5, we outline the primary contributions of our research and set the stage for future research directions.

2 Related Work

In this section we provide an overview of the production planning and control method DBR, including a comprehensive description of the forward scheduling approach and the applied planning parameters. Subsequently, we outline the SBM concept, as well as other related strategies, aimed at minimizing computational effort in simulation studies.

2.1 Drum-Buffer-Rope

The production planning and control method DBR is introduced by Goldratt and Cox [2]. Its fundamental concept revolves around scheduling the bottleneck workstation, thereby aligning the entire production system with the throughput of this critical constraint. In the DBR framework, the gross requirements are batched into lot sizes by the Master Production Schedule (MPS) to generate production orders, often utilizing Periodic Order Quantity lot-sizing policy [12]. After establishing the lot size, the bottleneck must be scheduled. This scheduling effectively requires two main buffers, which act as time horizons analogous to planned lead times in MRP [13]. However, DBR distinguishes itself by setting these two planning parameters at a system-wide level, applying uniformly across all materials concerning a particular production path, which reduces the number of planning parameters in comparison to MRP [12]. The two planning parameters are the Shipping-Buffer, which functions to prevent delays in completion, and the CCR (capacity-constrained resource)-Buffer, aiming to protect the bottleneck from starvation [14].

In DBR framework, a critical decision is whether to choose forward or backward scheduling for the bottleneck workstation. Forward scheduling determines when production at the bottleneck will be completed, identifying tardiness through planned end dates that fall after due dates. Whereas, backward scheduling determines when production should start, identifying tardiness through planned start dates that are set in the past. Although both approaches have benefits, as discussed by Mleczko [15], this study focuses on forward scheduling, primarily due to its avoidance of planning start dates in the past.

For applying forward scheduling, let's define I as the set of all production orders, where each order $i \in I$ is associated with a due date d_i. A scheduling window is established by identifying production orders whose due dates d_i, after subtracting the sum of the Shipping-Buffer S and CCR-Buffer C, fall beyond the current time t. The scheduling is performed iteratively, triggered whenever a new production order is within the scheduling window or when a production order is completed at the bottleneck workstation. Consequently, the subset of production orders to be scheduled at time t, denoted as $O(t)$, can be formulated as:

$$O(t) = \{i \in I | d_i - (S + C) \leq t \wedge i \notin R(t)\} \tag{1}$$

Here, $R(t)$ represents the set of production orders that have already been released for production. By excluding these released production orders from $O(t)$, a 'frozen zone' is created, effectively minimizing system 'nervousness' that arises from production orders with imminent due dates [16]. Once the set of production orders $O(t)$ for scheduling is established, they are sorted in ascending order according to their due dates d_i, thereby implementing an Earliest-Due-Date (EDD) approach for order release. When a production order is scheduled (at the moment when $i \in O(t)$), the earliest plan start date at the bottleneck workstation a_i is calculated by adding the CCR-Buffer C to the current time t. Given this earliest plan start date at the bottleneck workstation a_i and the mean expected planned processing time p_i, the plan start date s_i and plan end date e_i of each production order i at the bottleneck workstation can be calculated as follows:

$$s_i = \max(e_{i-1}, a_i) \tag{2}$$

$$e_i = s_i + p_i \tag{3}$$

Here e_{i-1} represents the plan end date of the preceding production order. Thus, the plan start date s_i is determined by either the plan arrival time at the bottleneck workstation a_i or the completion time of the preceding production order, whichever is later. After processing a production order at the bottleneck workstation, the deviation is assessed by comparing the actual end date $\hat{e}_0$ with the plan end date e_0 of this completed production order, calculating the deviation $\Delta e = \hat{e}_0 - e_0$. . The completed order is designated as index 0, subsequent production orders are systematically labeled with increasing indexes. If the production order is completed ahead of schedule (i.e., $\Delta e < 0$), forward rescheduling is initiated, pushing subsequent orders ahead but not past the earliest plan start date at the bottleneck workstation a_i. Conversely, in the case of a delay (i.e., $\Delta e >$ 0), backward rescheduling is performed, where any bottleneck idle time gaps between production orders are considered. As illustrated in Eq. (2), these gaps arise when the earliest planned start date of the scheduled production order, falls later than the planned end date of the preceding scheduled order. These gaps primarily emerge in periods when the gap between required and available capacity is large. The cumulative gap G_i, representing the sum of individual bottleneck idle time gaps $g_k = s_k - e_{k-1}$ preceding production order i, is calculated as:

$$G_i = \sum_{k=1}^{i} g_k \tag{4}$$

For both rescheduling scenarios the adjusted start date s_i' for each production order i has to be calculated, specifically for forward scheduling as follows:

$$s_i' = \max(s_i - \Delta e, a_i) \quad \textit{if } \Delta e < 0 \tag{5}$$

whereas, in case of backward scheduling, at first, the adjusted shift for each production order $\Delta e_i'$, considering idle time gaps at the bottleneck workstation, has to be calculated as:

$$\Delta e_i' = \max(\Delta e - G_i, 0) \quad \text{if } \Delta e > 0 \tag{6}$$

Based on the adjusted shift for each production order $\Delta e_i'$ the adjusted start date s_i' at backward scheduling is then:

$$s_i' = s_i + \Delta e_i' \tag{7}$$

For both forward and backward rescheduling the adjusted plan end date $e\prime_i$ for production order i can be calculated based on Eq. (3) substituting s_i with the adjusted start date s_i'. To summarize, only two planning parameters are essential for scheduling the bottleneck workstation and defining DBR: the CCR-Buffer, and the Shipping-Buffer. The following Table 1 outlines all the notations used.

Table 1. Notations for Forward Scheduling Description.

Notation	Description
I	Set of production orders
$O(t)$	Set of production orders considered for scheduling at time t
$R(t)$	Set of released production orders at time t
S	Shipping-Buffer
C	CCR-Buffer
t	Current time
d_i	Due date of produciton order i
a_i	Earliest plan start date of production order i at bottleneck workstation
s_i	Plan start date of production order i
s'_i	Adjusted plan start date after rescheduling for production order i
p_i	Plan process time of prodcution order i at bottleneck workstation
e'_i	Adjusted plan end date after rescheduling for production order i
e_0	Plan end date of completed production order at bottleneck workstation
$\hat{e}_0$	workstation
Δe	Deviation between plan end date and realized end date
$\Delta e'_i$	Adjusted shift for production order i
g_k	Size of the kth bottleneck idle time gap
G_i	Cumulative bottleneck idle time gap up to prodution order i

To improve performance, the original form of DBR dispatches production orders to the bottleneck workstation based on the Earliest-Constraint-Date (ECD), and subsequently on an EDD basis. Performance can be further enhanced by employing more sophisticated or varied combinations of sequencing and dispatching rules, as demonstrated by Thürer and Stevenson [17]. In addition, the incorporation of Buffer Management based on real-time information, further supports production control, as discussed in Woo et al. [18].

2.2 Efficient Simulation-Based Parameter Optimization

The concept of SBM is utilized in simheuristics to minimize costs in simulation experiments with limited resources [11]. This approach aims to explore all parameter sets within a predefined solution space efficiently. The SBM method evaluates the quality of simulation iterations by comparing the average costs against a dynamically adjusted percentile of all known iterations, skipping further replications if costs exceed a certain threshold. Key to this method is the initialization of the parameters lower bound (lb), upper bound (ub), percentile step, as well as the criteria for the minimum number of replications and iterations required to start SBM. Seiringer et al. [11] previously successfully applied the SBM concept in an MRP planned production system. Building on this, our current study integrates this SBM with an agent-based discrete event simulation model of a forward-scheduled DBR production system, further demonstrating SBM's broad applicability and effectiveness.

3 Simulation Model

In the following section, we initially describe the production system and its modeling within a simulation framework. Subsequently, we detail the conducted numerical study, outlining the tested production system environments, planning parameters, and settings for SBM.

3.1 Production System and Simulation Model

In this study, we investigate a multi-item multi-stage make-to-order flow shop production system with limited capacity and stochastics in demand, processing as well as customer-required lead times. Our observed production system is modeled as an agent-based simulation using AnyLogic and consists of five workstations {W1, ..., W5}, with workstation four identified as the bottleneck, and three products {P1, P2, P3}. Figure 1 illustrates the modeled production system, while Fig. 2 highlights the Bill of Materials (BoM), detailing the required workstations for producing the respective components and products. As shown in Fig. 2, the component C2 is initially produced using workstations {W1, W2, W3} along with the raw material R, which is always available and not planned. At the bottleneck workstation W4, component C2 is split into components C1/1 and C1/2. Finally, at workstation W5, component C1/1 is used to manufacture product P1, while C1/2 is utilized to produce either product P2 or P3. Each level of the BoM requires one specific component.

A real-world scenario similar to our observed setup is the automotive assembly line, where engine installation acts as a critical bottleneck due to the complex and time-consuming tasks of fitting the engine and connecting various systems. Such an assembly line is typically organized as a multi-item multi-stage flow shop where vehicles move from one station to the next, each dedicated to a specific part of the assembly process. Additionally, a comparable formulation is found in electronics manufacturing, which also follows a multi-item multi-stage flow shop configuration and faces significant bottlenecks in the assembly of complex multi-component devices such as smartphones.

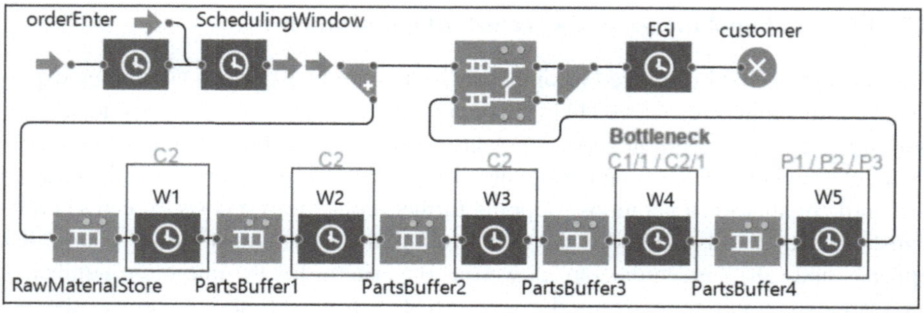

Fig. 1. Production System.

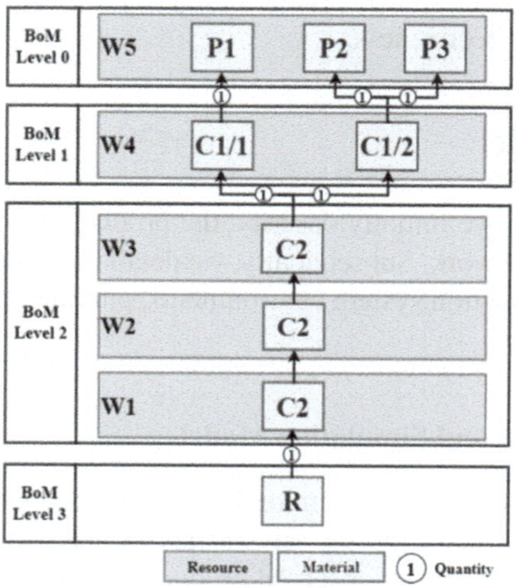

Fig. 2. Bill of Materials.

Each customer order is comprised of one three product type, selected through a random categorical distribution with probabilities ∈ {0.25, 0.50, 0.25}, reflecting the varying likelihoods of each product type being chosen. These customer orders have a discrete uniformly distributed lot size within the range of ∈ {1, 2}, from which a single production order is generated. The inter-arrival time of customer orders follows a lognormal distribution with a coefficient of variation (CV) of 0.3. The customer-required due date combines a deterministic part of 6 time units and a variable part that is exponentially distributed, with a mean 16 time units.

As we observe a make-to-order production system, these customer orders are directly transferred to production orders. The expected mean planned processing time at the initial three workstations {W1, W2, W3} is consistently set at 0.65 time units, as only one component, C1, is produced, as shown in Fig. 2. However, the expected mean planned processing time at the bottleneck workstation W4 is 0.70 time units for component C1/1 and 0.75 time units for component C1/2. Moreover, at the final workstation W5, the mean planned processing time is 0.70 time units for product P1, 0.60 time units for product P2, and 0.75 time units for product P3. All these times refer to produce one product respectively component. The available capacity at each workstation is 1440 time units per day. We calculate the shop load by integrating the inter-arrival times, expended mean planned processing times, and constrained capacity. This shop load represents the planned utilization, excluding the stochastic variability introduced by stochastic processing times. To ensure efficiency, we maintain the shop load at each non-bottleneck workstation at 10% lower than at the bottleneck workstation. For example, while a 95% planned utilization is maintained at W4, the other workstations operate at a planned utilization of 85%. These information about applied distributions and parametrization regarding the simulation model can be found in Table 3.

Scheduling for the bottleneck workstation, and consequently the entire production system, is conducted using the forward approach outlined in Sect. 2.1. During the simulation process at the workstations, the processing time is determined by a lognormal distribution, based on the above stated mean expected planned processing time and coefficient of variation of this expected planned processing time (CV PPT) 0.3, while also accounting for lot size. His CV PPT is varied only at the bottleneck workstation during the numerical study to investigate different behaviors in stochastic production systems. Within the system, production orders are dispatched up to the bottleneck using an Earliest-Constraint-Date (ECD), and post-bottleneck using an EDD, in alignment with the original version of DBR [2]. Finished goods are held in the Finished Goods Inventory (FGI) until reaching the customer-required due date. If tardiness occurs, delivery is executed immediately. Building on the previous description, the distributions were chosen for their non-negativity and widespread application in prior research. These values were calibrated based on preliminary studies to ensure they adequately challenge the production system without causing overload.

3.2 Numerical Study

In the following we describe the setup of the experiments for the performed numerical study with our developed simulation model. Using 20 replications per iteration effectively covers a meaningful range of stochastic effects. Without SBM all 20 replications must be simulated. To test how many replications can be saved using SBM for DBR, a first starting point are the evaluated SBM settings described in Seiringer et al. [11]. The settings S1, S2, S3 and S4 and the corresponding values are: with lower bound (lb) and upper bound (ub) combinations of (S1) lb = 0.05, ub = 0.4; (S2) lb = 0.1, ub = 0.5; (S3) lb = 0.02, ub = 0.8; and (S4) lb = 0.02, ub = 0.2. The most stringent setting is (S4), while the least restrictive setting is (S3), a SBM initialization phase of 5 iterations with 20 replications and a minimum number of 3 replications per iteration before SBM is applied to skip a running replication are used.

In addition to SBM, the planning parameter ranges for the DBR model, along with different shop loads and process uncertainties, i.e. CV PPT, for the bottleneck workstation are investigated as detailed in Table 2. Moreover, the applied distributions and parameterization concerning the simulation model are shown in Table 3. At the numerical study the shop load is adapted by changing the inter-arrival time of customer orders. Three different levels (low, medium, high) are investigated for both the shop load and the process uncertainty. This leads to 9 different production system environments. The tested ranges for the CCR-Buffer are from 1 to 12 time units, and for the Shipping-Buffer, they are from 1 to 24 time units, each with a step size of 1 time unit. Both ranges were set based on preliminary studies. In total 288 different parameter combinations are tested for DBR at the full factorial enumeration and DBR in combination with S1, S2, S3 and S4 for each of the 9 production system environments. This leads to 5760 individual replications for the full factorial enumeration at each production system environment and also potentially 5760 for each SMB setting {S1, ..., S4}. If SBM is enabled for the DBR model the 5760 should not be consumed to find the optimal solution. Hence, in total 259 200 potential individual replications are examined (9 production system environments x 5760 individual replications x 5 methods). Each replication is run for 8760 h simulation time (365 days) with a warm-up phase of 760 h (32 days). Each experiment is run subsequently on the same hardware a computer with an Intel i5-10500 CPU (3.1 GHz) and 32 Gb RAM. The parameter combinations are generated using R, then exported to a H2 database and subsequently queried by the Anylogic DBR model. The results are afterwards also stored into the same H2 database.

Table 2. DBR Parameter Ranges.

	Min	Max	Step Size	Iterations
Shop Load	0.85	0.95	0.05	3
CV PPT	0.30	0.90	0.30	3
Different production system environments				9
CCR-Buffer C [TU]	2	12	1	12
Shipping-Buffer S [TU]	2	24	1	24
Total Iterations per production system environment (no/SBM)				288
Individual Replications per production system environment (no/SBM)				5760

Table 3. Applied Distribution and Parametrization.

	Distribution Type	Parametrization
Product Type	Categorical	{P1 = 0.25, P2 = 0.50, P3 = 0.25}
Size of Customer Order	Discrete Uniform	U(1,2)
Inter-Arrival Time Customer Orders	Lognormal	Log(μ = cal. based on shop load {0.85, 0.90, 0.95}, σ =0.3)
Customer-Required Due Date Fix Part	Deterministic	6
Customer-Required Due Date Variable Part	Exponential	Exp(λ = 1/16)
Planned Processing Time {W1, W2, W3}	Lognormal	Log(μ = 0.65, σ = 0.3)
Planned Processing Time W4	Lognormal	Log(μ = {C1/1 = 0.70, C1/2 = 0.75}, σ = {0.3, 0.6, 0.9})
Planned Processing Time W5	Lognormal	Log(μ = {P1 = 0.70, P2 = 0.60, P3 = 0.75}, σ = 0.3)

4 Simulation Results and Discussion

To analyze the production system performance, overall costs are selected, which encompass the combined expenses of Work-in-Process (WIP) costs, FGI costs, and tardiness costs. The costing value for WIP is set at 0.5, half of the FGI costing value, which is 1. This distinction in costing values is logical, as storing finished goods typically incurs higher costs than holding WIP. The tardiness costs are set to 19, which represent a target service level of 95%. A service level of 0.95 indicates a 95% likelihood that the available inventory will meet customer demand, thus avoiding any stockouts. First, our analysis explores the impact of varying shop loads and process uncertainties on overall costs and parameterization. Subsequently, we examine the potential for reducing simulation budgets through the application of SBM.

4.1 Effects of Different Shop Load and Process Uncertainty

Figure 3 illustrates the minimum overall costs per time unit, expressed in cost units (CU) for the production system at different shop loads of 0.85, 0.90 and 0.95 and CV PTT at the bottleneck workstation W4, which is show in Fig. 1.

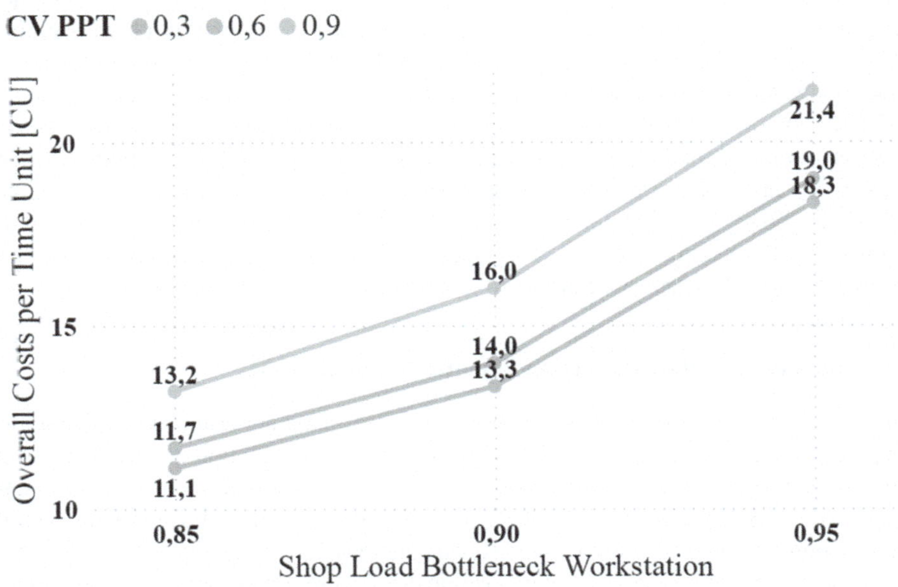

Fig. 3. Min. Overall Costs per Planned Utilization & CV PPT.

An increase in both shop load and process uncertainty, i.e. CV PPT, results in higher overall costs. This rise in costs can be attributed to two main factors. Firstly, at higher levels of shop load, the bottleneck workstation experiences fewer idle periods, thus limiting its ability to compensate for delays. Secondly, higher processing uncertainty results in larger discrepancies between the planned and actual end dates at the bottleneck workstation, which leads to larger rescheduling as indicated by Eq. (4) and Eq. (8). To discuss RQ1 in Table 3 the DBR parameterizations of the identified minimum overall costs, are presented for a comprehensive analysis. Whereby, C denotes the CCR-Buffer and S the Shipping-Buffer.

Table 4. Optimal DBR Planning Parameters.

		Shop Load		
		0.85	**0.9**	**0.95**
CV PPT	**0.30**	C = 6 S = 7	C = 7 S = 8	C = 8 S = 15
	0.60	C = 6 S = 8	C = 7 S = 9	C = 8 S = 15
	0.90	C = 6 S = 9	C = 7 S = 11	C = 8 S = 19

**all values expressed in Time Units [TU]*

As demonstrated in Table 4, variations in the production system environment have a marginal effect on the CCR-Buffer, with only a slightly longer buffer required at higher shop load. This necessity is because higher shop loads at the bottleneck workstation also increase the utilization of the preceding workstations, given that the inter-arrival time is adjusted in our model. Consequently, this increased utilization requires a longer CCR-Buffer to prevent the bottleneck from starvation. Whereas, a slight increase in process uncertainty at the bottleneck workstation necessitates a longer Shipping-Buffer. This buffer extension compensates for the growing variances between planned and actual completion times. Moreover, an increase in either of the two buffers enlarges the scheduling window, as specified in Eq. (1). This enlargement enhances the scheduling flexibility by considering a broader set of production orders for production release.

4.2 Simulation Budget Reduction Potential Through SBM

The intention of SBM is, to safe simulation time by skipping not promising replications. Skipping not promising solutions should not lead to worser solution quality. The saved simulation budget can then, for example, be used to test additional parameter combinations. In our DBR investigations the best solution is represented by the minimum overall costs within the selected result subset. To validate the effectiveness of SBM within the simulation model, it is essential to demonstrate that the minimum overall costs identified by SBM are equivalent to those found in the model without SBM. The data shown in Table 5 demonstrates the ability of SBM to identify the same minimum overall costs as those found by a full factorial (FF) enumeration for each production system environment. Since identical costs are observed for each SBM setting {S1, …, S4}, only a single representative value is displayed for SBM.

As seen in Table 5 the minimum overall costs could be found at each SBM setting for each production system environment. Pertaining to RQ3, this indicates that the

Table 5. Min. Overall Costs for Full Factorial Enumeration and SBM.

		Shop Load		
		0.85	**0.9**	**0.95**
CV PPT	**0.30**	FF = 11.1138 SBM = 11.1138	FF = 13.3347 SBM = 13.3347	FF = 18.3174 SBM = 18.3174
	0.60	FF = 11.6603 SBM = 11.6603	FF = 13.9776 SBM = 13.9776	FF = 18.9818 SBM = 18.9818
	0.90	FF = 13.2072 SBM = 13.2072	FF = 16.0121 SBM = 16.0121	FF = 21.3783 SBM = 21.3783

*all values expressed in Costs per Time Unit [CU]

likelihood of missing an optimal solution when using SBM, as opposed to conducting a full factorial enumeration, is negligible. With the established accuracy of the SBM implementation, we can analyze its performance relative to a full factorial enumeration at different production system environments using the data in following Table 6 and thereby comprehensively answer RQ4. This Table 6 details the number of replications used, with the improvement delta between a full factorial enumeration and with SBM application indicated in brackets. For instance, the 5760 replications in Table 6 concerning the full factorial (FF) enumeration correspond to 288 parameters, each with 20 replications. Δ1 denotes the relative difference in the number of replications between the full factorial enumeration and SBM setting S1 to S4, calculated as (SBM replications – full factorial enumeration replications) divided by a full factorial enumeration replications. Δ2 represents the average improvement delta across all four SBM settings for the identical shop load but different process uncertainty. Whereas, Δ3 indicates the average improvement delta for the same SBM setting across all three levels of process uncertainty. At every SBM setting, a significant performance advantage is evident compared to the full factorial enumeration. However, as either shop load or process uncertainty increases, the advantage in terms of replication savings lessens, which is observable when comparing Δ2 across identical shop loads and identical process uncertainties. This trend of reduced replication savings under SBM settings with increased shop loads or process uncertainty can be linked to the fact that higher shop loads induce more pronounced variations between replications. This leads to a greater probability of experiencing elevated costs and consequently requires the continuation of the current iteration without skipping. This observation answers RQ2, as different production system environments have an influence on the SBM performance. Across all SBM settings, setting S4, which is the most stringent, realizes the most substantial savings. S4 uniformly surpasses the other settings (S1, S2, and S3) within the same production system environments and maintains the superior average for varying process uncertainty Δ3, demonstrating the optimal delta values of −0.68, −0.61, and −0.47. These impressive results demonstrate that SBM with S4 not only yields the same minimum overall costs but also significantly reduces simulation budget.

Table 6. SBM Setting Performance at Different Production System Environments.

Shop Load 85%

		FF	S1 (Δ1)	S2 (Δ1)	S3 (Δ1)	S4 (Δ1)	Avg(Δ2)
CV PPT	**0.3**	5760	1991 (-0,65)	2279 (-0,6)	2716 (-0,53)	1690 (-0,71)	-0.62
	0.6	5760	2096 (-0,64)	2391 (-0,58)	2804 (-0,51)	1762 (-0,69)	-0.61
	0.9	5760	2443 (-0,58)	2730 (-0,53)	3061 (-0,47)	2102 (-0,64)	-0.56
		Avg(Δ3)	-0.62	-0.57	-0.5	-0.68	

Shop Load 90%

		FF	S1 (Δ1)	S2 (Δ1)	S3 (Δ1)	S4 (Δ1)	Avg(Δ2)
CV PPT	**0.3**	5760	2307 (-0,6)	2587 (-0,55)	2913 (-0,49)	1952 (-0,66)	-0.58
	0.6	5760	2494 (-0,57)	2766 (-0,52)	3046 (-0,47)	2129 (-0,63)	-0.55
	0.9	5760	2950 (-0,49)	3189 (-0,45)	3349 (-0,42)	2617 (-0,55)	-0.48
		Avg(Δ3)	-0.55	-0.51	-0.46	-0.61	

Shop Load 95%

		FF	S1 (Δ1)	S2 (Δ1)	S3 (Δ1)	S4 (Δ1)	Avg(Δ2)
CV PPT	**0.3**	5760	3202 (-0,44)	3401 (-0,41)	3453 (-0,4)	2890 (-0,5)	-0.44
	0.6	5760	3309 (-0,43)	3501 (-0,39)	3538 (-0,39)	3019 (-0,48)	-0.42
	0.9	5760	3458 (-0,4)	3595 (-0,38)	3640 (-0,37)	3204 (-0,44)	-0.40
		Avg(Δ3)	-0.42	-0.39	-0.39	-0.47	

5 Conclusion

In this publication, we integrated SBM into the production planning and control method DBR to reduce simulation budget for method parameterization. Moreover, our analysis examined the impact of different production system environments, i.e., varying shop loads and process uncertainty, on both the performance and parameterization of DBR and the efficacy of SBM. At our numerical study we simulated a multi-stage multi-item DBR planned and controlled production system with limited capacity. Findings reveal that higher shop load or process uncertainty increases overall costs and the Shipping-Buffer, due to growing discrepancies in planned versus actual completion times. However, only at higher shop load a slight increase of the CCR-Buffer is required. Additionally, increasing either buffer widens the scheduling window, enhancing production system flexibility. SBM consistently matched the minimum overall costs of a full factorial enumeration across all SBM settings and production system environments, also offering notable simulation budget savings. Yet, these savings lessen with increased shop load or process uncertainty due to rising cost variations, prompting continued iteration. The most substantial simulation budget reduction was noted with the strictest SBM setting, independent of the production system environment. Future research on DBR should explore more complex production systems, with varied buffers for divergent production paths. For SBM, advancements should aim at refining the algorithm's precision by identifying parameters with significant performance impact during runtime and adjusting the step size of them dynamically, i.e., smaller in promising areas of the search space and larger otherwise.

Acknowledgments. This research was funded in whole or in part by the Austrian Science Fund (FWF) [P32954-G] G and the Funding Agency of the State of Upper Austria with the project FO999905125. For open access purposes, the author has applied a CC BY public copyright license to any author accepted manuscript version arising from this submission.

References

1. Goldratt, E.M.: Theory of constraints. North River Croton-on-Hudson (1990)
2. Goldratt, E.M., Cox, J.: The Goal: Excellence in Manufacturing (1984)
3. Altendorfer, K., Felberbauer, T., Gruber, D., Hubl, A.: Application of a generic simulation model to optimize production and workforce planning at an automotive supplier. In: 2013 Winter Simulations Conference (WSC), pp. 2689–2697. IEEE (2013). https://doi.org/10.1109/WSC.2013.6721640
4. Konopka, J.M.: Capacity utilization bottleneck efficiency system-CUBES. IEEE Trans. Compon. Packag. Manuf. Technol. A, **18**, 484–491 (1995). https://doi.org/10.1109/95.465141
5. Kefeli, A., Uzsoy, R.: Identifying potential bottlenecks in production systems using dual prices from a mathematical programming model. Int. J. Prod. Res. **54**, 2000–2018 (2016). https://doi.org/10.1080/00207543.2015.1076182
6. Chakravorty, S.S., Atwater, J.B.: Bottleneck management: theory and practice. Prod. Plan. Control **17**, 441–447 (2006). https://doi.org/10.1080/09537280600682752
7. Jodlbauer, H., Huber, A.: Service-level performance of MRP, kanban, CONWIP and DBR due to parameter stability and environmental robustness. Int. J. Prod. Res. **46**, 2179–2195 (2008). https://doi.org/10.1080/00207540600609297
8. Hubl, A., Altendorfer, K., Jodlbauer, H., Gansterer, M., Hartl, R.F.: Flexible model for analyzing production systems with discrete event simulation. In: 2011 Winter Simulation Conference (WSC), pp. 1554–1565. IEEE (2011). https://doi.org/10.1109/WSC.2011.6147873
9. Chiabaut, N., Buisson, C.: Replications in stochastic traffic flow models: incremental method to determine sufficient number of runs. In: Appert-Rolland, C., Chevoir, F., Gondret, P., Lassarre, S., Lebacque, J.-P., Schreckenberg, M. (eds.) Traffic and Granular Flow 2007, pp. 35–44. Springer, Heidelberg (2009). https://doi.org/10.1007/978-3-540-77074-9_3
10. Weiss, S., Schwarz, J.A., Stolletz, R.: The buffer allocation problem in production lines: formulations, solution methods, and instances. IISE Trans. **51**, 456–485 (2019). https://doi.org/10.1080/24725854.2018.1442031
11. Seiringer, W., Altendorfer, K., Castaneda, J., Gayan, L., Juan, A.A.: Potential of simulation effort reduction by intelligent simulation budget management for multi-item and multi-stage production systems. In: 2022 Winter Simulation Conference (WSC), pp. 1864–1875. IEEE (2022). https://doi.org/10.1109/WSC57314.2022.10015506
12. Steele, D.C., Philipoom, P.R., Malhotra, M.K., Fry, T.D.: Comparisons between drum–buffer–rope and material requirements planning: a case study. Int. J. Prod. Res. **43**, 3181–3208 (2005). https://doi.org/10.1080/00207540500076704
13. Hopp, W.J., Spearman, M.L.: Factory Physics. Foundations of manufacturing management. Waveland Press, Long Grove, Illinois (2011)

14. Ye, T., Han, W.: Determination of buffer sizes for drum–buffer–rope (DBR)-controlled production systems. Int. J. Prod. Res. **46**, 2827–2844 (2008). https://doi.org/10.1080/00207540600922948
15. Mleczko, J.: Improving the efficiency of moving bottleneck by dynamic classification of tasks. In: Xie, Y. (ed.) 4th International Conference on Advanced Computer Theory and Engineering, ICACTE 2011, pp. 457–460. ASME Press (2011). https://doi.org/10.1115/1.859933.paper104
16. Li, Q., Disney, S.M.: Revisiting rescheduling: MRP nervousness and the bullwhip effect. Int. J. Prod. Res. **55**, 1992–2012 (2017). https://doi.org/10.1080/00207543.2016.1261196
17. Thürer, M., Stevenson, M.: On the beat of the drum: improving the flow shop performance of the Drum–Buffer–Rope scheduling mechanism. Int. J. Prod. Res. **56**, 3294–3305 (2018). https://doi.org/10.1080/00207543.2017.1401245
18. Woo, K., Park, S., Fujimura, S.: Real-time buffer management method for DBR scheduling. IJMTM **16**, 42 (2009). https://doi.org/10.1504/IJMTM.2009.021503

A New Characterization of the Kendall Space of Planar Shapes

Ximo Gual-Arnau[1(✉)] and Lluïsa Gual-Vayà[2]

[1] Departament de Matemàtiques, Institute of New Imaging Technologies, Universitat Jaume I, Castelló, Spain
gual@uji.es

[2] Departament de Matemàtiques, Institute of Mathematics and Applications, Universitat Jaume I, Castelló, Spain
al395358@uji.es
http://gual.uji.es

Abstract. Shape analysis is utilized across a wide spectrum of scientific fields dealing with large datasets, often comprising 2D images, aiding in the study and decision-making regarding morphology, reconstruction, recognition, and object classification. In this context, the Kendall shape space serves as a common mathematical framework for analyzing and comparing geometric shapes. In this work, by embedding this space into a Euclidean space $\mathbb{R}^N$, we provide a novel characterization thereof, from which we define a new extrinsic sample mean. Having this new mean defined provides a reference point when applying kernel methods based on this embedding.

Keywords: 2D shapes · Kernel method · Embedding · Kendall space · Mean shape · Shape analysis

1 Introduction

The entertainment industry, particularly in areas such as film production, television, video games, and virtual reality, significantly benefits from animation technologies, human action and object recognition, as well as facial recognition. Hand-drawn 2D animation remains a fundamental component of the global entertainment industry. However, even with computer support, certain production processes still require substantial time and labor. Employing Kendall Space for shape similarity measurement and matching helps mitigate these high costs [7]. On the other hand, the study of human gait has recently gained momentum due to its potential application as biometrics for person identification. Techniques developed within the 2D Kendall space aid in leveraging the silhouette deformations of an individual as discriminant and recognition features [9]. Furthermore, object recognition stands as a primary objective for many image analysis and computer vision applications [10]. Specifically, 2D facial landmark sequences treated in Kendall space are utilized for the recognition of macro- and micro-facial expressions [8].

A. A. Juan et al. (Eds.): DSA ISC 2024, LNCS 14778, pp. 235–243, 2025.
https://doi.org/10.1007/978-3-031-78238-1_22

Therefore, in various scientific applications, planar shapes are often represented by a finite set of points delineating their contours. In this study, we propose a methodology for characterizing each planar shape based on a finite set of points, referred to as landmarks, which describe its contour. We operate within Kendall's shape space framework [2], where establishing a point-wise correspondence between shapes is essential. In cases where points of scientific significance, such as anatomical landmarks, are unavailable, we resort to mathematically describing a subset of points and deriving the remaining points, termed pseudo-landmarks, based on these descriptions.

Each shape is thus delineated by k landmarks in $\mathbb{R}^2$, representing corresponding points across all shapes. Given that the Kendall planar space Σ_k^2 is a Riemannian manifold, with an interpretation as the complex projective space having a maximal sectional curvature of 4 ($\mathbb{CP}^{k-2}(4)$), various metrics can be defined on Σ_k^2. Leveraging these metrics, along with the database of shapes and the distances induced by them, enables the application of diverse techniques for shape analysis. Some of these metrics are intrinsic to the manifold, while others are derived from an embedding of Σ_k^2 into another space.

In particular, in this paper, we will use an extrinsic distance in Σ_k^2, induced by the Euclidean distance of the embedding i of Σ_k^2 in $\mathbb{R}^{(k-1)^2}$ [4]. We provide a novel characterization of Σ_k^2 and we define a sample mean from the extrinsic sample mean defined in $\mathbb{R}^{(k-1)^2}$ using Euclidean distance, and we obtain the corresponding shape through the inverse function i^{-1}.

Although from a geometric standpoint it may seem more natural to work with the intrinsic geometry of a manifold, when it comes to applying machine learning algorithms, they have been developed for the linear case [3]. Particularly for data in a Euclidean space, machine learning algorithms are much simpler to implement. However, most real-world problems are nonlinear. In fact, the Riemannian manifolds that define the Kendall shape space are not linear. Therefore, with our approach, we embed the 2D Kendall space into $\mathbb{R}^{(k-1)^2}$, which allows us to linearize the problem and apply these algorithms effectively. Now, it is important that the behavior of intrinsic and extrinsic distances between planar shapes is similar, as well as that of intrinsic and extrinsic means defined from these distances, as we will see happens with our approach.

The paper is organized as follows. In Sect. 2, we provide a novel characterization of the Kendall space of planar shapes, from which we define an extrinsic sample mean of planar shapes. In Sect. 3, we present the example through which we will compare the Euclidean and intrinsic metrics. Section 4 showcases the results based on this example. Finally, in Sect. 5, we present the study's conclusions.

2 Methods

2.1 2D Kendall Space Distances

Initially, a $2D$ shape is represented by a configuration matrix X of size $k \times 2$, consisting of Cartesian coordinates of the k landmarks outlining a $2D$ planar domain. However, as the object's shape encompasses all geometric information invariant to translations, rotations, and changes of scale (similarity transformations), we eliminate translations and scale changes from X by multiplying it with the Helmert submatrix, H [2], and normalizing it by its Frobenius norm. Thus, the pre-shape of the configuration matrix X is given by a $(k-1) \times 2$ matrix:

$$Z = \frac{HX}{||HX||} \tag{1}$$

The pre-shape space S_2^k defined as the set of all possible pre-shapes, is a hypersphere of unit radius in $\mathbb{R}^{2(k-1)}$. Finally, to eliminate rotations, the $2D$ shape is obtained by taking the quotient of Z_X over all possible rotations of the plane. Thus, the $2D$ shape space, Σ_2^k, is the quotient space of S_2^k under rotations, i.e.,

$$\Sigma_2^k = S_2^k / SO(2).$$

Therefore, if π denotes the natural projection to the equivalence class, $\pi : S_2^k \to \Sigma_2^k$, a shape $[X]$ is an orbit associated with the action of the rotation group $SO(2)$ on the pre-shape, [2].

On the other hand, given a configuration matrix X representing a shape $[X]$, another matrix belonging to the same equivalence class $[X]$ is given by $H^T Z$.

The space of planar shapes Σ_2^k is a smooth manifold of dimension $2k-4$, isometric to the complex projective space $\mathbb{CP}^{k-2}$ (without singular part).

In this paper, we will consider the extrinsic distance in the space Σ_2^k as defined in [4]. Later, we will characterize the image $i(\Sigma_k^2)$ and use this distance to define an extrinsic mean and perform clustering of 2D shapes.

The extrinsic distance under consideration is defined by the embedding $i : \Sigma_2^k \to \mathbb{R}^{(k-1)^2}$. This embedding is characterized as follows: given a configuration matrix X and its corresponding pre-shape matrix Z, represented as,

$$Z = \begin{bmatrix} x_1 & y_1 \\ x_2 & y_2 \\ \vdots & \vdots \\ x_{k-1} & y_{k-1} \end{bmatrix} = \begin{bmatrix} v_1 \\ v_2 \\ \vdots \\ v_{k-1} \end{bmatrix},$$

the embedding i is defined by the map

$$[X] \mapsto i([X]) := (\langle v_i, v_i \rangle, \langle v_i, v_j \rangle, \det(v_i, v_j)) \in \mathbb{R}^{(k-1)^2}, \tag{2}$$

where $i < j$ with $i, j \in \{1, 2, \cdots, k-1\}$, and $\langle v_i, v_j \rangle$ denotes the usual scalar product of v_i and v_j. Therefore,

$$i([X]) := \left(x_i^2 + y_i^2, \quad x_i x_j + y_i y_j, \quad x_i y_j - x_j y_i\right) \in \mathbb{R}^{(k-1)^2}. \tag{3}$$

The map i is injective, meaning that, $[X_1] = [X_2]$ if and only if $i([X_1]) = i([X_2])$, and it does not depend on the choice of the representative of $[X]$ [4].

We define the extrinsic distance between two shapes $[X_1]$ and $[X_2]$ as given by the equation:

$$d_e([X_1],[X_2]) := \|i([X_1]) - i([X_2])\|, \tag{4}$$

where $\|\cdot\|$ denotes the norm in $\mathbb{R}^{(k-1)^2}$. Additionally, as demonstrated in [4], the distance d_e, with a slight modification in the definition of $i([X])$, coincides with the full Procrustes distance.

Finally, an embedding of the Euclidean space $\mathbb{R}^{(k-1)^2}$ into a Reproducing Kernel Hilbert Space (RKHS) can be considered to utilize the distance defined by the Gaussian kernel instead of the Euclidean distance. The Gaussian kernel $K : \Sigma_2^k \times \Sigma_2^k \longrightarrow \mathbb{R}$ is defined as:

$$K([X_1],[X_2]) = \exp\left(-\gamma d_e^2([X_1],[X_2])\right), \tag{5}$$

which is a positive definite kernel for all $\gamma > 0$.

This kernel is similar to the Procrustes Gaussian kernel defined the 2D Kendall shape space [5] and the Projection Gaussian kernel [6], considering the 2D Kendall shape space as a Grassmannian manifold.

2.2 A New Extrinsic Mean Shape

In this section, we begin by characterizing the submanifold $i(\Sigma_2^k)$ in $\mathbb{R}^{(k-1)^2}$, and subsequently, we define the extrinsic mean as the projection onto $i(\Sigma_2^k)$ of the usual arithmetic mean in $\mathbb{R}^{(k-1)^2}$.

Proposition 1. *Let $M \subset \mathbb{R}^{(k-1)^2}$ be the submanifold defined by the vectors:*

$$(a_i, b_{ij}, c_{ij}) \in \mathbb{R}^{(k-1)^2}. \tag{6}$$

where $i < j$ with $i, j \in \{1, 2, \cdots, k-1\}$, such that:

1. $\sum_{i=1}^{k-1} a_i = 1$, *and* $a_i > 0$ *for* $i = 1, \ldots, k-1$.
2. $b_{1j}^2 + c_{1j}^2 = a_1 a_j, \quad j = 2, \ldots, k-1$.
3. $a_1 b_{ij} = b_{1j} b_{1i} + c_{1j} c_{1i}, \quad i = 2, \ldots, k-1, \quad i < j$.
4. $a_1 c_{ij} = c_{1j} b_{1i} - b_{1j} c_{1i}, \quad i = 2, \ldots, k-1, \quad i < j$.

Then, $M = i(\Sigma_2^k)$.

Proof. Firstly, we observe that the dimension of M is $dim(M) = 2k-4$, therefore coinciding with the dimension of Σ_2^k. We need to demonstrate that $Im(i) = M$.

Let

$$\begin{aligned} V &= \left(x_i^2 + y_i^2, \quad x_i x_j + y_i y_j, \quad x_i y_j - x_j y_i\right) \\ &= (\langle v_i, v_i \rangle, \langle v_i, v_j \rangle, \det(v_i, v_j)) \in Im(i). \end{aligned} \tag{7}$$

Let us define $a_i = x_i^2 + y_i^2$. Then, $a_i \geq 0$, and since none of the Helmertized landmark coordinates coincide with the origin, we have that $a_i > 0$.

Moreover, since the pre-shape $Z \in S_2^k$, we have $\sum_{i=1}^{k-1} a_i = 1$.

Let

$$b_{ij} = \langle v_i, v_j \rangle = \langle (x_i, y_i), (x_j, y_j) \rangle = ||v_i|| \, ||v_j|| \cos\theta_{ij}. \tag{8}$$

Then,

$$c_{ij} = \det(v_i, v_j) = \langle (x_i, y_i), (y_j, -x_j) \rangle = ||v_i|| \, ||v_j|| \sin\theta_{ij}. \tag{9}$$

Therefore, $b_{ij}^2 + c_{ij}^2 = ||v_i||^2 \, ||v_j||^2 = a_i a_j$ and, in particular, $b_{1j}^2 + c_{1j}^2 = a_1 a_j$. Finally, since $\cos\theta_{ij} = \cos(\theta_{1j} - \theta_{1i})$, we obtain

$$b_{ij} = \sqrt{a_i}\sqrt{a_j}\cos(\theta_{1j} - \theta_{1i}), \quad c_{ij} = \sqrt{a_i}\sqrt{a_j}\sin(\theta_{1j} - \theta_{1i}).$$

From Eqs. 8 and 9, and using the expression for the cosine of the difference of angles, we obtain

$$b_{ij} = \sqrt{a_i}\sqrt{a_j}\left(\frac{b_{1j}}{\sqrt{a_1}\sqrt{a_j}}\frac{b_{1i}}{\sqrt{a_1}\sqrt{a_i}} + \frac{c_{1j}}{\sqrt{a_1}\sqrt{a_j}}\frac{c_{1i}}{\sqrt{a_1}\sqrt{a_i}}\right),$$

and then condition *3*. Similarly, using the expression for the sine of the difference of angles, we obtain condition *4*.

Therefore, $V \in M$, and we have demonstrated that $Im(i) \subset M$. Now, let $V = (a_i, b_{ij}, c_{ij}) \in M$, which satisfies conditions *1–4*. We look for $[X] \in \Sigma_2^k$ such that $i([X]) = V$. We suppose that the pre-shape of X is defined as

$$Z = \begin{bmatrix} \sqrt{a_1} & 0 \\ \frac{b_{12}}{\sqrt{a_1}} & \frac{c_{12}}{\sqrt{a_1}} \\ \vdots & \vdots \\ \frac{b_{1k-1}}{\sqrt{a_1}} & \frac{c_{1k-1}}{\sqrt{a_1}} \end{bmatrix}. \tag{10}$$

Then $||Z|| = 1$ and $i([X]) = V$, so $M \subset Im(i)$. □

Consider a sample of shapes $[X_1], \ldots, [X_n] \in \Sigma_2^k$, and denote $i([X_1]) = V_1, \ldots, i([X_n]) = V_n \in M \in \mathbb{R}^{(k-1)^2}$. The Euclidean sample mean $\bar{V} = \frac{\sum_{i=1}^n V_i}{n}$ does not belong, in general, to M. To define a sample mean shape in Σ_2^k, we consider first the following minimization problem:

Minimize $d_E^2(\bar{V}, M)$; that is, $\min_{W \in M} d_E^2(\bar{V}, W)$, where d_E denotes the Euclidean distance in $\mathbb{R}^{(k-1)^2}$.

If $\bar{V} = (\bar{a}_i, \bar{b}_{ij}, \bar{c}_{ij})$ and we denote $W = (a_i, b_{ij}, c_{ij})$, we want to minimize the function

$$f(a_i, b_{ij}, c_{ij}) = \sum_{i=1}^{k-1}(a_i - \bar{a}_i)^2 + \sum_{\substack{i=1 \\ i<j}}^{k-1}(b_{ij} - \bar{b}_{ij})^2 + \sum_{\substack{i=1 \\ i<j}}^{k-1}(c_{ij} - \bar{c}_{ij})^2,$$

under the conditions *1)–4)* of Proposition 1.

Once we have obtained $W \in M$ which gives the minimum, the sample mean shape will be the shape $[X] \in \Sigma_2^k$ such that $i([X]) = W$.

Example.- Triangles

In Fig. 1, we illustrate the centred triangles corresponding to the vertex triangles: $T_1 = \{(1,0), (1,1), (0,1)\}$ (in green) and $T_2 = \{(3,0), (3,1), (1,1)\}$ (in red), along with the sample mean triangle of T_1 and T_2 (in blue).

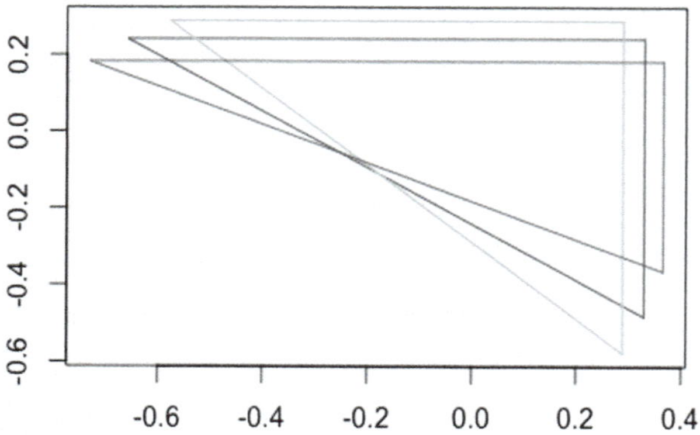

Fig. 1. Two triangles and their mean in blue. (Color figure online)

3 Material

In this section, we work with the data initially considered in [1] and processed in [2], consisting of 13 landmarks taken from near midsagittal 2D slices from magnetic resonance (MR) brain scans of a group of 14 control volunteers and a group of 14 schizophrenia patients. Specifically, in [1], a study was conducted with the aim of investigating whether there were differences in the intrinsic average shape of the brain between the two groups and in the variability of shapes. Identifying such shape differences between the control group and the group of schizophrenia patients provided valuable information to better understand the disease.

We will not conduct a detailed study to analyze any shape differences in the brain between the two groups, either in extrinsic average shape or in shape variability. Instead, we will simply calculate the extrinsic means of each group and compute the distances from the control group to both means and from the schizophrenia patient group to both means. This is done to observe whether the distances from the control group to its mean are greater than the distances to the mean of the schizophrenia patient group, and vice versa. This observation would suggest that the two groups have distinct characteristics in terms of the brain images they represent.

Additionally, we will compare the distances obtained with the intrinsic Riemann distances of the Kendall space of planar shapes. The intrinsic distances and the intrinsic mean have been obtained using the `riemdist` and `frechet` commands in the `shapes` library of `R`, respectively.

4 Results

We calculate the intrinsic sample mean of the landmarks of the 14 volunteer individuals $M_{i,v}$ and the intrinsic sample mean of the landmarks of the 14 individuals comprising the schizophrenia patient group $M_{i,s}$. The distance between these two means is $d_1 = d_i(M_{i,v}, M_{i,s}) = 0.0382$. Next, we compute the Euclidean means of the same two groups (14 volunteers and 14 patients), denoted as $M_{E,v}$ and $M_{E,s}$, respectively. The Euclidean distance between these two sample means is $d_2 = d_E(M_{E,v}, M_{E,s}) = 0.0388$. Therefore, the two distances d_1 and d_2 between the means obtained with the intrinsic and Euclidean metrics are very similar.

We also calculate the distance from each volunteer to the mean of the volunteer individuals and the distance from each volunteer to the mean of schizophrenia patients. We do this using both the intrinsic and Euclidean distances. The results are identical in both cases: Out of the 14 volunteer individuals, 11 of them are closer to the mean of the volunteer individuals, while the 3rd, 6th, and 11th individuals are closer to the mean of schizophrenia patients. The same results are obtained with both distances.

In Table 1, we list the volunteers whose difference between the distance to the mean of schizophrenia patients and the distance to the mean of the volunteer group is greater and lesser in both positive values (11 individuals) and negative values (3 individuals). We observe a very similar behavior in both distances; 3 out of the 4 individuals in the table coincide. In the first row of the table, we have the distances with the intrinsic metric, with the corresponding individual in parentheses, and in the second row, the same is considered but with the Euclidean distance.

Table 1. Difference between the distances to the means of the individuals in the volunteer group.

Difference	Greater and >0	Lower and >0	Greater and <0	Lower and <0
Intrinsic	0.0242 (12)	0.0035 (5)	−0.0009 (11)	−0.0084 (3)
Euclidean	0.0227 (12)	0.0030 (5)	−0.0002 (6)	−0.0076 (3)

Finally, we repeat the same calculations but considering the distances from schizophrenia patients to their respective means. The results are also the same: out of the 14 patients, 12 are closer to the patient mean, and 2 (the 5th and 13th patients) are closer to the volunteer mean. These results are also consistent across both metrics.

In Table 2, we list the patients whose difference between the distance to the mean of the control group and the distance to the mean of schizophrenia patients is greater or smaller in both positive (12 individuals) and negative values (2 individuals). In this case, we also observe that the behavior of the two distances is similar.

As we can see in Fig. 2, at first glance, the landmarks of the means of the volunteer group and the patient group are not significantly different; however, the distances from each group to its mean are generally much smaller than to the mean of the other group.

An interesting potential of the extrinsic mean is that when classifying individuals as healthy or with schizophrenia based on this extrinsic mean, classifi-

Table 2. Difference between the distances to the means of the individuals in the schizophrenia patient group.

Difference	Greater and >0	Lower and >0	Greater and <0	Lower and <0
Intrinsic	0.0290 (8)	0.0016 (12)	−0.0051 (5)	−0.0134 (13)
Euclidean	0.0294 (8)	0.0005 (2)	−0.0060 (5)	−0.0143 (13)

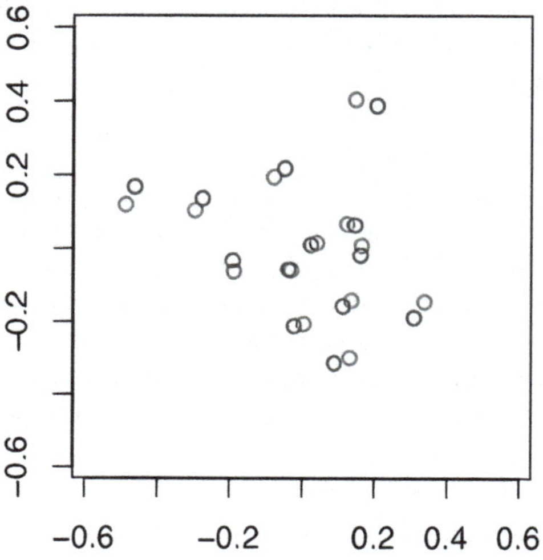

Fig. 2. Means of the volunteer group in red and patient group in blue. (Color figure online)

cation methods that rely on the linearity of space can be used. For example, some machine learning algorithms, which are not viable with intrinsic mean and distances.

5 Conclusions

The Kendall space of planar shapes is a Riemannian manifold, and as such, it is often advantageous to work with the intrinsic metric that defines the manifold. However, when dealing with powerful machine learning techniques in this space, it becomes interesting to have an embedding of it into a Euclidean space from which kernel methods can be applied. Whenever an embedding is defined, it is interesting to identify the image of the Kendall space under this embedding, which allows, for example, the definition of extrinsic sample means. In this work, starting from an embedding of the Kendall space into a Euclidean space, we characterize this space and define the corresponding mean. We also use a specific example to compare the intrinsic metrics and means with the Euclidean metrics and means, leading to the conclusion that the results are very similar.

These results open the door to the use of a new extrinsic sample mean, which can be employed, for example, to apply classification methods that rely on spatial linearity. Additionally, this work can serve as a guide for defining an extrinsic sample mean in the Kendall space of 3D shapes.

Acknowledgments. This research was partially supported by the Spanish Ministry of Science and Innovation (PID2020-115930GA-100 and PID2022-141699NB-I00). The second author has received funding through a research grant from the Institute of Mathematics and Applications (IMAC) at Jaume I University (UJI). The authors have no competing interests to declare that are relevant to the content of this article.

References

1. Bookstein, F.-L.: Biometrics, biomathematics and the morphometric synthesis. Biometrics Biomathematics Morphometric Synth. **58** (1996)
2. Dryden, I.-L., Mardia, K.-V.: Statistical Shape Analysis: with Applications in R. Wiley, Hoboken (2016)
3. Feragen, A., Lauze, F., Hauberg, S.: Geodesic exponential kernels: when curvature and linearity conflict. In: Proceedings of the IEEE Conference on Computer Vision and Pattern Recognition, pp. 3032–3034 (2015)
4 Gimeno, V., Gual-Arnau, X., Ibáñez, M.-V. and Simó, A.: A Gaussian kernel for Kendall's space of m-D shapes. Pattern Recogn. **144**(3) (2023)
5. Jayasumana, S., Salzmann, M., Li, H., Haramdi, M.: A framework for shape analysis via Hilbert space embedding. In: Proceedings of the IEEE International Conference on Computer Vision, pp. 1249–1256 (2013)
6. Jayasumana, S., Hartley, R., Salzmann, M., Li, H., Haramdi, M.: Kernel methods on Riemannian manifolds with Gaussian RBF kernels. IEEE Trans. Pattern Anal. Mach. Intell. **37**, 2464–2477 (2015)
7. Liu, S., Wang, X., Wu, Z., Seah, H.-S.: Shape correspondence based on Kendall shape space and RAG for 2D animation. Vis. Comput. **36**, 2457–2469 (2020)
8. Tanfous, A.B., Drira, H., Amor, B.B.: Sparse coding of shape trajectories for facial expression and action recognition. IEEE Trans. Pattern Anal. Mach. Intell. **42**(10), 2594–2607 (2020)
9. Veeraraghavan, A., Roy-Chowdhury, K., Chellappa, R.: Matching shape sequences in video with applications in human movement analysis. IEEE Trans. Pattern Anal. Mach. Intell. **27**(12), 1896–1909 (2005)
10. Zhang, J., Zhang, X., Krim, H., Walter, G.-G.: Object representation and recognition in shape spaces. Pattern Recogn. **36**, 1143–1154 (2003)

GIS Analysis to Support Low Carbon Heating

Paula Carroll and Peter Keenan(✉)

School of Business, University College Dublin, Dublin, Ireland
{paula.carroll,peter.keenan}@ucd.ie

Abstract. Decarbonising heat is a challenge due in part to the large number of actors such as individual homeowners. In this paper, we explore the use of Geographical Information Systems (GIS) to identify opportunities for heat decarbonisation. We demonstrate the approach for the residential sector in Ireland, where 64% of the population lives in urban locations and the potential for ground source heat pumps is limited by access to land. In contrast, access to the gas network limits opportunities for gas-fired central heating in rural locations. This paper explores how GIS approaches can be combined with spatial data to identify opportunities to decarbonise heat, providing useful insights and recommendations for policymakers on suitable locations for low-carbon heating technologies.

Keywords: Heat Pumps · GIS · Coefficient of Performance

1 Introduction

Large industrial or commercial entities and electricity producers are obliged to participate in the Emissions Trading System (ETS) to ensure these entities reduce their greenhouse gas (GHGs) emissions. Heat and transport are more challenging sectors to decarbonise due in part to the large number of actors such as individual homeowners and transport users. These sectors are not covered by the ETS and hence governments must find other ways through incentives and policies to reduce emissions and promote the adoption of low carbon alternatives. EU directives promote an increased replacement rate of old heating systems and an increased switch to solutions based on renewable energy [7]. 23% of the total energy used for heating and cooling in the EU in 2020 came from renewable sources [8]. Sweden supplied 66% of its heading needs from renewable sources through a combination of different renewable heat technologies. However, Ireland was ranked last with just over 5% of its heating needs served by renewable sources, and Ireland failed to meet its 2020 renewable heat target of 12%.

In this paper, we focus on the residential heating sector in Ireland. We combine statistical modelling with Geographic Information Systems (GIS) techniques to identify opportunities for heat decarbonisation. According to the 2022 census [4], 64% of the population lives in urban locations where the potential for

A. A. Juan et al. (Eds.): DSA ISC 2024, LNCS 14778, pp. 244–255, 2025.
https://doi.org/10.1007/978-3-031-78238-1_23

ground source heat pumps is limited by access to land. In contrast, 36% live in rural locations where access to the gas network limits opportunities for gas fired central heating. Gas is currently a lower cost and emissions option compared to other fossil fuels. We explore the distribution of residential building stock and heating source by location and couple these data with geothermal profiles and other GIS perspectives. We aim to identify opportunities to decarbonise heat in Ireland by providing a set of recommendations for the best locations of ground or air source heat pumps. In particular, we aim to answer the following research questions:

1. What fuels/heating types are used in Ireland by location?
2. What heating opportunities/challenges are identified using GIS?

2 Background/Literature Review

There are several technology options available to improve the sustainability of heating homes and buildings, many of which are at a high technology readiness level. District Heating Systems (DHSs) use some type of local fuel or heat resources to satisfy local heating demands. Heat energy is distributed through a network of pipes. The heat may be sourced by recovering waste heat from industrial or power generation processes, or can be generated from renewable sources such as biogases, or biofuels, or can be generated by industrial scale heat pumps. DHSs have been fully operational in countries with high renewable heat performance such as Sweden where DHSs have been in operation since the 1950s [19].

The development of DHSs is discussed in [11,12]. The current generation of DHSs are considered to be the fourth generation and are smart energy systems which aim to balance energy supply and conservation with space and hot water heat demand in an efficient way. The focus compared to earlier generations is to use innovative algorithms and technologies to minimise losses and facilitate low temperature heat sources to be integrated into the DHS. This requires institutional frameworks and policy supports such as cost incentives in order to be successful. Fifth generation DHSs using ultra-low temperatures are now under consideration [14]. These developments may allow low temperature heat sources to be leveraged to serve local heating needs, thus reducing the costs and inefficiencies of transporting energy to serve heating needs.

Heat pumps (HPs) are another high technology readiness level option. HPs are low carbon energy recovery systems that use electricity to exchange energy from air-, ground-, or water-sources, and transfer that energy as heat to a building's heating system. HPs can be used to provide space heating and/or hot water. Aerothermal, geothermal and hydrothermal energy are potential renewable energy sources to supply heat [1]. Heat pumps can be used to boost the heat delivered from a District Heating Network (DHN) at the heat demand location. Using renewably generated electricity to power the heat pumps allows heat to be provided in a sustainable and clean way. Heat energy production is defined as renewable if the average annual efficiency performance of heat produced to electricity consumed exceeds a certain threshold. EU directives set the threshold as $\frac{1/15}{\eta}$ where η is the ratio of total gross production of electricity to the primary energy consumption for the production [6,7]. With the current EU average renewable electricity mix, this threshold is 2.5. Hence heat pumps that achieve an average efficiency performance of at least 2.5 are deemed to be producing renewable heat.

The efficiency of a HP depends on the thermal efficiency of the homes and buildings where heat pumps are installed [2], and the patterns of usage and occupancy of the building by the occupants. The performance of heat pumps varies in different locations as a result of different local meteorological and geographical conditions. Ground source heat pumps are also affected by geological conditions and have significant space requirements for installation with land availability being affected by the population density. Similarly, water source heat pumps require access to a body of water and a network of collector pipework to extract the heat from the water body. Both ground and water HPs can have more consistent performance than air source HPs but their uptake is limited by the higher setup costs and space requirements. Heat pump location and performance is a spatially related issue and GIS can be used to explore and understand HP opportunities and to guide public policy.

HPs perform less well in winter when lower temperatures also mean that heating demand is highest; since the system must do more work to extract heat and transfer it to the building's heating system. Equation 1 shows how the Coefficient of Performance (COP) is calculated. COP is the ratio of the heat produced, Q, to the input energy, W.

$$COP = Q/W \tag{1}$$

The Seasonal Performance factor (SPF) is a weighted average of COP over a full heating season and gives an estimate of how the system performs over time. Renewable heating systems with an SPF of at least 2.5 are deemed to be renewable heat [6].

In general, Air Source Heat Pumps (ASHPs) are less efficient at producing heat than geo- or hydro-sources since air temperature varies more than ground or the body of water temperature, and can be much lower in winter. Ground and water source heat pumps benefit from the more favourable and less variable temperature conditions than air, so can achieve high efficiency. The ground and water temperatures in winter can be higher than the ambient air temperature meaning these HPs have less work to do to transfer heat to the building.

ASHPs require a smaller land footprint than ground source HPs (GSHPs) so may be more suited to urban areas. The outdoor unit is the size of an air conditioning unit and generally planning permission is not required. Since the ASHP outdoor unit is above ground, it is exposed to the ambient conditions and frost or ice can form on the external heat exchanger causing deterioration in performance. This can be reduced by locating the unit in a sunny or sheltered location. ASHPs can perform well in energy efficient buildings. The homes assessed in [2] are deemed to provide renewable heat with SPFs greater than 2.5.

HPs have a lifetime of 15–25 years, longer than that of conventional boilers. The plastic underground heat collectors of GSHPs are expected to last for up to 50 years [18]. Hence significant time on the HP system design is advised to ensure the system is correctly dimensioned to the user's needs and that adequate thermal comfort levels can be provided over the lifetime of the HP system.

Direct electric heating is a well established technology, storage and resistive electric heaters are used in residential buildings. They consume electricity directly to provide heat, either by storing heat, typically in ceramic materials, or providing instantaneous heat via heating elements. With a high penetration of electricity produced by renewable sources such as wind or solar, these options may provide a contribution to the overall heating balance.

Another low carbon heat option is to replace or convert fossil fuel boiler systems to run on bio or low carbon fuels. Some of these biogases could be blended with natural gases and delivered using existing gas network infrastructure in urban areas. Other options are at a lower technology readiness level with trials underway and regulations in development.

Finally, where the building fabric, or other limitations inhibit the use of HPs alone, hybrid HP systems may be considered. Such systems combine the use of a HP with a backup heating system such as a gas boiler which can be switched to at times of peak thermal demand [13]

There are multiple decisions to be made to decarbonise the heating sector. First the types of heating technologies to achieve climate action plans have to be supported by policy and incentives. Then citizens, and municipalities have to adopt the available heating services that serve their needs. In this paper, we focus on low carbon heating opportunities in Ireland.

The Government of Ireland prioritises "efficiency first" measures to improve the energy efficiency of buildings through retrofitting schemes for the existing building stock, thus lowering demand for energy for heating. It also aims to decarbonise heat by promoting the use of heat pumps and encouraging the development of district heating networks. Having failed to meet its 2020 renewable heat target, the most recent National Energy and Climate Plan aims for 600,000 Heat Pumps to be installed by 2030 of which 400,000 will be in retrofitted homes. The 2022 census shows that 71,000 households used heat pumps with 57,000 using air source heat pumps and 14,000 using ground source heat pumps, a large increase from a low base of 6,555 heat pumps in 2013. While residential energy use declined in 2022 compared to 2021, owing to increased fuel prices and milder weather, it still constituted 45.3% of heat demand [17], of which 94.3% came from fossil fuels.

The government of Ireland are considering multiple policies and options such as a renewable heat obligation for energy suppliers that will require a proportion of the energy supplied for heating to come from renewable sources. A national heat study was conducted to identify and assess options to decarbonise heating in Ireland [15,16].

Ireland lies at latitudes and longitudes (55.5, −10.25) - (51.5, −6) in the North Atlantic Ocean on the European continental shelf and benefits from the North Atlantic Current and Gulf Stream airflows. Ireland experiences a maritime temperate climate so has a milder climate than other places at similar latitudes, having mild winters and tepid summers. Hence, there is demand from the residential sector for space heating in winter and domestic hot water heating and little to no demand for space cooling. Heat demand is low density because of the dispersal patterns of the population of 5.15 million people in 2022, the nature of the housing stock, and Ireland's temperate climate. Extreme weather events are rare in Ireland. Electricity distribution system operators need to design their networks to meet the peak coincident demand of their consumers. The peak demand occurs in countries like Ireland during winter evenings, typically in December and January between 17:30 and 18:00. Additional demand from HPs will further increase peak demand particularly during extreme weather events or cold snaps [3].

Heating degree-days (HDD) and cooling degree-days (CDD) are weather-related energy consumption indicators. HDDs are the days on which the ambient air temperature dips below a threshold and so heating is needed. In the EU, the baseline temperature is 15.5 °C.

In this paper we focus on the role of HPs to serve Ireland's residential heating needs. HPs systems in countries like Ireland are mainly used for space heating, their performance during the heating season when they provide space and water heating is likely to be of more interest than that during the rest of the year when only hot water is needed. Hence we define the heating season performance factor (HSPF) as the SPF over the heating season or HDDs.

3 Methodology

We explore the research questions in Sect. 1 using GIS approaches and open-source data available from a combination of providers such as the Geological Survey of Ireland, the Central Statistics Office (CSO of Ireland), and the Climate Data Store.

ASHPs play a significant role decarbonising heating in Ireland [2]. A multiple linear regression model (MLR) to estimate the hourly COP as a function of weather and operating conditions is described in [2] as follows:

$$\begin{aligned} COP = 3:758 - 0:05 * flowtemp + 0:087 * outtemp - 0:034 * zone1temp \\ +0:030 * steadystate + 0:042 * Rain + 0:061 * dewptemp \end{aligned}$$

where $flowtemp$ is the ASHP outflow temperature, $outtemp$ is the outside ambient air temperature, $zone1temp$ is the living area room temperature, $steadystate$ is the time the AHSP is in steady state operation (in minutes), $Rain$ is the precipitation in mm, and $dewtemp$ is the dew point air temperature. All temperatures are in °C.

The MLR model is trained on data from a sample of retrofitted homes in Ireland with good overall model fit metrics reported in [2]. This gives us a useful estimate of the expected COP. As with all statistical modelling, training on additional data and additional explanatory variables could improve the model's ability to predict COP. In addition this simple linear model may not fully capture any nonlinear or interaction between explanatory variables. Nonlinear or machine learning models could also be used, but at the cost of explainability and simplicity of the models.

We simulate the performance of an ASHP over a ten-year time frame assuming similar ASHP systems and specify the operating conditions as: Outflow temperature of 37.54 °C, a living area thermal comfort of 21.16 °C, with the ASHP operating in steady state for 47.9 min during each heating hour when the system is on.

ERA-5 is a fifth-generation reanalysis provided by the European Centre for Medium Range Weather Forecasts-ECMWF [10]. We estimate the heating season performance of ASHPs over a ten-year time frame using the MLR model in [2], and using weather data from ERA-5. Hourly data of many atmospheric, land, and oceanic climate variables such as ambient air temperature at 2 m (2 m) above the land or water surface, precipitation, and wind speed from 1979 to the present at a spatial resolution of $0.25^\circ \times 0.25^\circ$ i.e., a $\sim$31 km grid. Reanalysis provides a picture of past weather and climate. Reanalysis blends observations with past short-range weather forecasts rerun with modern weather forecasting models. Reanalysis data is produced by running a numerical weather prediction model in hindcast mode, i.e., a projection back in time rather than a forecast prediction of future values. The reanalysis models assimilate historical observations and produce a time series of historical weather variable values. This affords us a greater understanding of the weather that prevails over a long period of time.

The ERA-5 data are available to download from [5]. We extracted a sample of weather variables (air temperature at 2 m, rain, dewpoint temperature) in .nc format from the ERA-5 reanalysis system for a ten-year period from 2012–2021 to cover Ireland by specifying latitudes and longitudes ('area': [55.5, -10.25, 51.5, -6.]). The Fuel Allowance is a social welfare payment to alleviate fuel poverty in Ireland which runs from the end of September to April each year. For the purpose of this study we assume Heating Degree days occur in months ['01', '02', '03', '04', '10', '11', '12', - i.e. Jan - Apr and Oct-Dec excluding summer months (May - Sept). We use the Python NetCdf package to extract the weather values.

We assume the ASHP is on during daytime hours 6.00–22.00 since generally heating is not required at night while the occupants sleep, well insulated buildings retain sufficient heat to meet night-time thermal comfort needs. The ASHP system turns on and off during the day as needs be to achieve the desired thermal comfort level. Hence, we select the ERA-5 2 m meter temperature, dew point temperature and total precipitation data for January - April and October to December from 2012–2021 from hours 06.00 to 22.00 for analysis. Temperatures in Kelvin are converted to Celsius, and precipitation is converted to mm for use in the MLR model. We apply the MLR model using Python to the average operating conditions and ten-year ERA-5 weather data by geographic locations to assess the potential ASHP performance by location. We then plot the averaged HSPF per location on a map of Ireland to visualise the ASHP performance by location using the ArcGis Pro software. This approach is replicable for other regions by selecting the corresponding ERA-5 data.

Our objective is to support decision making about suitable locations for ground or air source heat pumps to transform heating in Ireland from fossil fuel based to renewable heat sources. The GIS approach we apply to data for Ireland could be replicated in other locations.

4 Results, Analysis and Discussion

Figure 1 is a choropleth map answering our research question: what fuels/heating types are used in Ireland by location?

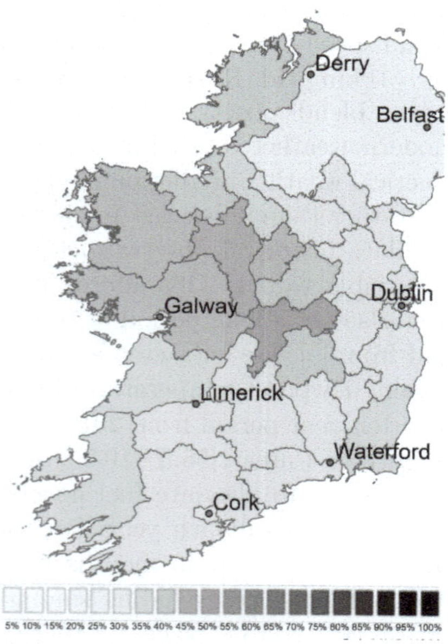

(a) Proportion of homes using solid fuel 2022

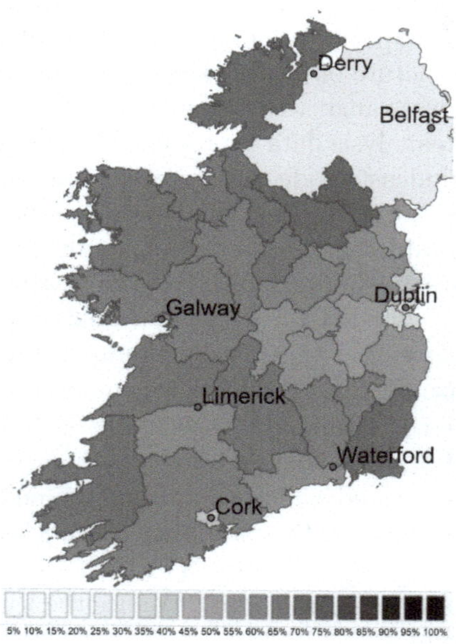

(b) Proportion of homes using oil 2022

Fig. 1. Heating type: solid fuel (left), oil (right)

Figure 1a shows the proportion of homes solid fuel, that is coal or peat, drawn from the 2022 census. Ireland has a long history of using peat for home heating. The Irish government is actively discouraging solid fuel use and is introducing regulations to restrict the types of fuels that can be commercially harvested and sold. Figure 1b shows the proportion of oil heating used by county. The remainder of heating in Ireland is mainly provided by the gas network. This distribution reflects the historical housing stock, the 2022 Census shows that nearly half of the new housing stock since the 2016 census used electricity as the main type of central heating, with Oil being much less common in newer properties.

We turn to our second research question on the heating decarbonisation challenges and opportunities identified using GIS. Figure 2 shows the distribution of the ERA5 average 2 m ambient air temperature during the heating days over Ireland from 2012–2021. This is the ambient outdoor air temperature and this determines the amount of work an ASHP would have to do to "lift" the ASHP heating system temperature to the home occupants desired thermal comfort level of approximately 20 °C for living space. A cooler temperature is recommended for sleeping areas. The average Heating Days temperature in Ireland over the ten-year period 2012–2021 is 8.23 °C.

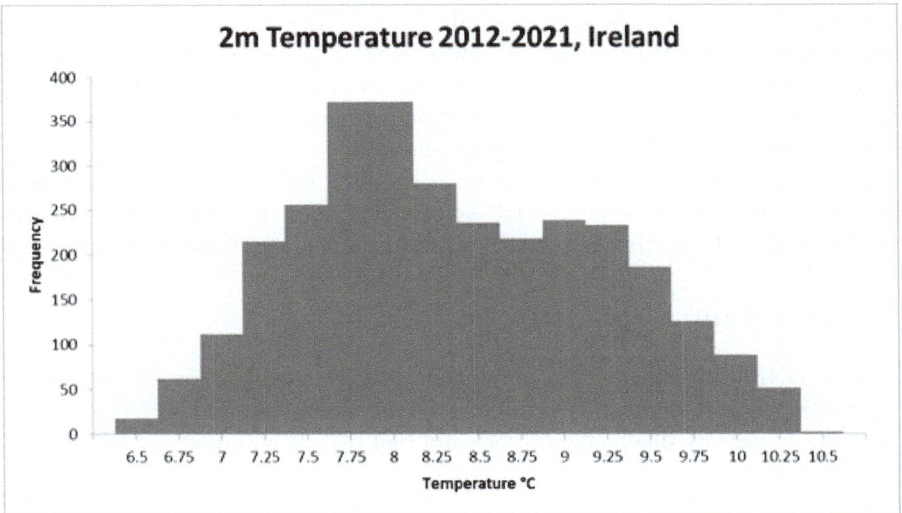

Fig. 2. 2 m Average Heating Day temperatures in Ireland over 2012–2021.

Figure 3 shows the 2 m temperatures during an extreme weather event that occurred in 2018 contrasted with the same day in 2019. Such events are rare in Ireland but are increasing in frequency. We see that the temperature values in 2019 are typical for Ireland, while in contrast low temperatures prevailed over Ireland during the extreme weather event in 2018. The impact of the extreme weather and estimated additional average electricity demand such events might create is described in [3]. ASHPs have to work harder in colder weather to achieve adequate thermal comfort levels. As GSHPs have their pipe network deep underground they are not so susceptible to extreme weather events.

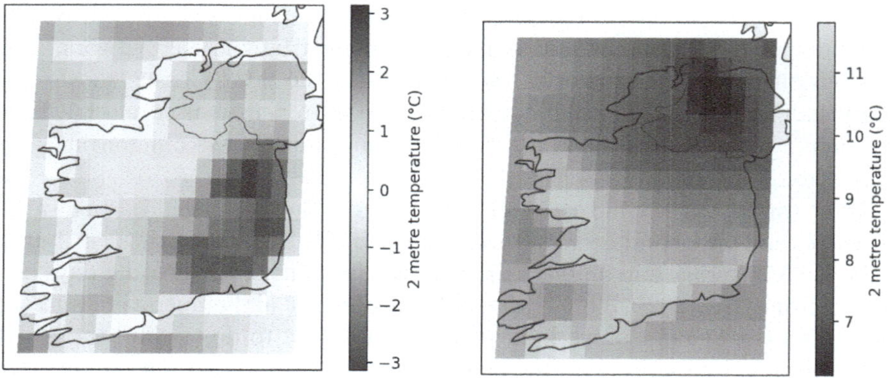

(a) Feb2018 (b) Feb2019

Fig. 3. 2 m temperature 28th February 2018 (left), 28th February 2019 (right)

Finally, Fig. 4 shows the average COP over the Heating Days, i.e., the HSPF, estimated with the MLR model described in [2]. The estimated HSPF exceeds the 2.5 renewable heat threshold in all locations. Warmer counties in the South East of the country yield the best ASHP performances, and we note that these areas currently use a significant proportion of oil.

We highlight that the results are for a sample deeply retrofitted home with high BER and a well managed ASHP system. We note that if the ASHP system is not operated efficiently these levels of performance could not be guaranteed.

5 Conclusion

HPs and DHNs are expected to play an important role in a low-carbon world. CO2 emissions from domestic heating can be reduced by around 50% (1–2 tonnes a year per house) by displacing oil, or solid fuel [18]. It is important for the success of heat decarbonisation policies that an appropriate mix of heating solutions be matched to the local requirements and renewable resource availability. Our ongoing analysis is based on multiple open data sources including the 2022 census of Ireland [4]. The ASPHs in Fig. 4 are estimates for a home with a high building energy efficiency. Further research will examine scenarios to understand the potential performance in buildings that are not so energy efficient, or when the ASHP is not operating at optimal conditions, or for future climate scenarios when the weather in Ireland may be warmer and wetter during the winter. Other further work will combine the spatial distribution shown in Fig. 4 with other GIS data, for instance, the data on the suitability of local geology for GSHPs. This would allow identification of the regions where heat pump installation would bring the greatest benefits. Other proposed work will look at the rural electrical distribution network and identify where a change of fuel towards electricity might place additional demands on the network and so require additional investment. Our work should allow the creation of a GIS based tool to allow decision

Fig. 4. Estimated ASHP Heating Days COP over 2012–2021.

makers understand the regions where government promotion of HPs should be concentrated to achieve maximum benefits.

Ireland is behind other jurisdictions in adopting the use of district heating and HP technologies. Ireland has yet to develop regulations and planning rules for geothermal energy. Ownership of geothermal energy has yet to be clarified, and there is no process for licensing the exploration for, or development and production of, geothermal energy resources. There is no mandatory national or local reporting on geothermal projects beyond requirements under existing environmental and planning regulations [9]. The recommendation on the suitability of biomass boilers and AHSPs as the most cost-effective heating solutions in Ireland also noted, that GSHPs may also be effective [15]. Our study demonstrates the renewable performance of ASHPs and supports the recommendation that ASHPs can contribute to the decarbonisation of heat in Ireland.

Further research is required to evaluate the performance of GSHPs and other heating options. An understanding of deeper geothermal resources has been hindered by a lack of data. There are currently no geothermal DH schemes in Ireland. Improvements in drilling could provide better data to highlight opportunities to harness recent geothermal technologies. GIS visualisations can help policy makers to identify challenges and opportunities and to quantify the parameters and constraints that should be used in optimisation models to select the best mix of heating solutions to serve heating needs.

References

1. Carroll, P., Chesser, M., Lyons, P.: Air source heat pumps field studies: a systematic literature review. Renew. Sustain. Energy Rev. **134**, 110275 (2020). https://doi.org/10.1016/j.rser.2020.110275
2. Chesser, M., Lyons, P., O'Reilly, P., Carroll, P.: Air source heat pump in-situ performance. Energy Build. **251**, 111365 (2021). https://doi.org/10.1016/j.enbuild.2021.111365
3. Chesser, M., O'Reilly, P., Lyons, P., Carroll, P.: The impact of extreme weather on peak electricity demand from homes heated by air source heat pumps. Energy Sources, Part B: Econ. Plan. Policy **16**(8), 707–718 (2021). https://doi.org/10.1080/15567249.2021.1945169
4. (CSO), C.S.O.: Census of population 2022 profile 2 - housing in Ireland. Technical report, Central Statistics Office (CSO) (2023). https://www.cso.ie/en/releasesandpublications/ep/p-cpp2/censusofpopulation2022profile2-housinginireland/occupieddwellings/
5. ERA-5: Climate data store. https://cds.climate.copernicus.eu/. Accessed 8 Apr 2024
6. European Commission: Commission decision of 1 march 2013 establishing the guidelines for member states on calculating renewable energy from heat pumps from different heat pump technologies pursuant to article 5 of directive 2009/28/ec of the european parliament. Official Journal of the European Union (2013). https://eur-lex.europa.eu/legal-content/EN/TXT/?uri=CELEX:32013D0114. Accessed 13 Feb 2024

7. European Parliament: Directive (EU) 2018/2001 of the European parliament and of the council of 11 December 2018 on the promotion of the use of energy from renewable sources. Off. J. Eur. Union **128**, 83–206 (2018). https://eur-lex.europa.eu/eli/dir/2018/2001/oj
8. Eurostat: Climate data store (2022). https://ec.europa.eu/eurostat/en/web/products-eurostat-news/-/edn-20220211-1. Accessed 9 Apr 2024
9. Government of Ireland: Geothermal energy in Ireland: A roadmap for a policy and regulatory framework. Technical report (2020). https://www.gov.ie/en/publication/abe7a-geothermal-energy-in-ireland-a-roadmap-for-a-policy-and-regulatory-framework/. Accessed 23 Feb 2024
10. Hersbach, H., et al.: The era5 global reanalysis. Q. J. R. Meteorol. Soc. **146**(730), 1999–2049 (2020). https://doi.org/10.1002/qj.3803
11. Lund, H., et al.: The status of 4th generation district heating: research and results. Energy **164**, 147–159 (2018). https://doi.org/10.1016/j.energy.2018.08.206
12. Mazhar, A.R., Liu, S., Shukla, A.: A state of art review on the district heating systems. Renew. Sustain. Energy Rev. **96**, 420–439 (2018). https://doi.org/10.1016/j.rser.2018.08.005
13. Saffari, M., Keogh, D., De Rosa, M., Finn, D.P.: Technical and economic assessment of a hybrid heat pump system as an energy retrofit measure in a residential building. Energy Build. **295**, 113256 (2023). https://doi.org/10.1016/j.enbuild.2023.113256
14. Saini, P., Huang, P., fiedler, F., Volkova, A., Zhang, X.: Techno-economic analysis of a 5th generation district heating system using thermo-hydraulic model: a multi-objective analysis for a case study in heating dominated climate. Energy Build. **296**, 113347 (2023). https://doi.org/10.1016/j.enbuild.2023.113347
15. SEAI: District heating and cooling - spatial analysis of infrastructure costs and potential in Ireland. Technical report, The Sustainable Energy Authority of Ireland (2022). https://www.seai.ie/publications/District-Heating-and-Cooling.pdf. Report 2 of the National Heat Study Accessed 29 Feb 2024
16. SEAI: Heating and cooling in Ireland today archetype profiles, spatial analysis, and energy efficiency potential. Technical report, The Sustainable Energy Authority of Ireland (2022). https://www.seai.ie/publications/District-Heating-and-Cooling.pdf. Report 2 of the National Heat Study Accessed 29 Feb 2024
17. SEAI: Energy in Ireland 2021 report. Dublin: Sustainable energy authority of Ireland. Technical report, The Sustainable Energy Authority of Ireland (2023). https://www.seai.ie/publications/Energy-in-Ireland-2023.pdf. Accessed 13 Feb 2024
18. Staffell, I., Brett, D., Brandon, N., Hawkes, A.: A review of domestic heat pumps. Energy Environ. Sci. **5**(11), 9291–9306 (2012). https://doi.org/10.1039/c2ee22653g
19. Werner, S.: District heating and cooling in Sweden. Energy **126**, 419–429 (2017). https://doi.org/10.1016/j.energy.2017.03.052

Enhancing Predictive Models in Insurance: A Feature Selection Analysis

Elnaz Ghorbani[1,3] and Raquel Soriano-Gonzalez[2,3](✉)

[1] Department of Computer Science, Universitat Oberta de Catalunya, 08018 Barcelona, Spain
eghorbanioskalaei@uoc.edu

[2] Department of Applied Statistics and Operations Research and Quality, Universitat Politècnica de València, Ferrandiz y Carbonell, s/n, 03802 Alcoy, Spain
rsorgon@epsa.upv.es

[3] Research Centre on Production Management and Engineering (CIGIP), Universitat Politècnica de València, Ferrandiz y Carbonell, s/n, 03802 Alcoy, Spain

Abstract. The effective management of large-scale datasets with diverse features is essential in data preprocessing for machine learning applications. Feature selection plays a critical role in ensuring the identification of the most relevant features, thereby improving model accuracy and insights. This paper investigates various feature selection techniques using a car insurance dataset, focusing on feature selection while considering an ordinal categorical feature as the target feature. Additionally, this study establishes a foundation for enhancing predictive models for customer classification by drawing insights from a real-world case study in the insurance industry. Four methods were compared: least absolute shrinkage and selection operator (LASSO), gradient boosting, stepwise regression, and univariate feature selection model. Results of validation indicate that LASSO outperforms other methods using the Bangdiwala statistic and its corresponding graph to evaluate the behavior of features in each model. This research underscores the significance of feature selection in refining insurance-related analytics and achieving a more comprehensive understanding of the field.

Keywords: Feature Selection · Supervised Learning · Insurance analytics

1 Introduction

Managing large-scale datasets with diverse features is one of the critical steps in data pre-processing in machine learning and data mining problems [1]. Effectively managing the feature selection procedure helps choose the most relevant and informative features, leading to more accurate models and insightful analyses. Through various approaches such as filter methods [2], wrapper methods [3], and embedded methods [4], feature selection aims to diminish the overfitting and model complexity. Filter methods evaluate the relevance of features

A. A. Juan et al. (Eds.): DSA ISC 2024, LNCS 14778, pp. 256–267, 2025.
https://doi.org/10.1007/978-3-031-78238-1_24

based on statistical measures or scores. By quantifying the relationship between each feature and the target variable of the dataset, filter methods provide valuable information about the importance of individual features [2]. On the other hand, wrapper methods assess subsets of features by training models on different combinations and selecting the best-performing subset. This approach is computationally intensive but can lead to the selection of highly relevant features that might have been overlooked by simpler filter methods [5]. Subsequently, embedded methods integrate the feature selection process directly into the model training procedure. Unlike filter and wrapper methods, which handle feature selection as a separate step following model training, embedded methods perform feature selection at the same time as the model training process. During the training phase, these methods automatically determine the relevance of features and adjust their weights accordingly, optimizing the model's performance in real-time [6]. Reducing dataset size can be accomplished through one of two approaches: either by reducing the number of features or by reducing the number of samples [7]. In this paper, the feature selection techniques are developed in order to reduce the size of our dataset by selecting the most influential features. Moreover, this paper exclusively concentrates on supervised learning methodologies since it relies on labeled data.

In this study, we investigate a real case study dataset of car insurance, an aspect of the automotive industry. A car insurance policy is an agreement between an insurance company and the vehicle owner, aiming to offer financial protection in the event of theft or damage to the vehicle [8]. Improving the quality of insurance data through effective pre-processing, including feature selection, is expected to enhance the estimation process by refining the dataset and focusing on the most relevant information. By identifying and employing only the most influential features, feature selection optimizes the predictive power of machine learning models, particularly in tasks like insurance reservation prediction, customer segmentation, pricing strategies, and fraud detection mechanisms [9]. As evidenced by previous research in the field, the strategic application of feature selection techniques enhances model performance and gathers innovation and optimization within the insurance sector. This paper highlights the impact of feature selection on data quality issues when applied to a car insurance case study dataset. It uses and compares different feature selection methodologies such as LASSO, gradient boosting, univariate feature selection, and stepwise regression. The added value brought by this study lies in the introduction of a target feature that is ordinal categorical, which significantly complicates the feature selection process. The focus lies on the critical role of feature selection in clarifying insurance-related analytics and unlocking deeper insights into this field. The remainder of this paper is organized as follows. Definitions of the materials and methods are presented in Sect. 2, whereas Sect. 3 discusses the results. Finally, the conclusion is presented in Sect. 4.

2 Materials and Methods

The dataset used in this study is provided by an insurance company. The insurance company provides a comprehensive database that includes policies for all clients across various types of automobile insurance. For this study, we have specifically selected the records of clients who hold a passenger car insurance policy. It is important to note that the database of unique passenger car clients is completely free of missing data. The dataset consists of 63 features and 36,767 rows, with each row representing the information of a single client. Table 1 shows the different features of the dataset with corresponding descriptions. This table organizes features into distinct groups based on their content, including identification information, information related to the policies, vehicle characteristics, and contract-related features. The target variable under consideration is the 'id13' feature, comprising twelve distinct categorical values representing the client's score. The target is defined as customers' credit, categorized as J, I, or H, to indicate those prone to fraud or payment default, making them undesirable clients for the insurance company. Conversely, customers classified as A, B, and C consistently fulfill their payment obligations and represent a benefit to the company. The remaining categories, D, E, and F, represent customers occupying an intermediate position and remain of interest to the insurance company. The distribution of categories in the target variable is uneven, with categories E, D, and F representing 30.19%, 18.22%, and 17.06% of clients, respectively, making them the most populous. Conversely, the categories indicating undesirable clients, namely H, I, and J, comprise only 2.54% of the total. This imbalance in the distribution of categories suggests that the dataset may present challenges when classifying clients according to the target variable. To initiate customer segmentation, the primary step involves feature selection, utilizing only the most informative features for analysis. In the next paragraphs, the four methods of feature selection are discussed. To begin with, LASSO stands out as a widely used technique in feature selection for its ability to identify essential features while reducing the number of irrelevant or redundant variables [10]. It achieves this by solving the following optimization problem:

$$\underset{\beta}{\text{minimize}} \left(\frac{1}{2n} \sum_{i=1}^{n} (y_i - \beta^T x_i)^2 + \lambda \sum_{j=1}^{p} |\beta_j| \right) \tag{1}$$

where n is the number of observations, p is the number of features, y_i is the target variable for the ith observation, x_i is the feature vector for the ith observation, β is the coefficient vector to be estimated, and λ is the regularization parameter. The first term represents the residual sum of squares (RSS), while the second term is the penalty term. This penalty helps make the coefficient vector sparse by setting less important feature coefficients to zero. As a result, LASSO selects a subset of the most relevant features and provides a simpler and more interpretable model. However, its lack of stability when there are many features relative to observations, and its tendency to arbitrarily select among highly correlated features, can lead to the erroneous exclusion of important variables from the model.

On the other hand, gradient boosting is another powerful technique for feature selection. Unlike LASSO, which directly penalizes the coefficients of the features, gradient boosting builds a predictive model by combining multiple weak learners, typically decision trees [11]. During the training process, gradient boosting sequentially adds trees to the ensemble, with each new tree focusing on the residual errors of the previous ones. This iterative process allows gradient boosting to identify and prioritize features that contribute the most to reducing prediction errors. By analyzing the feature importance scores obtained from the gradient boosting model, researchers can effectively select the most relevant features for the prediction task. Although feature importance scores can be helpful, interpreting them in complex models can be challenging. Additionally, this approach can be computationally intensive and prone to overfitting, especially with deep decision trees, requiring careful hyperparameter tuning [12].

Stepwise regression is a traditional statistical method used for feature selection, particularly in linear regression models. It operates by iteratively adding or removing features based on their statistical significance [13]. The stepwise procedure typically involves two main steps: forward selection and backward elimination. In forward selection, the algorithm starts with an empty model and iteratively adds the most statistically significant features until no additional features improve the model's performance. In backward elimination, the algorithm begins with a model containing all features and sequentially removes the least significant ones until no further removal enhances the model's performance. Although this technique is widely used and seems suitable, it suffers from overfitting the data, biased estimates, inflated type of error, and high computational time that it needs [14].

Univariate feature selection is a more straightforward approach that evaluates each feature individually based on statistical tests or scoring methods [15]. Common techniques include analysis of variance (ANOVA) [16], chi-square test [17], mutual information [18], and correlation coefficients [19]. Univariate feature selection ranks features according to their relevance to the target feature, allowing researchers to select the top-ranked features for further analysis. While univariate methods may overlook complex interactions between features, they are computationally efficient and provide insights into the individual contributions of each feature to the prediction task. Univariate methods are computationally efficient and offer insights into each feature's contribution to the prediction task. However, these techniques may lose the ability to capture nonlinear relationships between features and the target feature, which can limit their capacity to model the complexity of the problem. Additionally, by considering only one feature at a time, they may overlook interactions between multiple features that could be important for prediction [20].

Table 1. Car insurance dataset features.

Feature	Description	Type
Identification information		
id02	Activity type code	Categorical-Numeric
id11	Company Score	Numeric
id13	Policy holder's score	Categorical-Numeric
id16	Age of the policyholder	Numeric
id21	Postal code	Numeric
id24	Type of area where the insured lives	Categorical
id41	Premium amount	Numeric
id42	Commission amount	Numeric
id43	Total loss amount	Numeric
id57	Number of historical commercial actions	Numeric
id58_1, id58_2,id58_3	Links with the client (spouse, children, siblings)	Numeric
id59	Historical net premiums of linked clients	Numeric
id60	Historical commissions paid by linked clients	Numeric
id61	Historical cost of claims of linked clients	Numeric
AceptoCulpaSinAntecedentes	Policyholder contact status for incident data	Binary
ProvincialD	Province code of the policyholder	Numeric
ComarcaID	Region code of the policyholder	Numeric
NacionalidadID	Nationality code	Numeric
Edad	Age of insured when joining the company	Numeric
antiguedad_carnet	Age of the insured's driver's license when joining the company	Numeric
Conductor2ID	If second identifier exists	Binary
Information related to the policies		
id14	Score Mediation network	Categorical
id15	Policy contracting channel	Categorical
id17	Agency code	Numeric
id18	Subagency code	Numeric
id25	Number of policies of all lines in force of the client	Categorical- Numeric
id26	Maximum number of policies that a client has had in force	Categorical- Numeric
id27	No. of branches currently contracted by the client	Categorical- Numeric
id29	No. of policies canceled by the client in the current year	Categorical- Numeric
id30	Miscellaneous claims number	Numeric
id31	No. of claims during the reference year of Miscellaneous	Categorical- Numeric
id32	No. of historical auto accidents of the client	Numeric
id34	Sum of the number of years of each auto policy in force	Numeric
id36	Total historical customer claims frequency	Numeric
id37	Frequency of client's historical auto accidents	Numeric
id53	No. policies canceled by the company throughout the client's life	Categorical- Numeric
id54	Method of payment	Categorical
id55	Return or bonus	Numeric
DGT_001	Accidents with victims per year	Numeric
DGT_002	Accidents with animals per year	Numeric
IGN_001	Distance to the sea in meters	Numeric
Perm_dias_real	Permanence of the insured in the company	Numeric
Perm_dias_est	Permanence until expiration	Numeric
CoincideFechaFinAnulación	When a policy end date coincides with a cancellation date	Binary
id01.Propietario	If the owner is hiring	Binary
id01.ConductorID	If the hiring party is the driver	Binary
Prop.Conduc	If the owner is the dirver	Binary
Vehicle characteristics features		
id70	Power of the policy vehicle	Numeric
id71	Number of doors of the policy vehicle	Categorical- Numeric
id72	Type of vehicle by energy	Categorical
SINCOTotalVehiculos	Total number of different license plates	Numeric
SINCOMatriculaIncluida	License plates without registration included	Binary
ContrataLunas	If the glasses of the car are included in insurance or not	Binary
ScoreAda	Score of the vehicle based on brand type	Numeric
ValorVehiculoMercado	Value of the vehicle in market	Numeric
AnyoMatricula	Year in which the vehicle was registered	Numeric
Insurance contract features		
ContrataAsistencia	If the assistant is included in the policy or not	Binary
ContrataDanyosPropios	If the damage insurance is included in the policy or not	Binary
SINCOAnyos	Coverage periods and claims in the last 5 years from SINCO	Categorical- Numeric
SINCOSiniestrosMateriales	Number of accidents with material damage from SINCO	Categorical- Numeric
SINCOSiniestrosPersonales	Number of accidents with bodily injuries from SINCO	Categorical- Numeric

3 Results

In this section, the assumptions and results arising from the implementation of the four feature selection methods are presented. First, a backward stepwise algorithm is developed to choose a subset of valuable features by iteratively deleting less significant features from the model. For the LASSO method, a grid search is conducted to determine the optimal parameter λ minimizing the value of AIC (Akaike Information Criterion). On the other hand, the univariate feature selection method employs the F-statistic and associated p-values. This model selects all the features since no number is given as a threshold to identify the most influential features from the feature set. Finally, a gradient boosting classifier is initialized without specifying any hyperparameters. All the computations are performed in the R environment, utilizing various libraries including ordinalNetCV for LASSO [21], rms for stepwise model [22], ordinal library for univariate ordinal regression model [23], and xgboost [24].

As a visual tool to depict the relationship between the different feature selections obtained in each method a Venn diagram is employed, which is presented in Fig. 1. This diagram illustrates how the selected features overlap or differ among methods, providing a clear understanding of the consistency and diversity of the selected sets. From this diagram, it is observed that a set of 23 features is common in all four models used, a robust set of predictors with consistent importance across different methodologies.

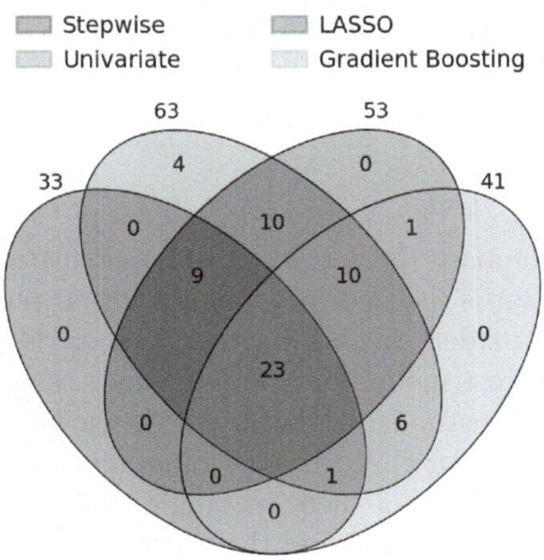

Fig. 1. Comparison of selected features by four feature selection methods.

To each of the selections obtained from the models, an ordinal Random Forest model is applied to verify the robustness of the techniques and assess their goodness of fit [25]. To compare the results of the four mentioned feature selection methods, a Bangdinwala test is utilized, a statistical method to assess agreement the test and train sets of the ordinal Random Forest model, when rating items on a categorical scale [26]. Through its agreement index, this method can measure the coordination between different feature selection methods and the target feature, especially when the latter is ordinal categorical. The Bangdiwala agreement coefficient varies between 1 and -1, where 1 indicates strong agreement, 0 indicates no deal better than chance, and -1 indicates strong disagreement.

The Bangdiwala method offers several significant advantages in feature selection evaluation, especially for ordinal categorical features. This agreement index evaluates concordance in terms of correct or incorrect test and train sets classification and considers the strength of agreement between classifications. In this study, we analyze the results of the Bangdiwala agreement index to determine which feature selection method produces models most consistent with the target feature. Table 2 shows the comparison results of the four feature selection methods. It is evident that although three of the models achieved similar results, the LASSO model performed the best in this statistic, indicating more excellent concordance with the ordinal categorical target feature. However, it is important to note that the univariate model cannot calculate this statistic for ordinal categorical target features, which is a limitation compared to other models.

Table 2. Comparison results of the four feature selection methods with Bangdinwala Statistic.

Models	B. Statistic	No. of features
Stepwise	0.618	33
LASSO	**0.671**	**53**
Univariate	-	63
Gradient Boosting	0.613	41

To evaluate and compare the results of different feature selection methods, the Bangdiwala diagram is utilized, providing a visual representation of agreement levels among the models. In this diagram, the x-axis represents predictions made by a Random Forest model for each customer category according to different variable selection methods, while the y-axis shows the actual distribution of customers in each category. Within the diagram, each category is symbolized by rectangles. The height of these rectangles reflects the number of actual customers in each category (y-axis) and the width represents the number of predicted customers (x-axis). A central black rectangle indicates high concordance between the predicted and actual customers in that category. The larger this rectangle, the better the concordance between the predictions and the actual data for that category of the target feature. The gray rectangles surrounding the black area denote moderate agreement, suggesting that although there is a correlation between the predictions and the actual data, there is a margin of error and some predictions do not perfectly match. The outer white area, on the other hand, shows areas of disagreement, indicating prediction outcomes that do not align with the observed reality. In Figs. 2, 3, 4, the different Bangdiwala diagrams for each feature selection method are displayed. For an effective comparison of variable selection methods, it is essential to consider several factors. Ideally, squares should appear in each category, showing an exact prediction of the number of customers compared to the actual data. A good indication of precise results is the placement of these squares along the diagonal, implying that the predictions coincide with the target categories. This evaluation must be extended to all categories of the target feature, as there may be variations in the level of agreement among them. Therefore, a comprehensive evaluation of all categories in the diagram is necessary.

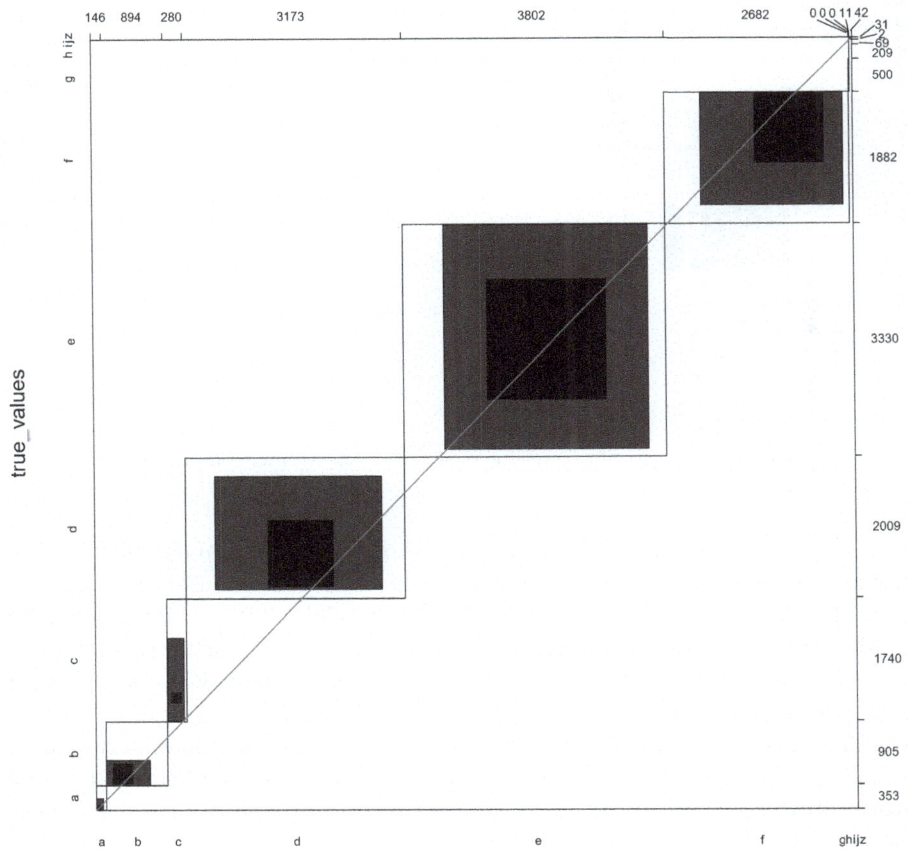

Fig. 2. Bangdiwala chart for evaluating the Lasso method.

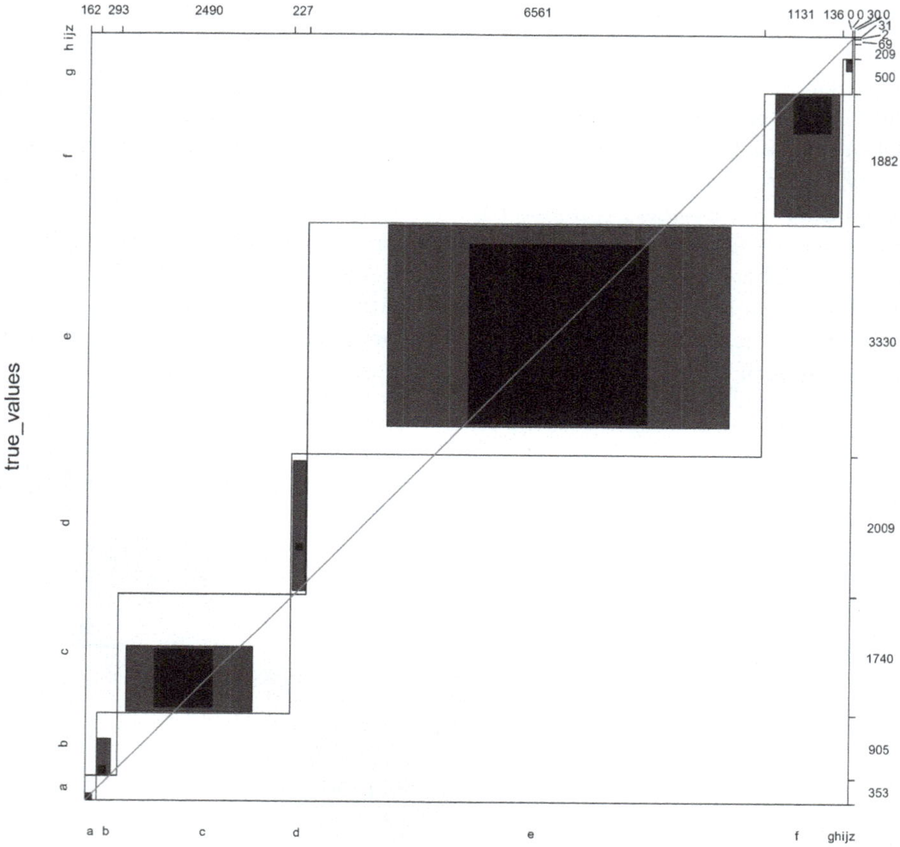

Fig. 3. Bangdiwala chart for evaluating the stepwise method.

In the analysis of specific models, the LASSO method stands out with five categories showing a high degree of agreement, reflected by large black squares along the diagonal. In contrast, the stepwise model achieves high agreement in only three categories, while the gradient boosting model achieves significant agreement in three categories, with two additional ones showing well-defined black squares, albeit not exactly on the diagonal, indicating certain deviations from the expected results of the target feature. This detailed evaluation reinforces the utility of the Bangdiwala agreement index and confirms that feature selection through LASSO provides the most optimal results for the ordinal categorical response variable.

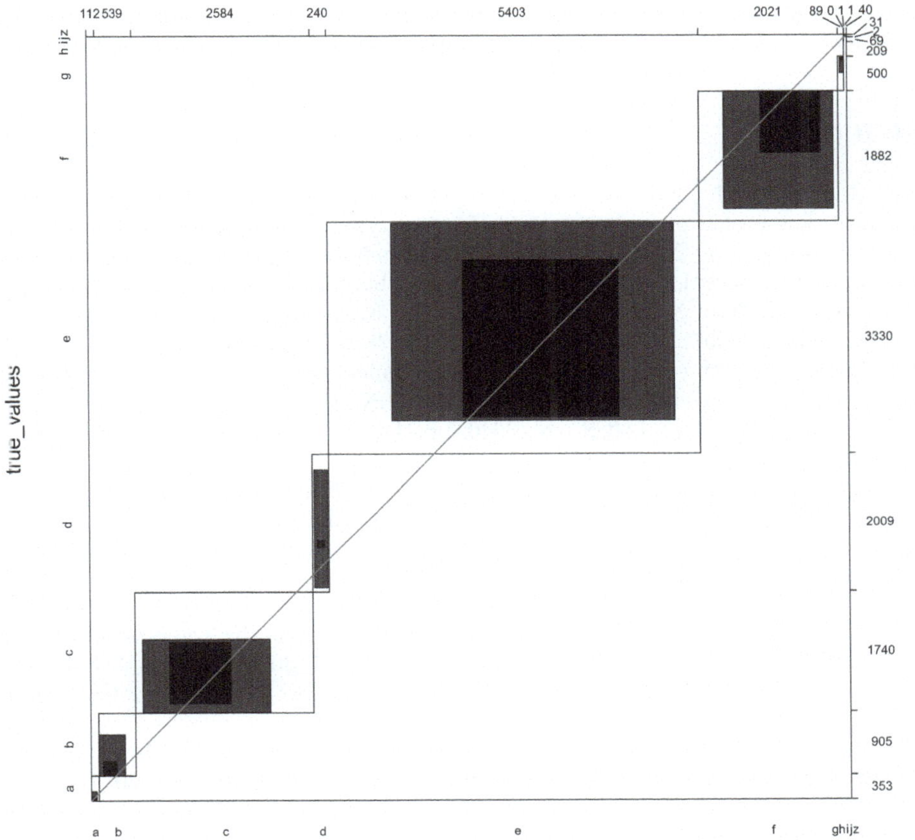

Fig. 4. Bangdiwala chart for evaluating the gradient boosting method.

4 Conclusion

In conclusion, this study highlights the critical relevance of feature selection in building predictive models for complex datasets, particularly in a real-world case study associated with car insurance. The target feature in this study is an ordinal categorical feature, adding complexity to feature selection and model interpretation. The ordinal nature of the target feature demands a careful approach to feature selection and result interpretation to ensure that the models are effective and provide relevant information for decision-making in the context of car insurance and other machine-learning applications. To address this challenge, four feature selection methods were compared. According to the Bangdiwala test LASSO feature selection method performs effectively in comparison with three others including stepwise, univariate, and gradient boosting methods. However, LASSO is a powerful method, addressing its limitations such as multicollinearity and instability due to high feature-to-observation ratios could be addressed in future work. One potential subject for improvement could be exploring the use of elastic net regularization, which offers a hybrid approach combining LASSO and Ridge regression, thus potentially mitigating the shortcomings of LASSO.

References

1. Li, J., Cheng, K., Wang, S., Morstatter, F., Trevino, R.P., Tang, J., Liu, H.: Feature selection: a data perspective. ACM Comput. Surv. (CSUR) **50**(6), 1–45 (2017)
2. Sánchez-Maroño, N., Alonso-Betanzos, A., Tombilla-Sanromán, M.: Filter methods for feature selection – a comparative study. In: Yin, H., Tino, P., Corchado, E., Byrne, W., Yao, X. (eds.) IDEAL 2007. LNCS, vol. 4881, pp. 178–187. Springer, Heidelberg (2007). https://doi.org/10.1007/978-3-540-77226-2_19
3. Talavera, L.: An evaluation of filter and wrapper methods for feature selection in categorical clustering. In: Famili, A.F., Kok, J.N., Peña, J.M., Siebes, A., Feelders, A. (eds.) IDA 2005. LNCS, vol. 3646, pp. 440–451. Springer, Heidelberg (2005). https://doi.org/10.1007/11552253_40
4. Lal, T.N., Chapelle, O., Weston, J., Elisseeff, A.: Embedded methods. In: Guyon, I., Nikravesh, M., Gunn, S., Zadeh, L.A. (eds.) Feature Extraction. Studies in Fuzziness and Soft Computing, vol. 207, pp. 137–165. Springer, Heidelberg (2006). https://doi.org/10.1007/978-3-540-35488-8_6
5. Sánchez-Maroño, N., Alonso-Betanzos, A., Calvo-Estévez, R.M.: A wrapper method for feature selection in multiple classes datasets. In: Cabestany, J., Sandoval, F., Prieto, A., Corchado, J.M. (eds.) IWANN 2009. LNCS, vol. 5517, pp. 456–463. Springer, Heidelberg (2009). https://doi.org/10.1007/978-3-642-02478-8_57
6. Liu, H., Zhou, M.C., Liu, Q.: An embedded feature selection method for imbalanced data classification. IEEE/CAA J. Automatica Sinica **6**(3), 703–715 (2019)
7. Jović, A., Brkić, K., Bogunović, N.: A review of feature selection methods with applications. In: 2015 38th International Convention on Information and Communication Technology, Electronics and Microelectronics (MIPRO), pp. 1200–1205. IEEE (2015)
8. Subudhi, S., Panigrahi, S.: Detection of automobile insurance fraud using feature selection and data mining techniques. Int. J. Rough Sets Data Anal. (IJRSDA) **5**(3), 1–20 (2018)
9. Taha, A., Cosgrave, B., Mckeever, S.: Using feature selection with machine learning for generation of insurance insights. Appl. Sci. **12**(6), 3209 (2022)
10. Freijeiro-González, L., Febrero-Bande, M., González-Manteiga, W.: A critical review of lasso and its derivatives for variable selection under dependence among covariates. Int. Stat. Rev. **90**(1), 118–145 (2022)
11. Upadhyay, D., Manero, J., Zaman, M., Sampalli, S.: Gradient boosting feature selection with machine learning classifiers for intrusion detection on power grids. IEEE Trans. Netw. Serv. Manage. **18**(1), 1104–1116 (2020)
12. Chong, I.-G., Jun, C.-H.: Performance of some variable selection methods when multicollinearity is present. Chemom. Intell. Lab. Syst. **78**(1–2), 103–112 (2005)
13. Zogala-Siudem, B., Jaroszewicz, S.: Fast stepwise regression based on multidimensional indexes. Inf. Sci. **549**, 288–309 (2021)
14. Harrell, F.E.: Regression Modeling Strategies: With Applications to Linear Models, Logistic and Ordinal Regression, and Survival Analysis. SSS, Springer, Cham (2015). https://doi.org/10.1007/978-3-319-19425-7

15. Jain, S., Saha, A.: Rank-based univariate feature selection methods on machine learning classifiers for code smell detection. Evol. Intel. **15**(1), 609–638 (2022)
16. Shakeela, S., Sai Shankar, N., Mohan Reddy, P., Kavya Tulasi, T., Mahesh Koneru, M.: Optimal ensemble learning based on distinctive feature selection by univariate ANOVA-F statistics for IDS. Int. J. Electron. Telecommun. 267–275 (2021)
17. Sikri, A., Singh, N.P., Dalal, S.: Chi-square method of feature selection: impact of pre-processing of data. Int. J. Intell. Syst. Appl. Eng. **11**(3s), 241–248 (2023)
18. Bennasar, M., Hicks, Y., Setchi, R.: Feature selection using joint mutual information maximisation. Expert Syst. Appl. **42**(22), 8520–8532 (2015)
19. Zhou, H., Wang, X., Zhu, R.: Feature selection based on mutual information with correlation coefficient. Appl. Intell. **52**(5), 5457–5474 (2022)
20. Solorio-Fernández, S., Ariel Carrasco-Ochoa, J., Martínez-Trinidad, J.F.: A review of unsupervised feature selection methods. Artif. Intell. Rev. **53**(2), 907–948 (2020)
21. Wurm, M.J., Rathouz, P.J., Hanlon, B.M.: Regularized ordinal regression and the ordinalnet R package. J. Stat. Softw. **99**(6) (2021)
22. Harrell, F.E.: Regression modeling strategies. R package version, pp. 6–2 (2012)
23. Christensen, R.H.B.: ordinalregression models for ordinal data. R package version 10(2019), 54 (2019)
24 Chen, T., et al.: XGBoost: extreme gradient boosting. R package version 0.4-2, 1(4), 1–4 (2015)
25. Hornung, R.: Ordinalforest: Ordinal forests: Prediction and variable ranking with ordinal target variables. R package version 2 (2018)
26. Bangdiwala, S.I.: The agreement chart. Institute of statistics mimeo series# 1859 (1988)

Spatial Interpolation Model With Covariates Using Thin Plate Splines

David Hervás[1(✉)], Patricia Carracedo[1], and Guillermo Franco[2]

[1] Department of Statistics and OR, Universitat Politècnica de València, 03801 Alcoy, Spain
{daherma,pcarracedo}@eio.upv.es
[2] Global Head of Catastrophe Risk Research, Guy Carpenter, 1166 Avenue of the Americas, New York, NY 10036, USA
guillermo.e.franco@guycarp.com

Abstract. Earthquakes have historically inflicted significant losses, with the period from 2000 to 2019 witnessing approximately 721,000 fatalities and US$636 billion in economic damages worldwide. Catastrophic earthquakes can wreak havoc on multiple structures and infrastructure systems simultaneously, leading to profound economic repercussions. While some losses can be mitigated through enhanced construction practices or transferred via the (re)insurance market, rapid response in certain scenarios is critical. Parametric insurance emerges as a strategic tool in such instances, designed to provide predetermined payouts for specific earthquake scenarios, often utilizing publicly available data for near-real-time responsiveness. However, parametric solutions are susceptible to basis risk, prompting efforts to minimize this discrepancy through evolutionary computation, machine learning, and biased-randomized algorithms. This study aims to help insurers by predicting losses produced by earthquake in Chile rapidly and with minimal uncertainty. Through a thin plate spline (TPS) advanced interpolation method this research seeks to advance understanding and address key challenges in parametric insurance design and implementation.

Keywords: Thin plate spline · Interpolation · Forecast · Earthquake · Chile

1 Introduction

Throughout history, earthquakes have caused numerous losses. In the period between 2000 and 2019, earthquakes caused approximately 721,000 deaths and US$636 billion in economic losses worldwide [5,6]. A catastrophic earthquake may simultaneously affect multiple structures and infrastructure systems, triggering great economic impacts [10]. Some of these losses can be mitigated with better construction practices and some can be transferred via the (re)insurance market [9].

A. A. Juan et al. (Eds.): DSA ISC 2024, LNCS 14778, pp. 268–275, 2025.
https://doi.org/10.1007/978-3-031-78238-1_25

In scenarios where having a rapid response of an insurance policy matters, such as when a government needs cash to relocate victims or when a corporation needs to guarantee its existence throughout the earthquake recovery process, using a form of insurance known as "parametric insurance" is an advantageous strategy. Parametric insurance is designed to pay a pre-determined amount for specific pre-agreed scenarios, sometimes arranged in a table that specifies the amount payable for an earthquake occurring within a certain geographic area and of a certain magnitude [8]. The parameters required to determine the appropriate payout can be obtained from public sources in near-real-time [18], which offers parametric insurance the ability to respond very fast, often within just a few days of the event.

The parametric mechanism is designed to pay fast but not accurately. Parametric solutions suffer from basis risk, which is the difference between the actual payout paid by the policy and the payout that should take place based on the damages experienced. Therefore, much of the effort to develop better parametric solutions has revolved around reducing this basis risk. Techniques proposed in the literature to minimize basis risk have relied on evolutionary computation [7], machine learning [1], and biased-randomized algorithms [2].

All the methods cited rely on measuring basis risk with the aid of a catastrophe risk model, a Monte Carlo-type numerical simulation that yields an estimation of loss for a large sample of synthetic but realistic earthquake events [11]. Models of this kind provide valuable information for decision-making for entities in the public and private sectors, contributing to better preparation and response to these catastrophic events. Machine learning models have generated great interest of their own in seismology and actuarial fields and offer new techniques for loss earthquake estimation [5,13].

In this paper, we retake the problem of minimization of basis risk using a novel approach that consists of approximating a catastrophe model simulation using just the four parameters typically employed to trigger a parametric policy, that is the longitude and latitude of the earthquake's epicenter, its focal depth, and its magnitude. We consider the framework for the design of parametric policies established in [8] and we tackle that main challenges that were identified in that work, namely (1) the incompleteness of samples for all combinations of locations and magnitudes, and (2) the lack of smoothness in the results produced by stochastic catastrophe models.

In this work, we use spatial interpolation to predict the payout of a hypothetical parametric policy for Chile for earthquakes at any arbitrary locations, based on the limited sample provided by the synthetic event catalog of a catastrophe model. This is a common statistical technique for the visualization of spatial data in geography and earth and environmental sciences [15]. Although kriging is more popular and more widely used for stochastic interpolation [15], other deterministic techniques such as Thin Plate Splines (TPS) have powerful and attractive features for the visualization of complex predictor-response relationships. Specifically, for the purpose of this work, which is to estimate losses rapidly and with minimal uncertainty, a deterministic methodology is attractive as it results in a unique solution.

This paper is structured as follows. Section 2 starts by describing the basis of the database. This section ends with an exposition of the TPS interpolation method. Section 3 presents the main results of this statistical technique and finally, in Sect. 4, the most important conclusions of the study are shown.

2 Data and Methodology

In this section we will begin by detailing the database used and, then the statistical techniques applied to the spatial data.

2.1 Data

A disadvantage of working with earthquakes is that the frequency of occurrence of these events is low and the historical record is insufficient to obtain meaningful statistics. To overcome this shortcoming, as mentioned above, catastrophe risk models exist which through simulation, are able to generate stochastic events in harmony with the seismic setting of interest [2]. Specifically, in this study, these models have provided all the necessary data, such as the characteristics of the earthquakes and the losses they have caused, to apply the spatial interpolation technique presented here.

The particular building portfolio includes 129753 observations corresponding to fundamental earthquakes characteristics in Chile such as location, magnitude, loss, depth and frequency. Figure 1 shows the logarithm of loss produced by earthquakes occurred in that specific location of Chile. It should be noted that, the losses in a given area is the total losses generated by the earthquake in that location.

2.2 Methodology

TPS is an interpolation approach that has been developed to investigate a frequently occurring problem in geosciences, the modelling of scattered data [12]. Specifically, TPS are a type of smoothing spline used for the multidimensional visualization of complex relationships between continuous explanatory variables and explained variable. Thin plate splines are depicted as a flexible surface which is warped to fit the data. Each continuous variable is plotted on an individual x-axis, creating a bivariate, two-dimensional surface. The outcome is then plotted on a y-axis, continuously across that bivariate surface in three-dimensions [19]. Specifically, in this work, earthquake losses in Chile were modeled by TPS, considering as explanatory variables the depth and magnitude. To do this, the package fields [16] of statistical software R [17] was used. The Tps function in R fits a TPS surface to irregularly spaced data. The model is fitted using a generalized additive model which equation is denoted by:

$$y = \beta_0 + f(x) + e \tag{1}$$

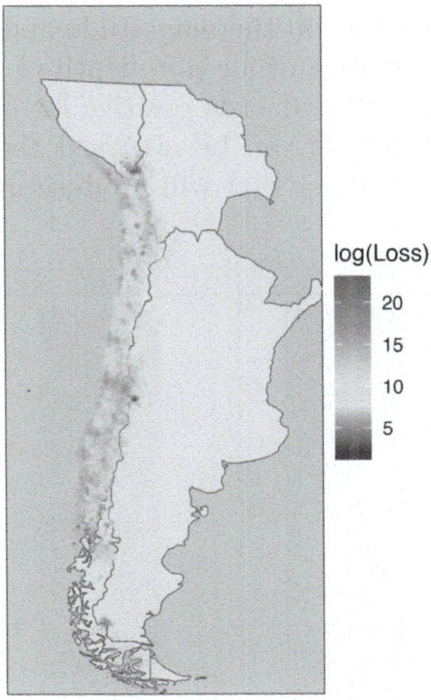

Fig. 1. Spatial distribution of logarithm of loss in Chile.

where y is the explained variable, β_0 is a constant. $f(x)$ is a d dimensional surface representing the fit of explanatory variables to the explained variable by means of splines. This function also works for just a single dimension and is a special case of a spatial process estimate (Kriging). A fast version of this function uses a compactly supported Wendland covariance and sparse linear algebra for handling larger data sets. e is the error term which provides a smoothing function based on a penalized least squares method. It is described as the tension needed to double a thin plate. For this reason, if e increases, the smoothness of the spline will increase. e parameter is chosen as a default by generalized cross-validation [4]. There are many advantages of using TPS compared to other techniques. Some of them are described below. One of the most important advantages of using TPS is that, a priori, no knowledge of the functional form of the data or the relationship of interest is required. In addition, unlike high degree splines, the determination of the number of nodes and their location is obtained automatically from the smoothing function in TPS. Another advantage is that it is a very powerful tool for visualizing complex predictor-response relationships, and heat maps can be added to further emphasize surface curvature. At the end, in order to measure the effect of continuous predictors, it also allows optimal control of continuous confounders [3,14].

3 Results

The results of prediction of TPS model are shown in Fig. 2 by means of contour plot for Chile. It shows the prediction of losses generated by the earthquake in each geographic unit depending on the magnitude and amplitude of the event. As expected, both variables, magnitude and depth of the earthquake, have an effect on losses. Specifically, this effect is positive for magnitude of the impact, while it is negative for depth. It can be observed the earthquakes of greater magnitude and shallower are those that will produce more losses.

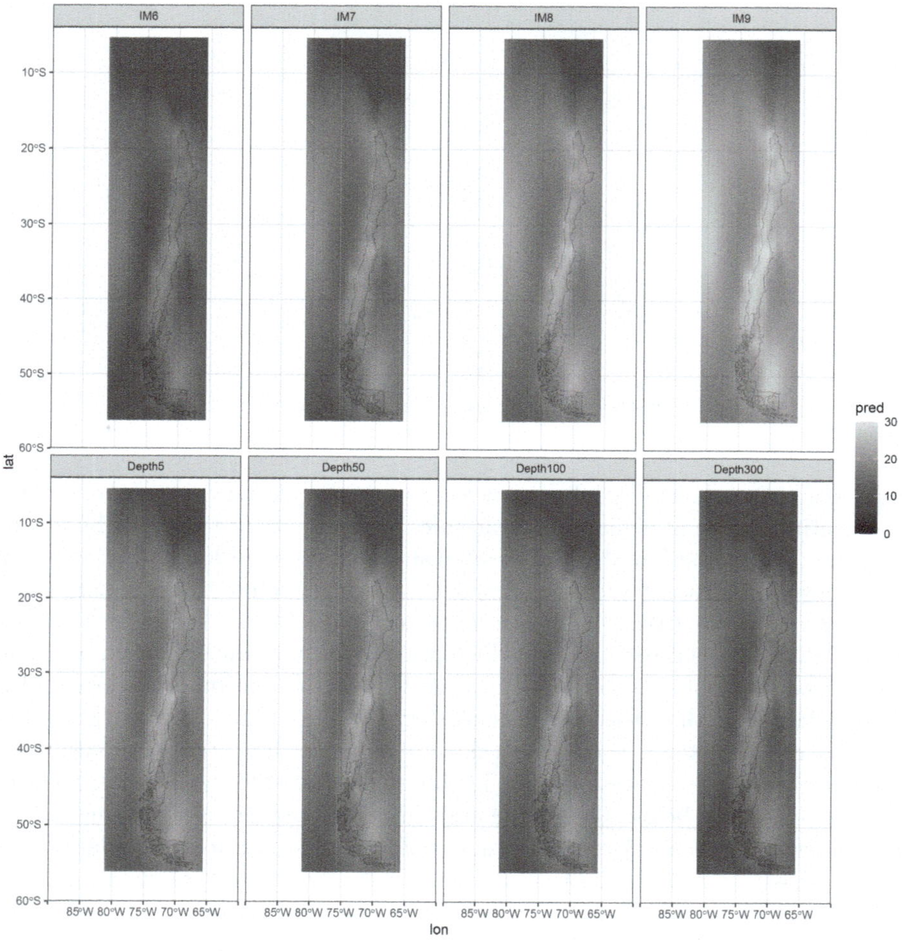

Fig. 2. Spatial prediction of logarithm of loss in Chile using TPS for the following magnitudes: 6, 7, 8, and 9 (at a fixed depth of 50 km) and for the following depths: 5 km, 50 km, 100 km and 300 km (at a fixed magnitude of 7.5).

In order to validate the model, the results of residuals of TPS model are shown in Fig. 3. It can be seen that, the model fits very well when there are many losses, while a little worse when there are few losses. This is in line with the objective of the insurance companies as they want to predict earthquakes more accurately, particularly in terms of anticipating high losses.

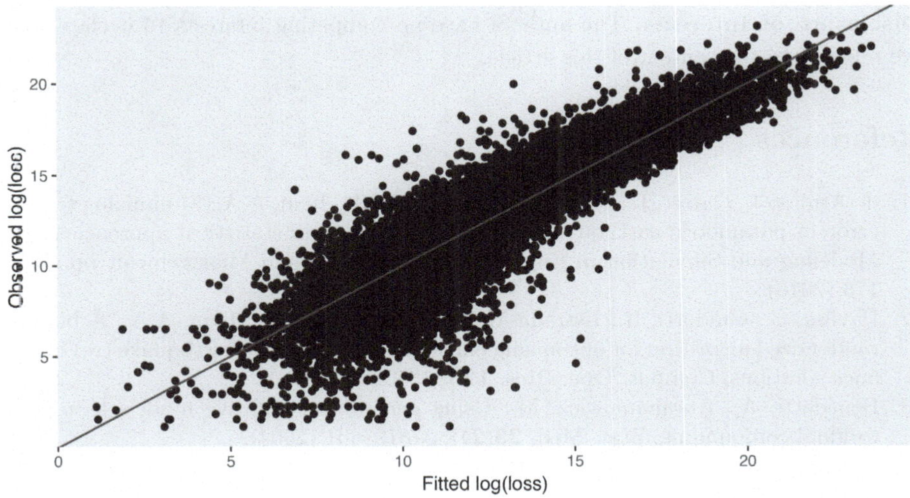

Fig. 3. Observed values versus fitted values.

4 Conclusion and Future Work

Predicting losses caused by earthquakes using statistical techniques is essential for effective disaster planning and response. These statistical methods enable governments and insurers to anticipate the economic and social impact of earthquakes, allowing for proper preparation and efficient resource allocation. They rely on historical seismic event analysis, probabilistic risk modeling, and geospatial data to accurately estimate potential losses in affected areas. The ability to predict these losses helps mitigate the impact of earthquakes, facilitating a swift and effective response to protect lives and aid in the recovery of affected communities.

In order to predict values of a variable of interest between known spatial points, the Univariate kriging methods including ordinary kriging and universal kriging spatial interpolation methods are the most popular and widely used [15]. Despite these stochastic interpolation techniques, there are others such as TPS which is deterministic. The latter provides certain advantages such as its operational simplicity, that is, it does not require more effort and experience. Other important add value is that complex predictor-response relationships can be perfectly visualized, adding heat maps if necessary. For this reason, this paper propose a spatial interpolation model with covariates using TPS in order to predict the caused by earthquakes in Chile. The TPS model is a good model for predicting losses from the Chile earthquake. The model fits very well when there are many losses which fulfills the insurance company's objective. As a future line of work, it is proposed to carry out a stratified sampling in order to obtain more high-magnitude earthquakes, since they are less frequent but produce more losses. In turn, adjust the model to better predict low losses from earthquakes.

Disclosure of Interests. The authors have no competing interests to declare that are relevant to the content of this article.

References

1. de Armas, J., Calvet, L., Franco, G., Lopeman, M., Juan, Á.A.: Minimizing trigger error in parametric earthquake catastrophe bonds via statistical approaches. In: Modeling and Simulation in Engineering, Economics and Management, pp. 167–175 (2016)
2. Bayliss, C., Guidotti, R., Estrada-Moreno, A., Franco, G., Juan, A.A.: A biased-randomized algorithm for optimizing efficiency in parametric earthquake (re) insurance solutions. Comput. Oper. Res. **123**, 105033 (2020)
3. Benedetti, A., Abrahamowicz, M.: Using generalized additive models to reduce residual confounding. Stat. Med. **23**(24), 3781–3801 (2004)
4. Bookstein, F.L.: Principal warps: thin-plate splines and the decomposition of deformations. IEEE Trans. Pattern Anal. Mach. Intell. **11**(6), 567–585 (1989)
5. Chen, W., Zhang, L.: An automated machine learning approach for earthquake casualty rate and economic loss prediction. Reliab. Eng. Syst. Saf. **225**, 108645 (2022)
6. Cred, U.: Human Cost of Disasters. An Overview of the last 20 years: 2000–2019. Cred, UNISDR, Geneva (2020). https://www.emdat.be/publications. Accessed 07 Mar 2024
7. Franco, G.: Minimization of trigger error in cat-in-a-box parametric earthquake catastrophe bonds with an application to Costa Rica. Earthq. Spectra **26**, 983–998 (2010)
8. Franco, G.: Construction of customized payment tables for cat-in-a-box earthquake triggers as a basis risk reduction device. In: Proceedings of the International Conference on Structural Safety and Reliability (ICOSSAR), pp. 5455–5462 (2013)
9. Franco, G.: Earthquake mitigation strategies through insurance. Encycl. Earthq. Eng. 1–18 (2014)
10. Goda, K., Wenzel, F., Daniell, J., Beer, M., Kougioumtzoglou, I., Patelli, E.: Insurance and reinsurance models for earthquake. Encycl. Earthq. Eng. 1184–1206 (2015)
11. Grossi, P., Kunreuther, H., Windeler, D.: An introduction to catastrophe models and insurance. In: Catastrophe Modeling: A New Approach to Managing Risk, pp. 23–42. Springer (2005)
12. Keller, W., Borkowski, A.: Thin plate spline interpolation. J. Geodesy **93**, 1251–1269 (2019)
13. Kong, Q., Trugman, D.T., Ross, Z.E., Bianco, M.J., Meade, B.J., Gerstoft, P.: Machine learning in seismology: turning data into insights. Seismol. Res. Lett. **90**(1), 3–14 (2019)
14. Lamina, C., Sturm, G., Kollerits, B., Kronenberg, F.: Visualizing interaction effects: a proposal for presentation and interpretation. J. Clin. Epidemiol. **65**(8), 855–862 (2012)
15. Meng, Q., Liu, Z., Borders, B.E.: Assessment of regression kriging for spatial interpolation-comparisons of seven GIS interpolation methods. Cartogr. Geogr. Inf. Sci. **40**(1), 28–39 (2013)
16. Nychka, D, Furrer, R , Paige, J, Sain, S: fields: Tools for spatial data (2021). https://github.com/dnychka/fieldsRPackage. R package version 15.2

17. R Core Team: R: A Language and Environment for Statistical Computing. R Foundation for Statistical Computing, Vienna, Austria (2023). https://www.R-project.org
18. Wald, D.J., Franco, G.: Financial decision-making based on near-real-time earthquake information. In: 16th World Conference on Earthquake Engineering, pp. 1–13. Santiago de Chile, Chile (2017)
19. Wood, S.N.: Thin plate regression splines. J. R. Stat. Soc. Ser. B Stat Methodol. **65**(1), 95–114 (2003)

Engineering

The Trigger Arc TSP: Optimise the Picking Process in Warehouses with Compactable Storage Systems

Carmine Cerrone[1(✉)], Maria Truvolo[1], Raffaele Dragone[1], and Raffaele Battaglia[2]

[1] Department of Economics and Business Studies, University of Genoa, Genoa, Italy
carmine.cerrone@unige.it, {maria.truvolo,raffaele.dragone}@edu.unige.it
[2] FourClicks S.r.l. Innovative Startup, Torre Del Greco, Italy
raffaele.battaglia@fourclicks.it

Abstract. This work introduces the Trigger Arc Travelling Salesman Problem (TATSP), a variant of the classical TSP that models dynamic arc costs influenced by the solution's path. It is presented a real-world application that can be modelled through the TATSP, aiming to optimise operations in warehouses with compactable storage systems. An Integer Linear Programming model (ILP) is described and implemented. Finally, the experimental results obtained from a set of generated instances are discussed.

Keywords: Trigger Arc · Warehouse Storage · Travelling Salesman Problem

1 Introduction

In this work, we introduce the Trigger Arc Travelling Salesman Problem (TATSP), a new variant of the TSP, where the cost of arcs can vary in relation to other arcs traversed by the tour. Specifically, for each arc a_i, its cost $c(a_i)$ may change if, in the part of the tour preceding a_i, an arc a_j related to the arc a_i has been crossed. By defining this new problem, we aim to enable the modelling of real-world scenarios where the traversal of an arc by a vehicle can lead to modifications in the structure of the road graph. In recent years, there has been a surge of interest in tackling variations of classical optimisation problems, particularly those related to graph traversal with dynamic attributes. While traditional formulations of problems like the Hamiltonian Cycle [1] and Traveling Salesperson Problem [2] have focused on static or time-dependent graph structures [3], emerging applications in fields like logistics [4] demand more sophisticated approaches. These applications introduce complexities where edge costs not only vary with time but also dynamically change based on traversal history or system state. For example, Fontaine et al. [5] discuss temporal graphs in which edge availability is time-dependent, reflecting scenarios in which the presence of edges may change

A. A. Juan et al. (Eds.): DSA ISC 2024, LNCS 14778, pp. 279–289, 2025.
https://doi.org/10.1007/978-3-031-78238-1_26

over time. Carmesin et al. [6] explore variants, where edges become deleted or untraversable based on the vertices visited, adding another layer of complexity to the traversal problem. Zaki et al. [7] highlight scenarios in which edge costs dynamically alter not only with time, but also with the traversal history or state of the system, reflecting the evolving nature of real-world optimisation challenges. Bossek et al. [8] contribute by studying a variant of the TSP known as the node weight dependent TSP (W-TSP). In the W-TSP, additional weights on the nodes affect the cost of a tour, with the cost increasing according to the weight of the nodes already visited. In this work, Kovács et al. [9] analyse the optimal route constructed from edges that have good fitness value. The fitness value of an edge in the proposed algorithm depends on the quality of these container routes, suggesting that the algorithm evaluates and prioritises routes based on how effectively they facilitate container movement within the logistic system. Cheikhrouhou & Khoufi [10] gave a comprehensive survey on several applications for Multiple Traveling Salesman Problem (MTSP). MTSP has been adopted in different real-life applications to obtain optimized multiple vehicle routes, focusing on mobile robots and unmanned aerial vehicles. In Ambrosino and Cerrone [11] is presented a problem in which the arcs are interconnected through the presence of both positive and negative costs that compensate each other to achieve solutions as close to zero as possible. Di Placido et al. [12] introduce a variant of the TSP where the cost of the arcs depends on the point at which the route covers a target, additionally associating a reward in relation to how closely the route approaches each node. In this paper, we study the TATSP a new variant of the TSP, in which arc costs are intrinsically linked to arcs already visited. By exploring the dynamics of graph traversal, we aim to address real-world scenarios in which the cost of an arc depends on the arcs traversed so far. This work is structured as follows: Sect. 2 provides a formal description of the problem. Section 2.1 proposes a mathematical model capable of depicting the problem. Section 2.2 reports the scenario of a real problem that can be modelled through the TATSP. Section 3 presents the experimental results obtained on some test graphs. Finally, Sect. 4 contains the conclusions and future developments of this work.

2 Problem Definition

The Trigger Arc Travelling Salesman Problem is defined as the problem of identifying in a graph G a Hamiltonian cycle of minimum cost considering relations defined between arcs. In other words, we might say that for each arc a in the graph, relations with other arcs can be defined, which act as triggers. If the trigger arc is traversed, the cost of the arc a is reset. For this reason, only the last trigger arc encountered before reaching the arc a is considered. The key concept in this problem is the definition of a trigger arc. If the relationship (t, a) is defined for arcs t and a, then arc t will act as a trigger for arc a, meaning that traversing arc t will result in the assignment of a new cost to arc a if a has not yet been traversed. More formally, consider a directed graph $G = (N, A)$ with weights on the arcs. The node $0 \in N$ is designated as the starting node(depot). Let $c(a) \in \mathbb{R}^+ \ \forall a \in A$ be defined as the cost of traversing the arc a. For each arc $a = (h, k)$, a set of relations $R_a = \{(a_1, a)|a_1 \in A\}$ is associated. Finally, let $c(r) \in \mathbb{R}^+ \ \forall r \in R_a$, be the traversal cost of the arc a if the relation r is active. Let $T = (a_1, a_2, a_3, \ldots, a_{|N|})$ be the ordered sequence of arcs starting at node 0 representing a Hamiltonian cycle in G. The relation $r = (a_i, a_j)$ is active if and only if the arcs $a_i, a_j \in T$, and the arc a_i precedes the arc a_j in T and there is no active relation $r_1 = (a_{\hat{i}}, a_j) \in R_{a_j}$ such that a_i precedes $a_{\hat{i}}$ in T. It follows that for each arc a, at most one relation can be active. As a result, the traversal cost of the arc $a = (h, k)$ will be equal to $c(a)$ if there are no active relations in R_a or $c(r)$ if r is the only active relation in R_a. In the next Section, the graph G, the cost function $c()$, and the set of relationships R will be formally defined.

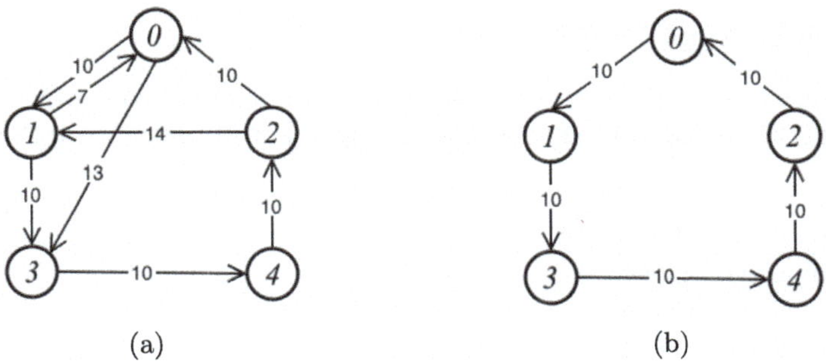

Fig. 1. TSP Example. Subfigure (a) represents the Input Graph. The subfigure (b) represents the Optimal Solution.

Taking into account the graph in Fig. 1(a) and solving the TSP, the cost of the optimal solution depicted in Fig. 1(b) is 50. Starting from the same graph, we introduce for the TATSP a relationship associated with the arc $(2, 1)$. Specifically, we have $R_{(2,1)} = \{r_1 = ((0, 3), (2, 1))\}$ with $c(r_1) = 4$. This relationship, when active, affects the weight of the arc $(2, 1)$, reducing it from 14 to 4. Consequently, we can identify a new optimal solution, represented by the tour in Fig. 2(a) with an objective function value of 44. The relationship r_1 is active in this solution because the trigger arc $(0, 3)$ is traversed before the arc $(2, 1)$. As a further example, associating another relationship r_2 with the arc $(2, 1)$, we get $R_{(2,1)} = \{r_1 = ((0, 3), (2, 1)), \ r_2 = ((4, 2), (2, 1))\}$ with costs $c(r_1) = 4$ and $c(r_2) = 7$. This results in an increased tour cost, since the cost of the arc $(2, 1)$ depends on the last trigger arc traversed by the solution before reaching $(2, 1)$. Specifically, the arc $(4, 2)$ is the last traversed before $(2, 1)$, affecting the cost of the acr to be set at 7, thus obtaining the solution depicted in Fig. 2(b) with a value of 47.

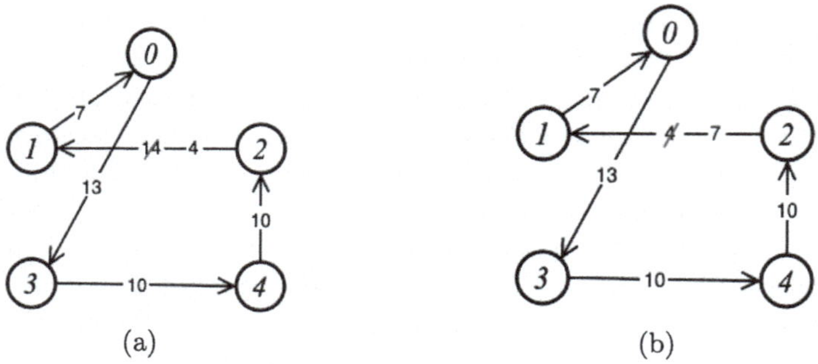

Fig. 2. TATSP Example. Solution (a): $R_{(2,1)} = \{r_1 = ((0,3),(2,1))\}$ with $c(r_1) = 4$. Solution (b): $R_{(2,1)} = \{r_1 = ((0,3),(2,1)),\ r_2 = ((4,2),(2,1))\}$ with $c(r_2) = 7$

2.1 Mathematical Model

This Section presents the Integer Linear Programming model capable of describing the TATSP. To formalise the problem, it is essential to define all the required inputs.

- $G = (N, A)$: directed graph.
- N: set of nodes.
- A: set of directed arcs.
- $c(a) \in \mathbb{R}$, $\forall a \in A$, is the cost of arc a.
- $R_a = (a_1, a) \in A \times \{a\}$, is the set of relationships associated with the arc a.
- $R = \cup_{a \in A} R_a$, is the set of all relationships.
- $c(r) \in \mathbb{R}$, $\forall r \in R$, represents the cost associated with the relationship r (if active).

Our Integer Linear Programming model is based on the following decision variables:

$$x_a = \begin{cases} 1, & \text{if arc } a \text{ is in the tour} \\ 0, & \text{otherwise} \end{cases} \qquad y_r = \begin{cases} 1, & \text{if the relationship } r \text{ is active} \\ 0, & \text{otherwise} \end{cases}$$

$$y_a = \begin{cases} 1, & \text{if no relationship is active in} R_a \\ & \text{and } x_a = 1 \\ 0, & \text{otherwise} \end{cases} \qquad \hat{y}_r = \begin{cases} 0 \text{ or } 1, & \text{if for } r = (a_1, a), \text{arc } a \\ & \text{precedes arc } a_1 \text{in tour} \\ 0, & \text{otherwise} \end{cases}$$

In the proposed mathematical model, the starting node of the tour is, by definition, node 0. The binary variables x_a and $y_a = y_{ij}$ are defined for every arc $a = (i, j) \in A$. Specifically, the variables x_a are used to represent the solution, indicating which arcs belong to the tour. The variables y_a inform us when there is no active relationship for arc a in the solution. This implies that for the arc, its base cost $c(a)$ must be paid. Variables y_r and $\hat{y}_r$ are defined for each relation $r = (a_1, a) \in R$. Specifically, y_r indicates whether the relation r is active, while $\hat{y}_r$ helps to understand when, despite the fact that both arcs of the relation are used in the solution, the trigger arc a_1 is visited after the base arc a, hence the relation r should not be considered active. Then, the problem can be formulated as follows:

$$MIN\ z = \sum_{r \in R} c(r) y_r + \sum_{a \in A} c(a) y_a \tag{1}$$

Subject to the constraints:

$$\sum_{(i,j) \in A} x_{ij} = |N| \tag{2}$$

$$u_i + 1 \leq u_j + M(1 - x_{ij}) \qquad \forall (i,j) \in A / j \neq 0 \tag{3}$$

$$\sum_{i=1, i \neq j} x_{ij} = 1 \qquad \forall\, j \in N \tag{4}$$

$$\sum_{j=1, j \neq i} x_{ij} = 1 \qquad \forall\, i \in N \tag{5}$$

$$y_a + \sum_{r \in R_a} y_r = x_{ij} \qquad \forall\, a = (i,j) \in A \tag{6}$$

$$y_r \leq x_{ij} \qquad \forall\, r = ((i,j),(h,k)) \in R \tag{7}$$

$$u_i + 1 \leq u_h + M(1 - y_r) \qquad \forall\, r = ((i,j),(h,k)) \in R \tag{8}$$

$$u_h + 1 \leq u_i + M(1 - \hat{y}_r) \qquad \forall\, r = ((i,j),(h,k)) \in R \tag{9}$$

$$x_{ij} \leq (1 - x_{hk}) + (1 - y_{hk}) + \hat{y}_r \qquad \forall\, r = ((i,j),(h,k)) \in R \tag{10}$$

$$u_{\hat{i}} - M(\hat{y}_{r_2}) \leq u_i + M(2 - y_{r_1} - x_{\hat{i}\hat{j}}) - 1 \qquad \begin{matrix} \forall\, r_1 = ((i,j),(h,k)) \in R \\ \forall\, r_2 = ((\hat{i},\hat{j}),(h,k)) \in R \end{matrix} \tag{11}$$

$$x_{ij}, y_a \in \{0,1\} \qquad \forall\, a = (i,j) \in A \tag{12}$$

$$y_r, \hat{y}_r \in \{0,1\} \qquad \forall\, r \in R \tag{13}$$

$$u_i \in \{0, |N|\} \qquad \forall\, i \in N \tag{14}$$

The objective function (1) is defined as minimising the total costs of the arcs, where the costs are composed of the sum of the costs of the activated relationships and the base costs of the arcs when no relationship is active for an arc. The constraint (2) ensures that a number of arcs equal to the number of nodes in the set $|N|$ are selected. The family of constraints (3) ensures that the variables u represent the respective order of visit for each node in the tour. Moreover, these constraints also guarantee the absence of subcycles in the final solution. The constraints (4, 5) ensure that each node has exactly one incoming and one outgoing arc. The constraints (6) defined for each arc $a \in A$ ensure that if the arc is not present in the solution ($x_a = 0$), then neither the basic cost of the arc is paid, imposing ($y_a = 0 \implies y_a c(a) = 0$), nor any cost associated with its relations ($y_r = 0 \implies y_r c(r) = 0$) $\forall r \in R_a$. On the contrary, if the arc is used, $x_a = 1$, then a single cost can be associated with the arc a: either the base cost associated with the variable y_a or one among the costs of the relations $y_r \forall r \in R_a$. The constraints (7) require that for each relation

$r = ((i,j),(h,k)) \in R$, the corresponding variable y_r be active if the tour traverses the trigger arc (i,j), i.e., if $x_{ij} = 1$. Constraints (8) are defined for each relation $r = ((i,j),(h,k))$ and ensure that if the arc (i,j) does not precede the arc (h,k) in the solution, then the variable $y_r = 0$. The constraints (9) ensure that if arc (h,k) does not precede arc (i,j) in the solution, then variable $\hat{y}_r = 0$, implying that variables $\hat{y}_r$ may only assume the value 1 if arc (h,k) precedes arc (i,j) in the solution. The family of constraints (10) ensures that for each relation $r = ((i,j),(h,k))$, the trigger arc (i,j) can only be used if at least one of the following three conditions is satisfied: (i) the arc (h,k) is not used, (ii) the variable y_{hk} associated with the base cost of arc (h,k) is not activated, (iii) if both arcs (i,j) and (h,k) are used in the solution, the arc (i,j) is traversed after the arc (h,k). Finally, the family of constraints (11) defined for each pair of relations $r_1 = ((i,j),(h,k)), r_2 = ((\hat{i},\hat{j}),(h,k))$ differing only by the trigger arc requires that the active relation is the one whose trigger arc is traversed last but before the arc (h,k) in the visit sequence.

2.2 Real Application

This Section presents a real-world application that leads to the definition of the problem related to optimising the picking phase of items within a warehouse equipped with a compactable storage system. A compact storage unit is a logistical solution that features a set of mobile shelving units. These shelving units move along specific rails, as shown in Fig. 3. This system allows the company to optimise the available space but also incurs additional costs, both in terms of the energy consumption needed to move the shelves and in terms of the operational times required for the shelf movements. Thanks to the integration of the compact storage system within a corporate information system, as envisaged by the dictates of Industry 4.0 applications [13], information on the status of the warehouse aisles has been made available. Using this information, received in real-time, it has become possible to accurately estimate the times related to picking operations, also considering the opening and closing times of the aisles. Consequently, the need to model this problem by taking advantage of the new data arose.

The presence of groups of compactable aisles in these warehouses requires, during the item-picking phase, opening the compact storage in the desired aisle, causing the closure of the other connected aisles. This procedure increases the costs associated with accessing the closed aisles. Since operational costs and access times directly depend on the number of aisles moved, accessing a specific aisle can lead to an increase or decrease in the costs of the connected aisles, depending on the new aisle configuration. For example, to access the aisle $e - f$ as illustrated in Fig. 3, it is necessary to move the shelves d and e to the right.

3 Computational Experiments

In this Section, we present the results of our computational tests. The experiments were carried out on an Apple Silicon M1 8-Core 3200 MHz CPU. The

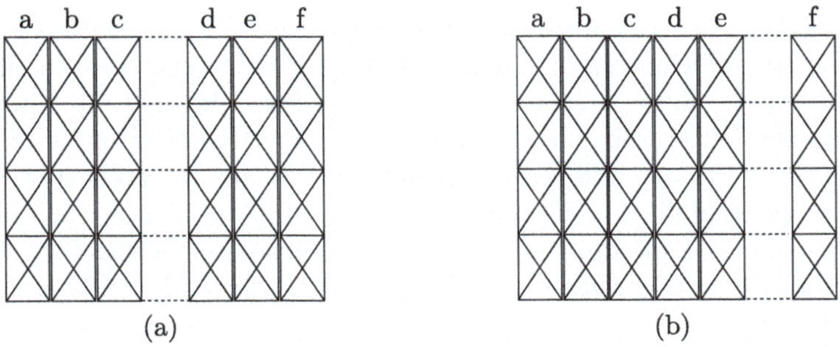

Fig. 3. Compactable storage system.

algorithm was implemented in the JAVA programming language using Cplex 22.11, with the default setting as the MILP solver, and a computation time limit of 1000 s.

3.1 Instances

To test the proposed mathematical model, a random generator is created that is used to produce 180 test instances. For each directed graph $G(N, A)$, nodes were placed within a square Euclidean space with a side of 5 km. The generated graphs were made complete by creating for each pair of nodes $i, j \in N$ their respective directed arcs (i, j) and (j, i). As the base cost for each arc $a = (i, j) \in A$, the Euclidean distance between the nodes is used, $c(a) = \text{EuclideanDistance}(i, j)$. The generation of each relation $r = (a_1, a_2) \in R$ present in the graph was created by randomly selecting pairs of arcs $a_1, a_2 \in A$. The cost of the relation $c(r)$ was generated by multiplying $c(a_2)$ by a random value in the range $[\frac{c(a_2)}{2}, 2c(a_2)]$. The instances were divided into three scenarios (*Balanced*, *Increase*, *Decrease*). For each scenario, 60 instances were created with a number of nodes n equal to $(10, 15, 20, 25)$ and with a number of relations per graph equal to $(n^2, 2n^2, 4n^2, 8n^2, 16n^2)$. For each scenario, with the number of nodes and the number of relations being equal, three different instances were created. All the computational results presented in this work, always report the average value obtained from the three created instances. Scenario *Balanced* is characterised by random costs for relations greater than or less than the base cost of the arc in the range $[\frac{c(a_2)}{2}, 2c(a_2)]$. The scenario *Increase* contains random costs for relations greater than the base cost of the arc in the range $[c(a_2), 2c(a_2)]$. In contrast, the scenario *Decrease* uses random costs for relations less than the base cost of the arc in the range $[\frac{c(a_2)}{2}, c(a_2)]$.

3.2 Results

In this Section, we analyse the computational results obtained for the three scenarios previously defined. To evaluate the impact of relationships on the cost of the tour, for each instance, the cost of the classic TSP executed without considering the relationships was calculated. This value was used in the tables to show the percentage decrease in the route costs. In all tables, where the model was unable to produce the optimal solution within the time limit of 1.000 seconds, the results are highlighted in bold. It can be observed that, with an increase in the number of nodes and the number of relationships, the model fails to finish within the time limit for any of the proposed scenarios. Tables 1 and 2 are related to the scenario *Balanced.* Table 1 shows, for each combination of the number of nodes and the number of relations, two key information: the average value of the objective function and the percentage decrease in the cost of the routes compared to the classic TSP. It is interesting to note that as the number of relations increases, the cost of the solutions tends to decrease. Despite the presence of relations that impose an increase in the cost of the arcs, the model manages to select paths capable of activating relations that result in a total decrease in the cost of the route. It also becomes interesting to observe that the percentages of reduction in the cost of the route are not proportional to the increase in the number of relations. Doubling the number of relations in the instances does not entail a corresponding doubling of the average savings in route costs. Table 2 is essential to verify the computational times associated with the proposed model and the respective MIP gap. For example, instances with 25 nodes containing $16(25^2) = 10,000$ relations produce a MIP gap of 36%.

Table 1. Average costs and % of improvement for the Balanced scenario.

	Average Objective Function					*% TSP Decrease*				
Nodes	(n^2)	$2(n^2)$	$4(n^2)$	$8(n^2)$	$16(n^2)$	(n^2)	$2(n^2)$	$4(n^2)$	$8(n^2)$	$16(n^2)$
10	14.801	14.778	14.243	14.419	14.191	3	3	6	5	7
15	16.874	16.687	16.555	16.036	**16.324**	2	3	3	6	**5**
20	19.308	19.072	19.277	**18.791**	**21.157**	3	4	3	**6**	**−6**
25	20.277	20.184	**20.261**	**20.287**	**24.530**	2	2	**2**	**2**	**−19**

Table 2. Average computational time and % of MIP Gap for the Balanced scenario.

	Average Computational Time					*% MIP Gap*				
Nodes	(n^2)	$2(n^2)$	$4(n^2)$	$8(n^2)$	$16(n^2)$	(n^2)	$2(n^2)$	$4(n^2)$	$8(n^2)$	$16(n^2)$
10	0,2	0,1	0,3	1,2	10,4	0	0	0	0	0
15	0,4	0,8	4,6	39,4	**719,2**	0	0	0	0	**3**
20	16,6	28,9	239,8	**900,5**	**1.000**	0	0	0	**3**	**28**
25	116,2	252,7	**675,5**	**774**	**1.000**	0	0	**5**	**11**	**36**

Tables 3 and 4 are related to the *Decrease* scenario. In this scenario, all relationships impose a decrease in the costs of the arcs. From Table 3, we can observe a significant decrease in the cost of the routes. In particular, for the scenario that contains 10 nodes and $16(10^2) = 1600$ relations, the solution improves by 23% compared to the TSP solution, as opposed to a 7% decrease in the *Balanced* scenario. Table 4 also shows that more instances reach the time limit, and the largest MIP gap reaches 42%.

Table 3. Average costs and % of improvement for the Decrease scenario.

	Average Objective Function					*% TSP Decrease*				
Nodes	(n^2)	2(n^2)	4(n^2)	8(n^2)	16(n^2)	(n^2)	2(n^2)	4(n^2)	8(n^2)	16(n^2)
10	13.949	13.452	13.131	12.475	11.723	8	12	14	18	23
15	16.393	16.203	15.522	14.594	**13.967**	4	5	9	15	**18**
20	19.101	18.436	**18.041**	**18.451**	**18.280**	4	8	**10**	**8**	**8**
25	20.044	**19.636**	**19.319**	**19.988**	**22.781**	3	**5**	**7**	**3**	**−10**

Table 4. Average computational time and % of MIP Gap for the Decrease scenario.

	Average Computational Time					*% MIP Gap*				
Nodes	(n^2)	2(n^2)	4(n^2)	8(n^2)	16(n^2)	(n^2)	2(n^2)	4(n^2)	8(n^2)	16(n^2)
10	0,2	0,1	0,4	1,5	8,6	0	0	0	0	0
15	0,6	3,4	12,7	90,5	**682**	0	0	0	0	**5**
20	29,2	56,9	**707**	**1.000**	**1.000**	0	0	**1**	**17**	**31**
25	185,5	**419,4**	**751,4**	**1.000**	**1.000**	0	**2**	**8**	**21**	**42**

Finally, Tables 5 and 6 relate to the *Increase* scenario. In this scenario, all relations impose an increase in the costs of the arcs. From Table 5, we note that for a number of relations equal to n^2 or $2(n^2)$, there is no real worsening of the solution. This indicates that it is possible to identify alternative solutions that avoid activating the relations. Obviously, as the number of relations in the instances increases, we observe a deterioration of the solution up to 11% in the case of instances with 10 nodes, obviously not considering the instances that reach the time limit. Table 6 also shows that fewer instances reach the time limit, and the largest MIP gap reaches only 17%.

Table 5. Average costs and % of improvement for the Increase scenario.

	Average Objective Function					*% TSP Decrease*				
Nodes	(n^2)	$2(n^2)$	$4(n^2)$	$8(n^2)$	$16(n^2)$	(n^2)	$2(n^2)$	$4(n^2)$	$8(n^2)$	$16(n^2)$
10	15.256	15.266	15.653	15.898	16.820	0	0	−3	−4	−11
15	17.132	17.132	17.381	17.423	18.489	0	0	−1	−2	−8
20	19.984	20.029	20.295	**20.421**	**22.349**	0	0	−2	**−2**	**−12**
25	20.716	**20.717**	**20.919**	**21.226**	**23.922**	0	**0**	**−1**	**−3**	**−16**

Table 6. Average computational time and % of MIP Gap for the Increase scenario.

	Average Computational Time					*% MIP Gap*				
Nodes	(n^2)	$2(n^2)$	$4(n^2)$	$8(n^2)$	$16(n^2)$	(n^2)	$2(n^2)$	$4(n^2)$	$8(n^2)$	$16(n^2)$
10	0,1	0,1	0,3	0,8	5,0	0	0	0	0	0
15	0,3	0,5	3,0	27,1	311,2	0	0	0	0	0
20	26,9	52,3	197	**603,9**	**1.000**	0	0	0	**2**	**15**
25	158,1	**411,6**	**560,8**	**674,9**	**726**	0	**1**	**3**	**6**	**17**

4 Conclusions

In this work, we have defined the Trigger Arc Travelling Salesman Problem (TATSP), a new variant of the classic TSP. In this problem, the cost of each arc present in the solution can vary in relation to the other preceding arcs in the optimal tour. Specifically, using a set R of relations between arcs, it is possible to model real-world scenarios where traversing an arc can imply modifying the costs of the arcs visited subsequently. In addition to defining the problem, an integer linear programming model capable of identifying the optimal solution for small instances was presented. As is evident from the experimental results, the proposed mathematical model can only generate exact solutions for very small instances. The term "small instance" refers to the number of nodes within the graph. When examining instances from the perspective of the number of relationships, it is observed that instances with $n = 25$ and n^2 relationships, although they have 625 relationships, are solved in an average time of 158 seconds. Furthermore, for instances involving 10.000 relationships, although the model does not produce the optimal solution, it manages to produce feasible solutions that are at most 17% worse than the optimal. This analysis suggests that the current model is most effective in scenarios with fewer relationships and can be used in other cases to validate the behaviour of future heuristic approaches on small instances. The most interesting future research directions seem to involve the definition of a new exact approach that reduces the number of constraints and variables of the model through the use of cutting techniques and column generation. Moreover, the implementation of heuristic techniques capable of quickly identifying good solutions to the problem could be interesting and useful.

References

1. Rahman, M.S., Kaykobad, M.: On Hamiltonian cycles and Hamiltonian paths. Inf. Process. Lett. **94**(1), 37–41 (2005)
2. Reinelt, G.: The Traveling Salesman: Computational Solutions for TSP Applications, vol. 840. Springer, Heidelberg (2003)
3. Wang, Y., Yuan, Y., Ma, Y., Wang, G.: Time-dependent graphs: definitions, applications, and algorithms. Data Sci. Eng. **4**, 352–366 (2019)
4. Larni-Fooeik, A., Ghasemi, N., Mohammadi, E.: Insights into the application of the traveling salesman problem to logistics without considering financial risk: a bibliometric study. Manage. Sci. Lett. **14**(3), 189–200 (2024)
5. Fontaine, R., Dibangoye, J., Solnon, C.: Exact and anytime approach for solving the time dependent traveling salesman problem with time windows. Eur. J. Oper. Res. (2023)
6. Carmesin, S., Woller, D., Parker, D., Kulich, M., Mansouri, M.: The Hamiltonian cycle and travelling salesperson problems with traversal-dependent edge deletion. J. Comput. Sci. **74**, 102156 (2023)
7. Zaki, A., Attia, M., Hegazy, D., Amin, S.: Comprehensive survey on dynamic graph models. Int. J. Adv. Comput. Sci. Appl. **7**(2) (2016)
8. Bossek, J., Casel, K., Kerschke, P., Neumann, F.: The node weight dependent traveling salesperson problem: approximation algorithms and randomized search heuristics. In: Proceedings of the 2020 Genetic and Evolutionary Computation Conference, pp. 1286–1294 (2020)
9. Kovács, L., Agárdi, A., Bányai, T.: Fitness landscape analysis and edge weighting-based optimization of vehicle routing problems. Processes **8**(11), 1363 (2020)
10. Cheikhrouhou, O., Khoufi, I.: A comprehensive survey on the multiple traveling salesman problem: applications, approaches and taxonomy. Comput. Sci. Rev. **40**, 100369 (2021)
11. Ambrosino, D., Cerrone, C.: The cost-balanced path problem: a mathematical formulation and complexity analysis. Mathematics **10**(5), 804 (2022)
12. Di Placido, A., Archetti, C., Cerrone, C., Golden, B.: The generalized close enough traveling salesman problem. Eur. J. Oper. Res. **310**(3), 974–991 (2023)
13. Lasi, H., Fettke, P., Kemper, H.G., Feld, T., Hoffmann, M.: Industry 4.0. Bus. Inf. Syst. Eng. **6**, 239–242 (2014)

Route Clustering Algorithms for Procurement Transport

Juan Moreno(✉), Josefa Mula, and Raul Poler

Universitat Politècnica de València, Research Centre on Production Management and Engineering (CIGIP), Alcoy, Alicante, Spain
{jmoreno,fmula,rpoler}@cigip.upv.es

Abstract. Efficiently distributing goods and materials is a vital aspect of logistics and supply chain management. In this paper, we develop a clustering mathematical programming model to consolidate the transportation logistics of an established set of routes. Each route has a determined weight and volume of transported goods per week. In this context, we study the impact of consolidating routes to procure the manufacturing facilities by leveraging the weight and volume of the established routes. To assist during the decision process, the model has different parameters that the decision maker can adjust in order to better model the supply chain under study. Furthermore, a branch and price algorithm is developed and tested to solve the real-world instances inspired in the automotive sector with up to 900 suppliers and 50 manufacturing facilities. Thus the decision tool has been validated in a first-tier supplier of the automotive industry.

Keywords: Edge clustering · Supply chain · Procurement Transport · Branch and Price

1 Introduction

Maximising economic potential and social benefits, and minimising environmental impacts are critical objectives in the efficient utilisation of productive and logistics resources in manufacturing supply chains. Procurement, as defined by [4], is the process by which goods and services are acquired in a supply chain, whose main aim is to source raw materials, components, or finished products for a factory or distribution centre from suppliers. Note that while the supplier "typically" assumes the responsibility for transport planning, there are cases, such as the automotive industry, in which manufacturers actively participate in making transport decisions due to the complex interactions between numerous manufacturing facilities and suppliers [10]. This indicates that transport decisions during procurement should be a collaborative effort made among all supply chain parties to ensure that the sourcing process is efficiently managed. Here we focus on optimising the tactical decisions involved in procurement transport planning by underscoring the importance of strategic collaboration by aggregating the transport resources of different routes to enhance supply chain efficiency.

A. A. Juan et al. (Eds.): DSA ISC 2024, LNCS 14778, pp. 290–305, 2025.
https://doi.org/10.1007/978-3-031-78238-1_27

According to [8], automatic algorithm-based procurement transport planning is preferable to manual planning based on planners' judgement and experience. The latter can be myopic and generate suboptimal plans. Such insight should be amplified when considering the optimisation of procurement transport problems when there are many nodes in the supply chain, where the coordination of the different stakeholders can reduce logistics, transportation costs and the environmental impact. One possible way to coordinate these logistic processes is to cluster the routes between suppliers and manufacturing facilities so that the routes in the same cluster share transportation resources. This is of outmost importance when routes are long and constrained by weight and volume because we can mix the required goods inside trucks to minimise the number of trucks. Hence this article aims to bridge this research gap by providing a novel mixed integer linear mathematical programming (MILP) model that clusters the routes between the suppliers and manufacturing facilities to minimise the procurement transport cost along the supply chain. Additionally, as a solution approach, branch and price algorithms is developed. Branch and price combine two methods: branch-and-bound and column generation (CG). It is often applied to problems like vehicle routing, crew scheduling, and cutting stock problems. CG is a technique to solve large MILP models with a determined decomposable structure. It was introduced in [11] to address a multi-commodity network flow problem. Then, [5] adapted it to LP problems with a decomposable structure. Furthermore, [12] demonstrated the efficiency of the CG technique to the cutting stock problem. For a detailed explanation about the inner workings of CG and branch and price techniques we refer readers to [14] and [6]. Here, a solution approach based on a branch and price algorithm using column generation is developed to solve real-world instances with up to 900 suppliers, 50 manufacturing facilities and 1,800 routes.

The main contributions of this paper are: (i) formulating a novel MILP model for the procurement transport supply chain route clustering, dubbed as SCRC, that integrates volume and weight distribution; (ii) developing a branch and price algorithm to find good quality solutions in a reasonable amount of time; (iii) validating and testing the SCRC model with a case study at a first-tier supplier of the automotive industrial sector. Readers are also referred to the literature reviews of [15] and [9], which focus on the integration of inventory decisions and transport decisions.

The rest of the paper is organised as follows. Section 2 presents the literature review. Section 3 describes the addressed problem. Section 4 introduces the mathematical formulation of the SCRC. Section 5 describes the branch and price procedure. Section 6 includes the computational experiments. Section 7 explains how the SCRC can be used as a decision support tool. Section 8 concludes the research and analyses future research lines.

2 Literature Review

Transport processes are, extensively, explored in academic research due to their critical role in supply chains. As highlighted by [13], transportation is essential for transferring raw materials and final products across different supply chain stages, including suppliers, warehouses, manufacturers, and retailers. [10] categorise transport processes in supply chains into two main types: the transport of materials from suppliers to manufacturing facilities, known as procurement transport; the distribution of finished goods from factories to consumers. This discussion focuses on addressing issues related to procurement transport.

Procurement transport problems can be analysed with two different dimensions: the configuration of the supply chain and the used transportation method. When considering supply chain configurations, [15] distinguish among four different scenarios: (i) from a single source to a single destination (SS); (ii) from a single source to multiple destinations (SM); (iii) from multiple sources to a single destination (MS); (iv) from multiple sources to multiple destinations (MM). The "source" reflects the number of considered supplier locations, whereas the "destination" indicates the number of manufacturing sites receiving supplies. Additionally, [8] highlight various transportation methods, including full-truckload (FTL), less-than-truckload (LTL) and milk run. Here the aggregation of transportation resources in the scientific literature is also a main concern.

[16] study the concept of delivery profiles, which determine the frequency at which deliveries are made and are set at a fixed rate. In this context, FTL transport occurs when the volume of goods exceeds the capacity of one vehicle and, therefore, direct shipment is a viable option. LTL is used when the volume of goods does not fill an entire truck. In this case, goods are shipped to a consolidation centre before being transported to their final destination. The proposed model is designed for the MS supply chain configuration and routes are consolidated by LTL. In [17], the proposed model considers demand uncertainty.

[7] introduce an MILP model to integrate production, procurement and transport planning decisions into manufacturing systems. The proposed model goes beyond traditional material requirement planning systems by considering the MS supply chain configuration in a unified optimisation framework. It takes into account the capacities of material and production resources, as well as procurement transport planning decisions. Routes can be consolidated using three modes of transport: FTL, LTL and milk run.

[3] address integrating production, inventory and procurement transport into a manufacturing context by defining this problem as an assembly routing problem (ARP). It introduces an MILP model for the ARP and proposes a three-phase decomposition matheuristic as a solution approach. Routes are consolidated using a milk-run scheme to procure products.

[1] study the optimisation of inbound transportation of an automotive factory from first-tier suppliers. The MS study uses FTL, LTL and courier express service (CES) as transport methods. Routes are consolidated using LTL as a transport method.

[2] examine a multiperiod tactical transportation plan by focusing on the inbound logistics for an automobile manufacturer. The study explores direct, indirect, and milk-run transportation strategies, each characterised by its unique cost structure. The objective is to determine the most efficient routes and strate-

Table 1. Comparison of procurement transport optimisation models

Article	SC configuration	Decision level	Transportation mode	Transportation capacity constraint	Volume and weight optimisation	Source aggregation	Destination aggregation	Modelling framework
[16]	MS	T	FTL|LTL	V		X		MILP
[17]	MS	O	FTL|LTL	V		X		MILP
[7]	MS	T	FTL|LTL|MR	V		X		MILP
[3]	MS	O|T	MR	V		X		MILP
[1]	MS	O	MR|FTL|LTL|CES	V|W	X	X		MILP
[2]	MR|FTL|LTL	T	MR	V				DA
SCEC	MM	T	FTL|LTL	V|W	X	X	X	MILP

Abbreviations. T (tactical), **O** Operational; **FTL** (full-truck load), **LTL** (less-than-truck load), **MR** (milk-run), **CES** (courier express service); **V** (volumetric), **W** (weight) **UH** Unit Holding; **DA** Decomposition Approach;

gies to optimise the transportation flow. The concept of clustering routes is introduced into an MS supply chain configuration.

Table 1 compares the supply chain configuration, transport modelling, and transport aggregation between the SCRC model proposal and those currently in the literature.

The main novelty of our proposal is the joint transport MM optimisation, i.e., between multiple sources and multiple destinations, while optimising the volume and weight mix using the FTL and LTL transport methods.

3 Problem Description

The objective of the proposed model is to reduce transportation costs in an MM supply chain by consolidating transport from different suppliers to various manufacturing plants from a tactical decision level. Figure 1 illustrates the considered supply chain, where circles are supply chain nodes and edges represent the established routes between them. Therefore, a node can be both a manufacturing facility and a supplier for other nodes, depending if the node has incoming and outcoming edges. Each route has a preset transport type, number of used trucks and volume and weight transported per week to procure the different manufacturing facilities. Depending on these parameters and route length, there is also a pre-established cost per week.

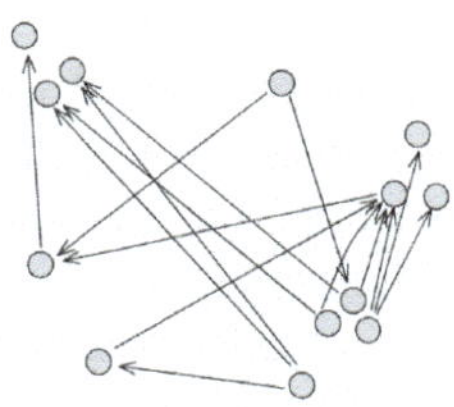

Fig. 1. Routes of the supply chain

If a route is in low demand, i.e., the number of trucks needed to transport goods per week is less than one, then the route is normally delegated to a logistics company with higher shipping costs. Otherwise, the employed transport type is FTL. Efficiency rarely reaches 100% due to varying demands or the need for urgent sourcing, which necessitate more trucks or sending partially filled trucks. Moreover, the minimum required number of trucks is determined by the route's total volume and weight, and the truck's maximum capacity in both volume and weight terms. Jointly managing the clustered routes offers the possibility of reducing the current costs. On the one hand, we eliminate the cost of delegating the routes in low demand because we consolidate all the routes in the cluster. On the other hand, we can reduce the number of trucks needed by mixing the routes dominated by weight and volume.

To optimise transportation decisions, we propose the concept of clusters by grouping routes for joint transportation management purposes. A cluster comprises both the receiving and sourcing nodes. We set both the receiving and sourcing central hubs nodes of the supply chain. The transport in the cluster can be separated into three stages: sourcing to the source central hub; transporting goods from the source central hub to the receiving central hub; distributing goods from the receiving central hub to the receiving nodes. The transportation cost associated with this cluster is derived in these three stages. Figure 2 illustrates the clustering process of a set of routes.

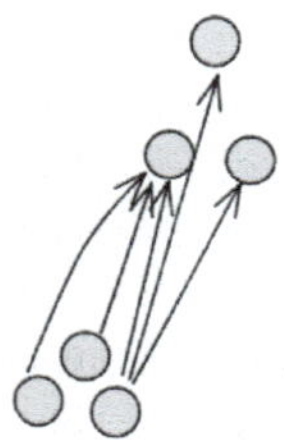

(a) Routes

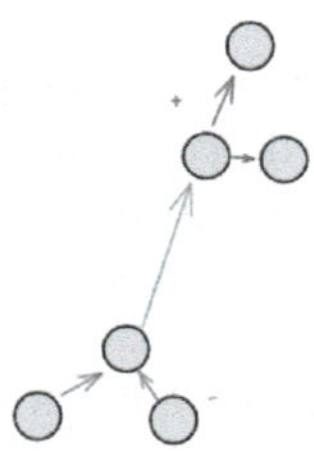

(b) Cluster

Fig. 2. Clustered routes example

Additionally, each route has a set of routes to which it can be clustered. This is because some routes carry materials that are incompatible for joint transport. Furthermore, the beginning or end of routes could be in different countries, and routes are not mixed because of customs.

The objective of the proposed optimisation method is to find new clusters to minimise the transportation costs. Not all routes need to be clustered. Therefore, a solution is determined by the sets of clusters and non-clustered routes. A possible solution for the supply chain example is illustrated in Fig. 3. Note that not all routes have been clustered.

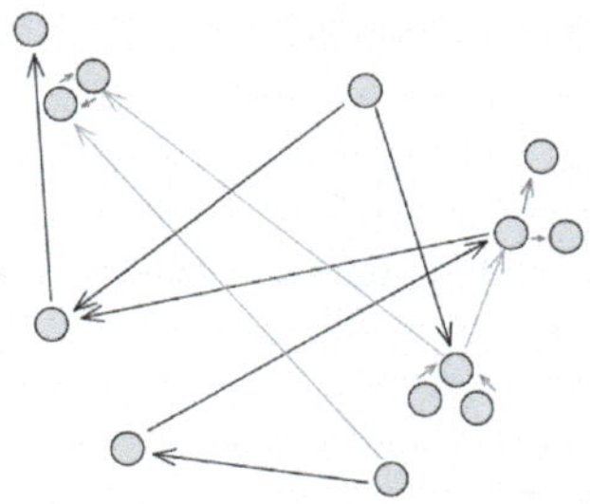

Fig. 3. Clustered supply chain

The problem can be stated as follows:
Initial information:

- Transport details of routes (weight, volume, current cost).
- Distance among all the nodes of the supply chain.
- Set of edges that can be grouped together.
- Cost per kilometre.
- Maximum weight and volume allowed per truck.
- Parameters concerning consolidation efficiency.

To determine:

- Set of clusters for transport consolidation.
- Set of unaltered routes.

With constraints:

- Adhering to the association graph.

Objective:

- Minimise transportation costs.

4 Mathematical Formulation

In this section, we describe the developed SCRC model. The nomenclature is provided in Table 2.

The objective is to minimise transportation costs, i.e., the cost of the non-altered routes and the cost of the different clusters.

$$\min z = \sum_{e \in E} Y_e c_e + \sum_{c} \gamma K_c + \gamma^+ K_c^- + \gamma^- K_c \tag{1}$$

Table 2. Nomenclature of SCRC

Indices	
V	Set of vertices of the graph (v =1, 2,·,V)
E	Set of edges of the graphs that represent the current routes (e =1, 2,·,E)
C	Set of edge cluster (e =1, 2,·,C)
Set of Indices	
$\Gamma(e)$	Set of edges that can be grouped with e
$\delta^+(e)$	Receiving node of the edge
$\delta^-(e)$	Source node of the edge
Parameters	
v_e	Volume transported on route e
w_e	Weight transported on route e
d_{ij}	Distance between nodes $i \in V$ and $j \in V$
c_e	Initial cost of route on edge e
max_v	Maximum volume carried by a truck
max_w	Maximum weight carried by a truck
α	Efficiency of transport where volume dominates the route
β	Efficiency of transport where weight dominates the route
γ	Efficiency of transport between hubs
γ^+	Efficiency of transport between receiving hubs and receiving nodes
γ^-	Efficiency of transport between sourcing nodes and sourcing hub
Decision Variables	
Y_e	Binary variable indicating if e is not clustered
X^c	Binary variable indicating if cluster c is used
X_{ce}	Binary variable indicating if edge e belongs to cluster c
X^-_{ce}	Binary variable indicating if node $\delta^-(e)$ is the sourcing hub for the sourcing nodes in cluster c
X^+_{ce}	Binary variable indicating if node $\delta^+(e)$ is the receiving hub for the receiving nodes
$X^e_{ce'e''}$	Binary variable indicating if the edge e is in c and the source node of e' is the sourcing hub for outgoing goods and receiving node of e'' if the is the receiving hub for incoming goods
D_{ce}	Distance between the sourcing and receiving hub of cluster c if e is in the cluster, 0 if not.
D^-_{ce}	Distance from $\delta^-(e)$ to the sourcing hub if e is in cluster c
D^+_{ce}	Distance from $\delta^+(e)$ to the receiving hub if e is in cluster c
K_c	Transportation cost between sourcing hub and receiving hub at cluster c
K^-_c	Transportation cost between sourcing nodes and sourcing hub at cluster c
K^+_c	Transportation cost between receiving hub and receiving nodes at cluster c

subject to:

$$Y_e + \sum_c X_{ce} = 1 \quad \forall e \in E \tag{2}$$

$$\sum_{e \in E} X_{ce} \leq M X^c \quad \forall c \in C \tag{3}$$

$$\sum_{e \in E} X_{ce} \geq 2X^c \quad \forall c \in C \tag{4}$$

$$\sum_{e \in E} X_{ce}^{-} = X^c \quad \forall c \in C \tag{5}$$

$$\sum_{e \in E} X_{ce}^{+} = X^c \quad \forall c \in C \tag{6}$$

$$\sum_{e' \in \Gamma(e)} X_{ce'}^{+} = X_{ce} \quad \forall c \in C \tag{7}$$

$$\sum_{e' \in \Gamma(e)} X_{ce'}^{-} = X_{ce} \quad \forall c \in C \tag{8}$$

$$\sum_{e', e'' \in \Gamma(e)} X_{ce'e''}^{e} = X_{ce} \quad \forall c \in C, e \in E \tag{9}$$

$$X_{ce'e''}^{e} + 2(1 - X_{ce}) + 1 \geq X_{ce'}^{-} + X_{ce''}^{+} \quad \forall c \in C, e \in E, e', e'' \in \Gamma(e) \tag{10}$$

$$\sum_{e' \in (E' \Gamma(e))} X_{ce'} \leq M(1 - X_{ce}) \quad \forall c \in C, e \in E \tag{11}$$

$$D_{ce}^{-} = \sum_{e' \in \Gamma(e)} X_{ce'}^{-} d_{\delta^{-}(e)\delta^{-}(e')} \quad \forall c \in C, e \in E \tag{12}$$

$$D_{ce}^{+} = \sum_{e' \in \Gamma(e)} X_{ce'}^{-} d_{\delta^{+}(e)\delta^{+}(e')} \quad \forall c \in C, e \in E \tag{13}$$

$$D_{ce} = \sum_{e', e'' \in \Gamma(e)} X_{ce'e''}^{e} d_{\delta^{-}(e')\delta^{+}(e'')} \quad \forall c \in C, e \in E \tag{14}$$

$$max_v K_{ce}^{-} \geq D_{ce}^{-} v_e \quad \forall c \in C, e \in E \tag{15}$$

$$max_w K_{ce}^{-} \geq D_{ce}^{-} w_e \quad \forall c \in C, e \in E \tag{16}$$

$$K_c^{-} = \sum_{e \in E} K_{ce}^{-} \quad \forall c \in C \tag{17}$$

$$max_v K_{ce}^{+} \geq D_{ce}^{+} v_e \quad \forall c \in C, e \in E \tag{18}$$

$$max_v K_{ce}^{+} \geq D_{ce}^{+} v_e \quad \forall c \in C, e \in E \tag{19}$$

$$K_c^{+} = \sum_{e \in E} K_{ce}^{+} \quad \forall c \in C \tag{20}$$

$$max_w K_c \geq (\sum_{e \in E} D_{ce} w_e)\alpha \quad \forall c \in C \tag{21}$$

$$max_v K_c \geq (\sum_{e \in E} D_{ce} v_e)\beta \quad \forall c \in C \tag{22}$$

Equation (2) determines if a route is clustered or not. If a cluster is formed, then it should have at least two routes (Eqs. (3–4)). For each created cluster, we should set a sourcing hub, and a receiving hub (Eqs. (5–8)). Equation (9–10) links the sourcing hub and the receiving hub for all the routes in the cluster. Equations (11) guarantees that only compatible edges can be set in a cluster. For each clustered route, Eq. (13) sets the distance from the outgoing node of the route to the sourcing hub. Equation (14) sets the distance from the sourcing hub to the receiving hub. Equation (13) establishes the distance from the receiving hub to the destination node of the route. Equations (18–19) determine the transportation cost from the outgoing node to the sourcing hub by taking into account the distance between them and the number of trucks needed, which depends on the volume and weight of the routes. Equation (20) determines the total transportation costs to the supplier hub. Similarly, Eqs. (18–20) determine the transportation costs of the cluster from the receiving hub to the destination nodes. Equations (21–22) determine the transportation cost from the sourcing hub to the receiving hub by taking into account the distance between them and the number of trucks needed, which depends on the volume and weight of routes and the efficiency parameters.

5 Branch and Price Solution Approach

There are two main resolution issues with the provided SCRC model. The first one is that a priori we do not know the exact number of needed clusters. Therefore, we could overestimate the number of clusters needed to find the optimal clustering of routes. The other is that the number of needed variables and constraints grows linearly in relation to the number of clusters set. Meanwhile, the decision space of the integer variables exponentially grows. This leads to an inefficient solution using current MILP solvers. Therefore, a column generation scheme based on branch and price is developed.

The key components of the branch and price algorithm are the master problem (MP), the restricted master problem (RMP), the pricing problem (PP) and the branching policy. The MP is a reformulation of a linear or integer programming model so that we can exploit the decomposable structure of the problem. However, the MP is typically too large to be solved directly because it involves a huge number of variables and columns that represent all the possible decision scenarios (e.g., all the possible routes in a routing problem or all the feasible clusters in our case). To manage this complexity, we consider only a smaller subset of variables, which leads to the formulation of the RMP to find new promising columns to be added to the RMP. The RMP only uses a subset of the variables and columns of the MP. Solving the RMP provides an optimum solution for the included subset of variables, but not necessarily for the entire set of original variables. Therefore, RMP integrality constraints are relaxed to find the current dual variables and new columns through the PP, also known as the subproblem, which is used to identify which variables (or columns) should be added to the RMP to potentially improve the current solution. The goal of the PP is to find variables with a negative reduced cost (in a minimisation problem) to be added to the RMP, which is likely to lead to a better objective value.

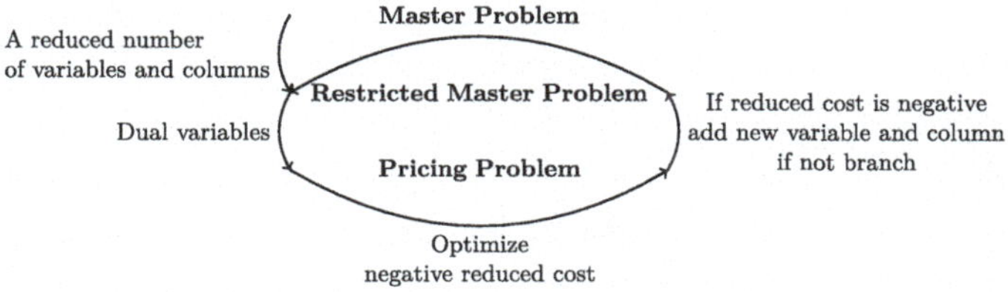

Fig. 4. Branch and Price

If there are no new columns with a negative reduced cost, then we have optimally solved the integrality relaxed MP. If all the variables associated with the solution are integers, then we have found an integer solution to the current branch. If not, we have to branch the current branch by adding further constraints to find integer solutions through the relaxed RMP. Figure 4 exemplifies the Branch & Price Algorithm.

5.1 Formulation of the Master Problem and the Reduced Master Problem

Here we firstly reformulate the model as a set-partitioning formulation (Dantzig-Wolfe decomposition). The key idea of the set-partitioning formulation for the SCRC is to enumerate all the feasible clusters with their associated costs. Let P be the set of all feasible clusters that satisfy the constraints imposed by the SCRC model. For each cluster $p \in P$, a binary variable λ_p is defined that equals to 1 if cluster p appears in the solution and 0 otherwise. Binary parameter a_e^p equals to 1 if cluster p contains the edge $e \in E$. Parameter c_p represents the cost associated with cluster c. In this case, we consider that the pre-established routes are clusters with pre-established cost and with $a_e^p = 1$ if e is the current route, and 0 otherwise. Therefore, the MP of our problem is:

$$\min \sum_{p \in P} c_p \lambda_p \tag{23}$$

subject to:

$$\sum a_e^p \lambda^p = 1 \quad \forall e \in E \tag{24}$$

$$\lambda_p \in \{0, 1\} \quad \forall p \in P \tag{25}$$

The objective function 23 is to minimise the cost of the clusters needed to transport goods. Meanwhile, 24 ensures that each route is only used in one cluster.

Here we employ the initial routes as the initial clusters used in the RMP with their associated costs.

$$\min \sum_{p \in P} c_p \lambda_p \tag{26}$$

subject to:

$$\sum a_e^p \lambda^p = 1 \quad \forall e \in E \tag{27}$$

$$0 \leq \lambda_p \leq 1 \quad \forall p \in P \tag{28}$$

Dual values are obtained by solving the RMP. Let $\pi_e, e \in E$, be dual variables associated with constraints (27). Then the reduced cost associated with cluster $p \in P$ is $c_p - \sum_e a_e^p \pi_e$. The feasible cluster with a negative reduced cost can improve the current solution to the RMP. For each iteration, we solve the subproblem that consists of finding a feasible cluster with the minimum negative reduced cost. The value of the variables associated with negative reduced cost (expressed as new columns) are added to the RMP. Then the RMP is solved to generate a new set of dual values. This process is repeated until no further variables with a negative reduced cost are discovered.

5.2 Pricing Subproblem

We solve the pricing problem using an MILP approach. The indices and parameters are the same as the model proposed before, but the decisions variables do not depend on the indices of clusters. Table 3 describes the new decision variables.

Table 3. Nomenclature of the pricing subproblem

Decision Variables	
X_e	Binary variable indicating if edge e belongs to the new cluster
X_e^-	Binary variable indicating if $\delta^-(e)$ is the sourcing hub for the suppliers in the new cluster
X_e^+	Binary variable indicating if $\delta^+(e)$ is the receiving hub for the nodes in the new cluster
$X_{e'e''}^e$	Binary variable indicating if edge e is in c and the source node of e' is the sourcing hub for outgoing goods and receiving node of e'' is receiving hub for incoming goods in the new cluster
D_e	Distance between the sourcing and receiving hub of the new cluster if e is in the cluster, 0 if not.
D_e^-	Distance from $\delta^-(e)$ to the sourcing hub if e is in the new cluster, 0 if not
D_e^+	Distance from $\delta^+(e)$ to the receiving hub if e is in the new cluster, 0 if not
K	Transportation cost between the sourcing nodes and the sourcing hub in the new cluster
K^-	Transportation cost between the sourcing hub and the receiving hub in the new cluster
K^+	Transportation cost between the receiving hub in the new cluster

The objective is to find new clusters with negative reduced costs. That is:

$$\min z = \gamma K + \gamma^- K^- + \gamma^+ K^+ - \sum_{e \in E} \pi_e * X_e \tag{29}$$

subject to:

$$\sum_{e \in E} X_e \geq 2 \tag{30}$$

$$\sum_{e \in E} X_e^- = 1 \tag{31}$$

$$\sum_{e \in E} X_e^+ = 1 \tag{32}$$

$$\sum_{e' \in \Gamma(e)} X_{e'}^+ = X_e quad \forall e \in E \tag{33}$$

$$\sum_{e' \in \Gamma(e)} X_{e'}^- = X_e \quad \forall e \in E \tag{34}$$

$$\sum_{e', e'' \in \Gamma(e)} X_{e'e''}^e = X_e \quad \forall e \in E \tag{35}$$

$$X_{e'e''}^e + 2(1 - X_e) + 1 \geq X_{e'}^- + X_{e''}^+ \quad \forall e \in E, e', e'' \in \Gamma(e) \tag{36}$$

$$\sum_{e' \in E' \Gamma(e)} X_{e'} \leq M(1 - X_e) \quad \forall e \in E \tag{37}$$

$$D_e^- = \sum_{e' \in \Gamma(e)} X_{e'}^- d_{\delta^-(e)\delta^-(e')} \quad \forall e \in E \tag{38}$$

$$D_e^+ = \sum_{e' \in \Gamma(e)} X_{e'}^- d_{\delta^+(e)\delta^+(e')} \quad \forall e \in E \tag{39}$$

$$D_e = \sum_{e', e'' \in \Gamma(e)} X_{ce'e''}^e d_{\delta^-(e')\delta^+(e'')} \quad \forall e \in E \tag{40}$$

$$max_v K_e^- \geq D_e^- v_e \quad \forall e \in E \tag{41}$$

$$max_w K_e^- \geq D_e^- w_e \quad \forall e \in E \tag{42}$$

$$K^- = \sum_{e \in E} K_e^- \tag{43}$$

$$max_v K_e^+ \geq D_e^+ v_e \quad \forall e \in E \tag{44}$$

$$max_v K_e^+ \geq D_e^+ v_e \quad \forall e \in E \tag{45}$$

$$K^+ = \sum_{e \in E} K_e^+ \tag{46}$$

$$max_w K \geq (\sum_{e \in E} D_e w_e)\alpha \tag{47}$$

$$max_v K \geq (\sum_{e \in E} D_e v_e)\beta \tag{48}$$

Similarly, (30–48) restrict the clusters used in the pricing problem to feasible ones.

5.3 Branching

Here, we use the depth-first branching rule, which once we can not find any new promising solutions through pricing problem we set the variable with the largest fractional to one. That is, we will fix a cluster that must be used. After fixing the variables, more columns are generated using the pricing algorithm. Repeating this branching strategy leads to an integer solution of the RMP. In our case, this method of branching does not guarantee to find an optimal solution, but it does guarantee a feasible one.

6 Computational Experiments

Here we compare both the performance of the SCRC model and the developed price and branch approach. For this purpose, 200 instances of different sizes were artificially generated and inspired in a first-tier supplier of the automotive sector. Depending on the total number of routes, we grouped instances into four main types: small, medium, large and extra larges. The routes in the instances were generated by randomly assigning a source (supplier) and a destination (manufacturing facility) of the supply chain under study. Furthermore, each route was randomly assigned to be constrained by volume or weight.

We used the Julia programming language and the JuMP package to run all the calculations. To solve both the RMP and PP, we have used the Gurobi commercial solver.

The Table 4 illustrates how we have grouped the instances.

Table 4. Nomenclature for instance types

Size	Number of Routes
Small	(0, 500]
Medium	(500,1000]
Large	(1000,1500]
Extra Large	(1500–2500]

Table 5 presents the mean objective value and resolution duration aggregated by the size of the instace for the different resolution procedures. For the SCRC MILP model we have set a time limit of 2 h. Meanwhile, for the branch and price resolution procedure, we have set a time limit of 30 min for each node iteration.

As we can observe, the branch and price algorithm outperforms the MILP model in both terms of solution quality and resolution time.

Table 5. Mean objective value and resolution duration per resolution procedure

Type	Resolution procedure	Transport Cost	Duration
Extra large	B&P	6864750.622806	1826 s
Extra large	MILP	7507608.574584	11520 s
Large	B&P	3811390.489451	1803 s
Large	MILP	4498435.223205	8820 s
Medium	B&P	2048727.759246	1435 s
Medium	MILP	2531413.865415	8010 s
Small	B&P	622577.984265	43 s
Small	MILP	673251.233579	5825 s

7 SCRC as A Decision Support Tool

The SCRC model can be used as a decision tool for clustering routes of various supply chains. Through the different efficiency parameters of the algorithm, decision makers can analyse the various clustering routes proposed by the algorithm. If we set all the efficiency parameters to 1, then we consider that clustering will not have any inefficiencies, which is rarely the case. Once we have identified where inefficiencies are from, for example, the sourcing to the main hub, then we can adjust this parameter and see how the system behaves. Figure illustrates the different savings proposed by algorithms when we increase the inefficiency of one of the parameters. Furthermore, if a decision maker thinks that a proposed cluster is not well suited for the supply chain being considered, additional constraints can be set in the mathematical model by selecting which routes can a route be clustered.

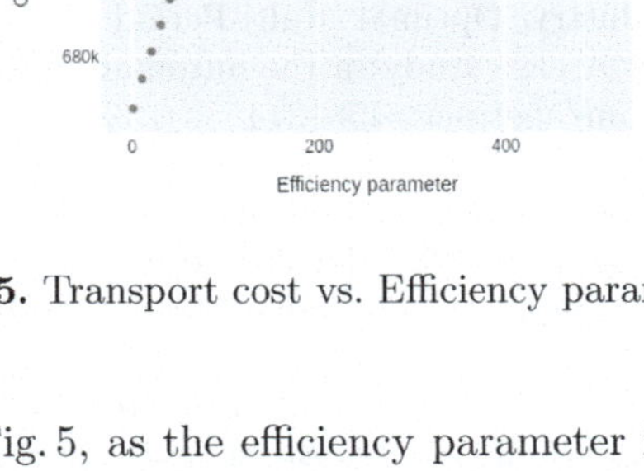

Fig. 5. Transport cost vs. Efficiency parameter

As we can see in Fig. 5, as the efficiency parameter increases the transport cost also increases. Furthermore, the transport cost can not overpass the initial cost of the problem, since if the new clusters are too costly it will choose to use the initial configuration.

SCRC model has been used as a decision tool for clustering routes at a first tier supplier of the automotive sector.

8 Conclusions and Future Research

In conclusion, this paper presents a pioneering effort in route clustering for a supply chain with multiple suppliers and manufacturing facilities. It introduces a mathematical model designed to consolidate the transport resource of large supply chains. Furthermore, the model has different parameters that the decision maker must decide. Not only the development of a mathematical model, but also the subsequent application of a branch and priced method to ascertain near-optimal solutions, underscore the robustness and innovative approach of this study. The decision tool, validated in a first-tier supplier of the automotive industry, provides a solid foundation for future explorations and applications in procurement transport planning.

Looking ahead, future research alternatives have been identified, and promise to further enhance the model's utility and applicability. These include exploring alternative metrics for evaluating the transport cost of each cluster by, for example, adding a vehicle routing problem for sourcing and delivery to the central hubs. Furthermore, it is possible to integrate the model to make operational decisions to dynamically adjust routes based on inventory levels and the demand of the manufacturing facilities. Additionally, it offers the possibility of addressing transport decisions about using reusable packaging in the supply. Regarding the resolution procedure, a future research potential is identified. For instance, using other branching strategies to ensure optimal solutions for medium-sized instances and applying metaheuristcs for the pricing problem for larger instances.

References

1. Baller, R., Fontaine, P., Minner, S., Lai, Z.: Optimizing automotive inbound logistics: a mixed-integer linear programming approach. Transp. Res. Part E: Logistics Transp. Rev. **163** (2022). https://doi.org/10.1016/j.tre.2022.102734
2. Bou-Zeid, I., Alpan, G., Penz, B., Serrano, C., Nguyen, A., Khatouf, M.A.: Optimal Multi-Period Tactical Transportation Plan for Inbound Logistics: A case study in the automotive industry. Optimal Multi-Period Tactical Transportation Plan for Inbound Logistics: A case study in the automotive industry. Technical report (2023), https://ssrn.com/abstract=4581244
3. Chitsaz, M., Cordeau, J.F., Jans, R.: A unified decomposition matheuristic for assembly, production, and inventory routing. INFORMS J. Comput. **31**(1), 134–152 (2019). https://doi.org/10.1287/ijoc.2018.0817
4. Chopra, S., Meindl, P.: Supply chain management : strategy, planning, and operation (2015)
5. Dantzig, G.B., Wolfe, P.: Decomposition principle for linear programs. Oper. Res. **8**(1), 101–111 (1960). https://EconPapers.repec.org/RePEc:inm:oropre:v:8:y:1960:i:1:p:101-111

6. Desrosiers, J., Lübbecke, M.E.: Branch-price-and-cut algorithms. In: Wiley Encyclopedia of Operations Research and Management Science. Wiley (2011). https://doi.org/10.1002/9780470400531.eorms0118
7. Díaz-Madroñero, M., Mula, J., Peidro, D.: A mathematical programming model for integrating production and procurement transport decisions. Appl. Math. Modell. **52**, 527–543 (2017). https://doi.org/10.1016/j.apm.2017.08.009
8. Díaz-Madroñero, M., Peidro, D., Mula, J.: A fuzzy optimization approach for procurement transport operational planning in an automobile supply chain. Appl. Math. Model. **38**(23), 5705–5725 (2014). https://doi.org/10.1016/j.apm.2014.04.053
9. Díaz-Madroñero, M., Peidro, D., Mula, J.: A review of tactical optimization models for integrated production and transport routing planning decisions (2015). https://doi.org/10.1016/j.cie.2015.06.010
10. Fleischmann, B., Kopfer, H., Sürie, C.: Transport planning for procurement and distribution. In: Stadtler, H., Kilger, C., Meyr, H. (eds.) Supply Chain Management and Advanced Planning. STBE, pp. 225–240. Springer, Heidelberg (2015). https://doi.org/10.1007/978-3-642-55309-7_12
11. Ford, L.R., Fulkerson, D.R.: Constructing maximal dynamic flows from static flows. Oper. Res. **6**(3), 419–433 (1958). http://www.jstor.org/stable/167028
12. Gilmore, P.C., Gomory, R.E.: A linear programming approach to the cutting-stock problem. Oper. Res. **9**, 849–859 (1961). https://api.semanticscholar.org/CorpusID:8079477
13. Langley, C.J., Novack, R.A., Gibson, G., Coyle, J.H.: Supply Chain Management: A Logistics Perspective. Cengage Learning, 11th edn. (2020)
14. Lübbecke, M.E., Desrosiers, J.: Selected topics in column generation. Oper. Res. **53**(6), 1007–1023 (2005). http://www.jstor.org/stable/25146936
15. Mosca, A., Vidyarthi, N., Satir, A.: Integrated transportation - inventory models: a review (2019). https://doi.org/10.1016/j.orp.2019.100101
16. Schöneberg, T., Koberstein, A., Suhl, L.: An optimization model for automated selection of economic and ecologic delivery profiles in area forwarding based inbound logistics networks. Flex. Serv. Manuf. J. **22**(3–4), 214–235 (2010). https://doi.org/10.1007/s10696-011-9084-5
17. Schöneberg, T., Koberstein, A., Suhl, L.: A stochastic programming approach to determine robust delivery profiles in area forwarding inbound logistics networks. OR Spectrum **35**(4), 807–834 (2013). https://doi.org/10.1007/s00291-013-0349-0

Capitalising Artificial Intelligence Capabilities to Foster Disruptive Events Anticipation and Proactive Resilience

Marco Arias-Vargas[1,2], Raquel Sanchis[1(✉)], and Raúl Poler[1]

[1] Research Centre On Production Management and Engineering (CIGIP), Universitat Politècnica de València, Valencia, Spain
rsanchis@cigip.upv.es

[2] Escuela de Ingeniería Industrial, Universidad de Costa Rica, San Pedro, Costa Rica

Abstract. Disruptive events keep happening, and the uncertainty about their consequences is growing. These events would affect many global supply chains since they have already proven to be fragile, so organisations within them need to be more resilient to face up to this turbulent context. The main objective of this work is to provide organisations with an easy-to-use tool to help them become more resilient by enhancing their anticipation capacity.

One of the main constituent capacities of resilience is anticipation. In this work, this is assessed through a developed web tool that monitors keywords and phrases in international news sources, analyses the sentiment of each news item based on artificial intelligence, guides preventive actions through access to scientific publications, and offers the option to ask a smart chatbot about the manner to receive fast ideas. The guidance provided through this work, and more specifically, through the developed web tool, would strengthen the organisations' anticipation capacity and enhance enterprise and supply chain resilience. Moreover, the main contribution of this work is to provide companies with a customising and straightforward anticipation and monitoring tool to help them be aware of and prepared for potential disruptive events.

The findings of this work are validated by eleven managers from different companies. All of them agree that the developed web tool is easy to implement and that it will help their organisations to increase resilience.

To the best of our knowledge, the specific combination of artificial intelligence techniques applied in this work has not been published before, giving this research a novel approach to sentiment analysis and decision-making.

Keywords: Resilience · news · sentiment analysis · chatbot · artificial intelligence · natural language processing

1 Introduction

The global business environment is getting more complex; circumstances are changing rapidly due to the fact that disruptive events (DEs) keep happening, and the uncertainty about their consequences is growing; while the business world has not fully recovered

A. A. Juan et al. (Eds.): DSA ISC 2024, LNCS 14778, pp. 306–319, 2025.
https://doi.org/10.1007/978-3-031-78238-1_28

from the pandemic and is still dealing with the Russia – Ukraine war and with international news that mention a potential financial recession and a potential mid-term crisis. The recent Israeli – Palestinian Conflict is worsening, adding uncertainty to organisations all over the world. Thus, organisations need to become more resilient to succeed. The stability index [1], designed to help supply chain (SC) professionals understand the stress created by DEs in the global SC community, shows that, in the last decade, this stress has jumped from medium to high level. Moreover, the SC Disruption Survey [2] states that 74% of the respondent organisations have been forced to rethink their global SC operations in the last two years due to several DEs. Organisations interacting in these global SCs need to be resilient to reduce their vulnerability, defined as an accelerating sensitivity to a harmful stressor, culminating in more harm than benefit from random events [3]. Enterprise resilience (ER) is defined as the capability to anticipate and be prepared to face DEs and, if their occurrence is inevitable, the capacity to recover as soon and efficiently as possible [4]. Moreover, SC resilience (SCR) is defined as the ability of an SC to maintain, resume, and recover operations in the aftermath of a severe DE [5]. To achieve resilience, organisations must develop preparedness, adaptive, and recovery capacities [6], and this can be achieved by implementing preventive and knowledge registration actions. The preparedness capacity is defined as the readiness capacity to face disruptions, assessing whether companies have the knowledge, means, and resources to anticipate different disruptions [7]. Focusing on this first capacity to start building resilience, it is necessary to lower the impact of DEs, applying preventive actions after a monitoring process warns about the probable appearance of a DE.

Artificial intelligence (AI) can be used to monitor and analyse news worldwide with robust natural language processing (NLP) algorithms. These techniques can be applied to measure sentiment analysis (SA) [8], allowing the reader to focus on negative, neutral, or positive news. With the correct news identified and analysed, managers can coordinate implementing the right preventive actions. Still, if the news findings are beyond the manager's knowledge, it would be advisable to read scientific publications from formal information sources, such as Google Scholar, and also combine that information with quick ideas gathered from consulting with chatbots powered by AI, such as ChatGPT [9].

Enterprise and SC resilience (SCR) has a substantial body of knowledge to guide organisations to enhance anticipation, preparedness, adaptation, and recovery strategies. Moreover, previous research studies have shown the importance of SA to identify patterns and have presented examples of specific applications related to resilience; this is a solid step to achieve resilience, but it is not enough. Generative AI is gaining popularity through organisations and SCs, but it needs to be focused and combined with other tools to build resilience. Combining powerful SA and generative AI techniques will empower organisations, making them more resilient by anticipating DEs. Additionally, this combination will help to close this gap. The main research question of this work is: Can a combination of news search, SA, and GPT, enhance anticipation capabilities and, thus, proactive ER?

This work proposes a monitoring and advisory tool powered by AI to reinforce an organisation's preparedness capacity and, in turn, ER. Section 2 presents a literature review regarding the relationship between resilience, anticipation, sentiment analysis, and AI models. Section 3 describes the proposed tool, from the process flow to technical features that explain its functionality. Section 4 explains the methodology applied to develop and validate the resilience-enhancing tool. In the last section, conclusions and future lines of research are outlined.

2 Literature Review

Facing DEs is one of the main challenges of every organisation. These events are defined as foreseeable or unforeseeable events that directly affect an enterprise's usual operation and its stability [7]. They can be generated within an organisation's supply chain (SC) or beyond that. Many of the most impactful DEs occur beyond the 'organisations' limits, but they impact them. These external events may be found in the news, and if so, their occurrence can be predicted and their impact reduced. In this sense, the authors in [10] present an example of the financial impact of predicting a DE. It is possible to find events using keywords and sentences to search among the news, and this news can be analysed with SA to narrow the search. Text SA is an important NLP task, as it analyses the text and tries to predict its emotion (positive, negative, or neutral) [8] There are many publications regarding SA, some with specific applications and others relating SA to Industry 4.0 (I4.0). The research presented in [11] proposes a green AI solution for an SA pipeline to understand better the changing public opinions on social media. The work performed by [12] links multimodal SA (text and images) with media information related to industrial products and services generated in I4.0, considering that their proposed model (TIBERT) does an outstanding job in emotion recognition in the industry, thus accelerating the development of I4.0.

Once organisations have access to news analysed through SA, it would be advisable for them to study the preventive actions they can implement to minimise the effect of a probable DE, which can be done using AI. In recent years, there have been many works regarding the use of AI to build enterprise and supply chain resilience (ESCR), such as the one presented by the authors in [13], who proposed an integrated AI-powered multicriteria decision model to evaluate AI techniques regarding their ability to support SCR strategies; the research developed in [14], where the authors concluded that there is a significant contribution in respect to SCR and the application of AI; and the work in [15] where a data mining approach for SC risk identification and prediction based on Twitter (now, X) was presented. Also, the authors in [16] who tested a model that combines vector autoregressive with SA through NLP deep learning models, and the work explained in [17] which showed a data-analytics approach based on different types of Support Vector Machines (SVM) that leverages system monitoring data for ER. Other publications show success factors organisations have applied when facing critical DEs. In [18] the authors present the results of a study in a group of manufacturing firms during the Covid-19 pandemic. The main finding is that 3D printing contributed positively to dealing with production disruptions due to its capacity to produce components and spare parts broken down in manufacturing facilities, thus emerging as a valid alternative when facing SC disruptions. Also, the work performed in [19] shows a structured literature review of risk impact on performance in SCs. One of the findings is that the most common types of DEs are internal supply issues and external disruptions, which negatively impact the SC performance. They analysed almost fifty scientific articles and exposed several risk mitigation strategies, such as collaboration, integration, visibility, and investment in resilience.

Many research works consider AI to be closely related to or even a part of I4.0; thus, considering the impact of I4.0 on ESCR is essential when analysing technologies to enhance resilience. The authors in [20] establish that I4.0 supports pre-disruption resilience measures, enabling more effective proactive risk management; they study the relationships between I4.0 enabler technologies (Cloud computing (CC), Internet of Things (IoT), Big data analytics (BDA), AI, Cyber-physical systems (CPS), Additive manufacturing (AM), and, Blockchain (BC)) and SC resilience antecedents, finding that BDA is the most topical and mature one. They also present an I4.0 SC resilience framework, which allows people to choose appropriate I4.0 technologies when noticing weaknesses in specific SC resilience antecedents. Additionally, in the work presented in [21], the authors analyse, through a bibliometric analysis, the interplay of I4.0 and SCR under investments constraints, concluding that companies can reduce the risk of disruptions and improve their ability to respond quickly to unexpected events by precisely investing in Digital SCR (DSCR), which they state is an intersected area among I4.0, SCR, and investment. The authors highlight that an organisational DSCR harnesses the power of I4.0 to increase the ability to digitally detect, avoid, manage, and recover from disruptions.

Moreover, authors in [22] focus on the Resilient SC4.0, which expresses the interoperability of I4.0 with a SC aiming to create resilience. They develop a research study to investigate how Key Performance Indicators (KPIs) for enhancing SCR are affected by I4.0 technologies, and, specifically regarding AI, they conclude that AI techniques enhance various elements of SCR, such as redundancy, flexibility, visibility, collaboration, agility, situation awareness, and velocity. Additionally, they analyse 17 KPIs for SCR and show the I4.0 technologies that have a direct or indirect impact on each KPI, where focusing on AI, they conclude that its methods can be used for proactive response to changes, highlighting that Machine Learning (ML) can improve various KPIs, such as time to recovery, by providing accurate information about the occurrence of a disruption in time.

Considering that there is an increasing number of scientific publications regarding strategies and ploys to minimise DEs and that organisations worldwide are learning from experiences, this growing body of knowledge can be helpful for others to anticipate and prevent DEs. This knowledge can be examined by managers searching scientific sources, and even more, it can be analysed and combined using generative AI, which can lead to new ideas and tactics. According to the authors in [23], AI chatbots are a driving force of I4.0. They define a chatbot as an AI agent that provides human interaction, using natural languages, with a computer system. They also state that according to I4.0 standards, person-based automation must be available and that because of the I4.0 strategy, a chatbot is the best interface candidate.

Currently, ChatGPT can help with these tasks since it interacts conversationally and can answer follow-up questions [9]. This chatbot is based on a generative pre-trained transformer (GPT), an autoregressive NLP model that uses deep learning neural networks to produce human-like text [8]. Moreover, transformers are deep feed-forward artificial neural networks with a self-attention mechanism, and they have been tremendously successful in NLP processing tasks and, more generally, at modelling sequential data [24]. Formal algorithms behind the transformers are not easily found in publications, but in [24], it is presented the pseudocode of the architectural components from which transformers are made, such as token embedding, positional embedding, attention, layer normalisation, and unembedding. These authors also describe prominent transformer architectures, such as Bidirectional Encoder Representations from Transformers (BERT) and GPT, and their specific large language models (LLMs), such as GPT-3, which is the base of ChatGPT (now using GPT-4).

All these techniques are being applied today in many organisations, and they can individually enhance ER. Still, as far as we know, combining news search, SA, and GPT has not been addressed.

3 Methodology

The steps taken to develop the SRA tool are shown in Fig. 1. The process started searching in the technical literature for applications of SA focused on enhancing resilience, learning from them, and using them as a basis to develop a new one using current techniques. Thus, step two was to evaluate different Python libraries and compare their techniques, for example, classical NPL models against transformer models. The third step was the combination of the selected techniques and, thus, their related libraries and the creation of an algorithm to combine the results from the chosen techniques. Step four was to include an easy way for the user to access Google Scholar and the creation of a chatbot that uses Chat GPT's capabilities. Step five tested the developed tool with different news sources and keywords. Step six was selecting the news sources, phrases, and words based on published keywords from related publications. Finally, step seven was to experiment with managers from different organizations, which also helped to define the phrases and keywords to include in the tool.

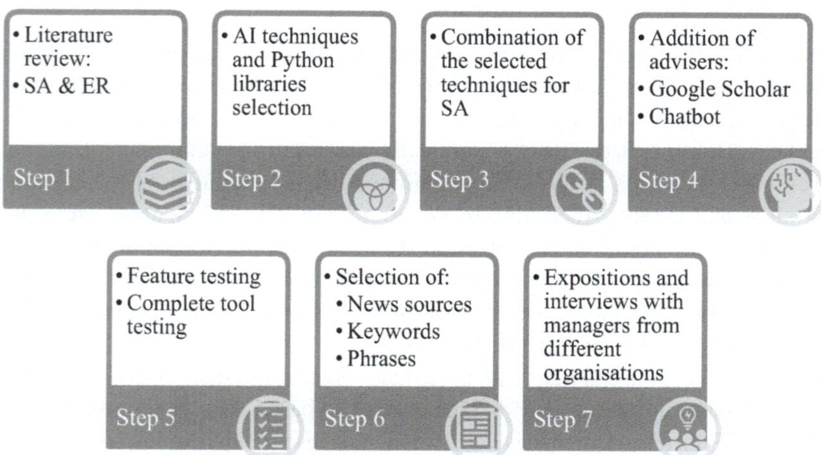

Fig. 1. Developing methodology for SRA tool

This tool has already been presented to eleven managers from different companies and industry sectors, and they all agreed that it would help their companies be more resilient. As an example, one of the managers said: "Yes, it helps to anticipate in an objective and informed manner. Effects from a global event can be linked to specific business issues. It helps to apply a risk management approach with easy-to-access information". Other respondent wrote: "it allows us to quickly find up to date relevant topics that might impact our business, so we can have a clear understanding of what is happening". The validation process had three phases; the first was the demonstration of the SRA Tool to each manager individually. In the second phase, each manager completed an online questionnaire about the tool's functionality. In the third phase, the research team interviewed each participant to gain better insights into their opinions and suggestions to improve SRA. Information on the participants is shown in Table 1.

Table 1. Profile of the participants

Job position	Company's function	Company size
Planning Manager	Food Distribution	Large
Logistics Director	Food Distribution	Large
General Manager	Third-Party Logistics	Medium
General Manager	Automotive distribution	Large
General Manager	Consulting services	Small
Financial Manager	Third-Party Logistics	Medium
General Manager	Medical instruments distribution	Medium
Logistics Manager	Manufacturing of cleaning supplies	Large
Operations Manager	Manufacturing of cleaning supplies	Large
Logistics Director	Auto parts distribution	Large
SC Director	Construction products distribution and retail	Large

Small: less than 51 workers. Medium 51 to 200 workers. Large: more than 200 workers.

All the participating companies are private and have operations at least in Costa Rica, the country where the validation process was implemented, and the size of the organisations is categorised based on Costa Rican standards.

91% of the participants recommended new keywords to include in the SRA, specific for each industry where they work. Moreover, 100% of the respondents showed their particular needs through customised recommendations to improve SRA, where the top request is to add flexibility to change and include new words and new news sources.

Based on the results, the interviewed managers consider the SRA tool helpful to increase resilience; additionally, all the participants show interest in customising SRA for their specific needs, which indicates their interest in the tool again. Their recommendations are already included in the current version of SRA, and now, each user can include specific words or phrases, according to their needs. This can be done before selecting the words or phrases to search for, typing text below the header "Enter new temporary keywords (optional", as shown in Fig. 2.

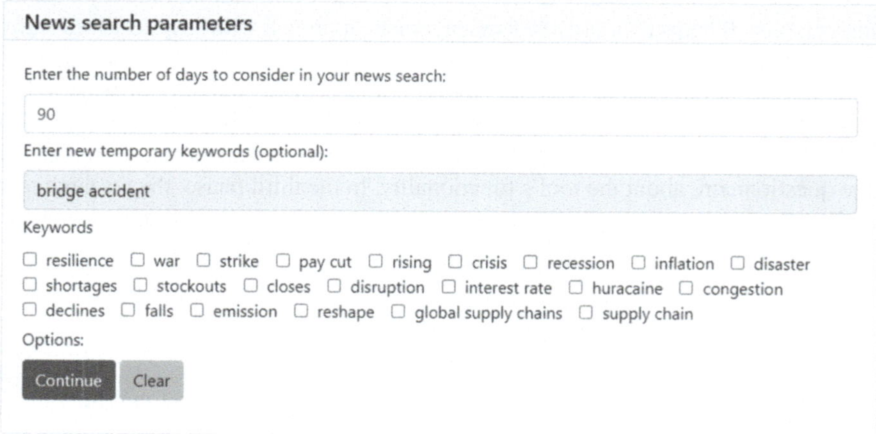

Fig. 2. Selecting the keywords (with new features)

4 The Smart Resilience Adviser Tool

The Smart Resilience Adviser (SRA) is a web tool designed to help managers anticipate potential DEs. It is capable of finding keywords and phrases in the news, analysing their sentiment, and offering the reader two options to study details about potential events found in the news: (i) an automatic search for related articles in Google Scholar and (ii) a guided chat conversation with an AI-empowered chatbot; ChatGPT. This tool was developed in Python using the PyWebIO library. A temporary evaluating version can be found at http://74.50.79.213:8080/, selecting user: Guest and password: 2. In Fig. 3, there is a business process model and notation (BPMN) diagram which explains the basics of the SRA Tool from the user login to the decision between reading Google Scholar articles or chatting with ChatGPT to get ideas to enhance resilience.

The SRA aims to simplify news monitoring and analysis, offering organisations an alternative to focus on their specific business requirements. The first step is the selection of keywords and news sources to monitor. Figure 4 shows an example of the current list of keywords available to analyse, which were selected based on expert opinion, but it can include more words and phrases since it is a variable list. After the user selects the corresponding keywords and phrases, the tool will display a similar screen for news sources. Words, phrases, and news sources can be fully customised for each organisation.

Once the user selects the words and phrases to search for and the news sources, the SRA shows and analyses each matching news with the corresponding keyword. This text analysis uses mainly four Python libraries: bs4, spaCy, NLTK, and transformers. Beautiful Soup (bs4) is a web scraping library for gathering data from web pages and different types of files, while spaCy is designed to build systems for information extraction or

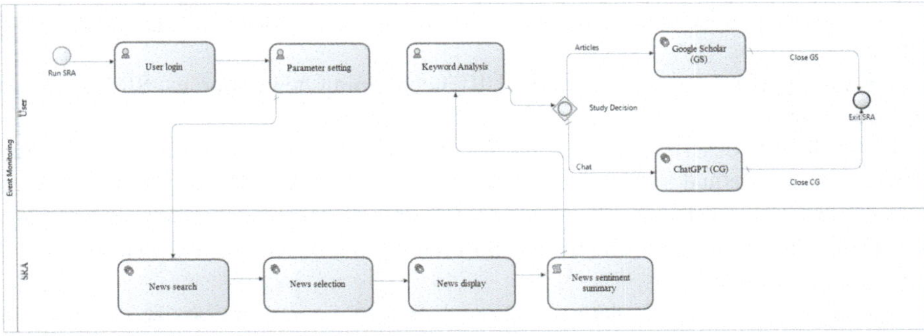

Fig. 3. SRA block diagram

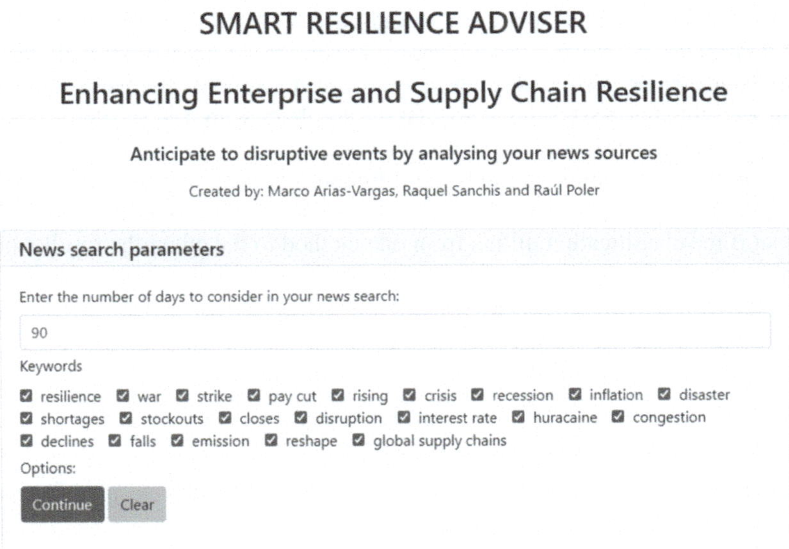

Fig. 4. Selecting the keywords (original screen)

NLP; it can process and recognise large volumes of text. The NLTK, known as a Natural Language Toolkit, is a platform for building programs to work with human language data, which contains several NLP libraries. The Transformers library offers thousands of pretrained models to perform tasks like text classification, information extraction, summarisation, and text generation.

When the SRA tool runs, it displays the news title, source, sentiment classification, publication date, and a link to read the entire news in a new tab. The SA options are "positive", "neutral", and "negative". This classification combines the results of two AI techniques: symbolic and statistical NLP, included in the NLTK library, and Transformers from the Transformers library. Complementing the aforementioned definition, Transformers are new deep learning neural networks particularly suitable for NLP [8].

Figure 5 shows a BPMN diagram with the SA evaluation process. The algorithm works as follows:

For each uniform resource locator (URL) containing news sources, SRA searches for each of the selected keywords in all the news available.

For each news description containing keywords, it counts the number of tokens, and if there are less or equal to 255, SRA applies SA with NLTK and Transformers. Tokens are a sequence of vocabulary elements or sub-words that represent a piece of text [24]. When SRA was developed, the maximum number of tokens in a text that the model could process was 255.

Specifically, NLTK calculates compound polarity scores to classify the news in positive, neutral, or negative sentiments. If this score is less or equal than -0.05, the news would be classified as "negative", but if this score is equal or greater than 0.05. The news would be positive; otherwise, the news would be classified as "neutral" [25]. The most extreme negative sentiment would have a -1 score in this model, and the most potent positive sentiment would have a score of 1. As a second classification method, Transformers also analyse each news based on the model named "distilbert-base-uncased-finetuned-sst-2-english", classifying them with a label among the same options; positive, neutral, or negative sentiment.

Since both classification methods are different, the next step that SRA does is to combine their results. If both methods give the same classification, that would be the result, but if the classification differs from one method to the other, the result would be "neutral".

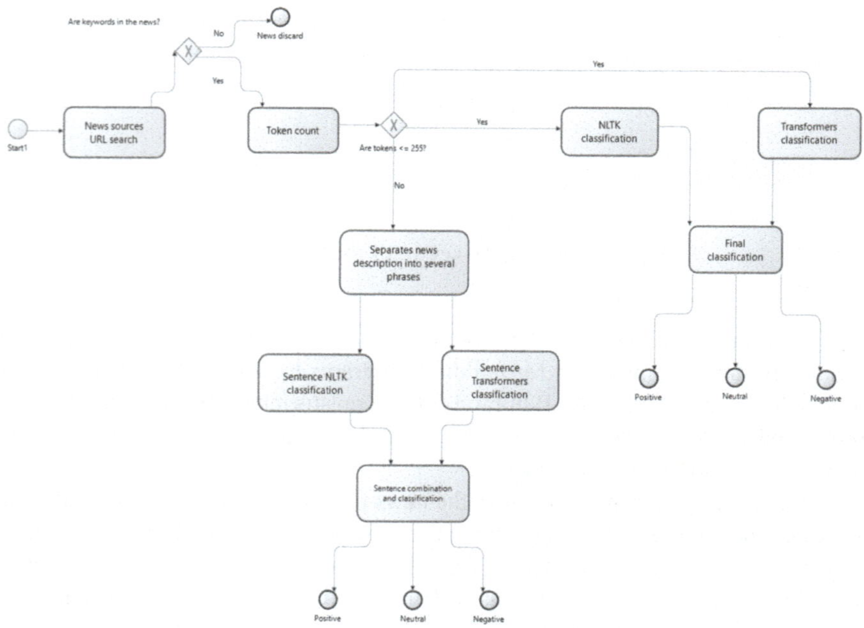

Fig. 5. SRA algorithm flowchart

If the number of tokens calculated in step 2 is higher than 255, SRA separates the description text to analyse the sentences separately, and then it combines the results to classify the news as positive, neutral, or negative. The classification process that SRA executes after the division of the news description text follows the same steps from 3 to 5. After reaching this point, SRA starts collecting descriptive statistics about the news sentiment classification.

Following the process, Fig. 6 shows an example of news from different sources (Financial Times and BBC) matching two keywords (crisis and war).

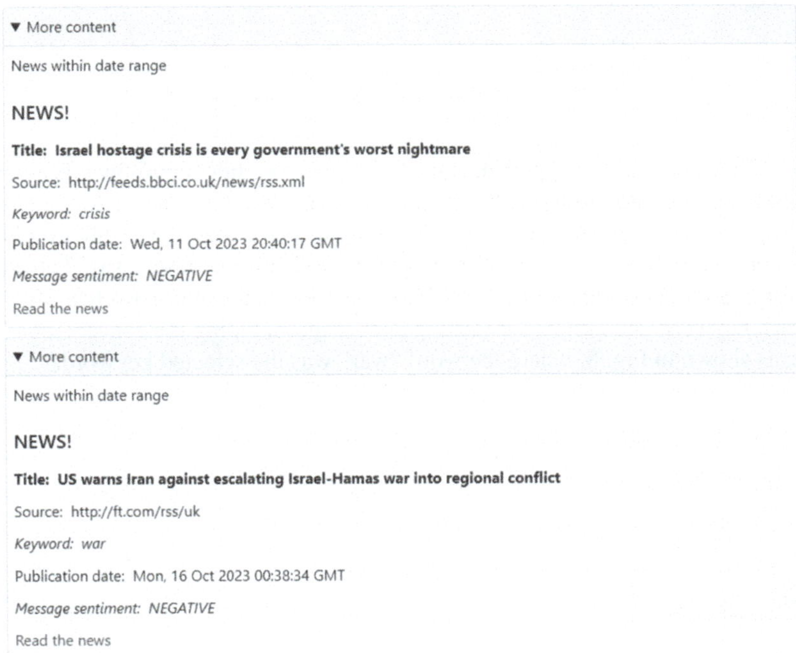

Fig. 6. Example of news analysed

After showing in the browser similar details for all the news found, the SRA tool displays a table summary with a row for each keyword found in the news. Every row includes descriptive statistics, such as the number of news containing each keyword and classified as positive, neutral, and negative. The last column of the table shows a link with a phrase that includes the row keyword, and by clicking on that link, the SRA tool opens the Google Scholar web page and searches for the mentioned phrase. The articles displayed after the search are options that the user can read and study to be more prepared to face a probable DE identified in the news by the keyword. Google Scholar can lead to hundreds or thousands of references depending on the keyword. Still, the goal here is to motivate practitioners to find and read academic publications, and if that becomes a continuous process, they will learn how to improve their searches. Figure 7 exhibits an example of the summary table.

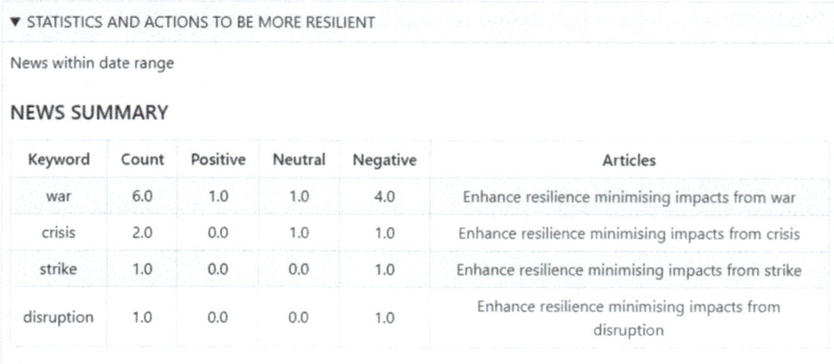

▾ STATISTICS AND ACTIONS TO BE MORE RESILIENT

News within date range

NEWS SUMMARY

Keyword	Count	Positive	Neutral	Negative	Articles
war	6.0	1.0	1.0	4.0	Enhance resilience minimising impacts from war
crisis	2.0	0.0	1.0	1.0	Enhance resilience minimising impacts from crisis
strike	1.0	0.0	0.0	1.0	Enhance resilience minimising impacts from strike
disruption	1.0	0.0	0.0	1.0	Enhance resilience minimising impacts from disruption

Fig. 7. News summary and link to academic articles

Since SRA is a tool to support managers, it seeks to guide them towards academic publications on the selected topic, but there is a chance that those users do not have the time or the habit to study this type of articles, and this is the reason why SRA tool offers a second option to obtain relevant information and to learn in a faster way. This second option is a chatbot powered by ChatGPT. If the user selects the option to chat, SRA takes the chosen keyword and guides the starting conversation, typing the conversation's first prompt, as shown in Fig. 8, where the word "war" was the selected keyword.

Fig. 8. Starting the chat

Once the "Continue" button is clicked, SRA connects with ChatGPT and displays the answers to the prompt, looking for ideas to minimise the effects of "keyword" in an SC. Figure 9 shows an example of the response received to the recommended prompt, which, in this case, is an interesting list of valid ideas to achieve a more resilient SC when facing a war as DE.

These answers can be what the user needs, but if they are not, the user can change the prompt and keep the conversation. To be more effective in this matter, we advise to follow the recommendations that Prompt Engineering states since it aims to design and refine prompts (questions or instructions) to elicit specific responses from AI models [26]. There are many private and open-source LLMs available. However, GPT was chosen because ChatGPT is recognised among the organisation's practitioners and because it continually improves, supported by an AI-leading company and its funding partners, such as Microsoft.

Question:	Enhance resilience minimising the effects of war in a Supply Chain
Analysis:	1. Establish a back-up supply chain: Develop a back-up supply chain in a different, safe region to ensure continuity of supply during war. 2. Develop alternative sources: Develop alternative sources for raw materials, components, and other supplies to reduce the dependency on a single supplier. 3. Increase inventory levels: Increase inventory levels of critical components and supplies to reduce the risk of supply disruption. 4. Strengthen relationships: Strengthen relationships with suppliers in different regions to ensure continuity of supply in case of war. 5. Utilise technology: Utilise technology such as cloud computing, artificial intelligence, and blockchain to keep track of supply chain operations and ensure visibility of supply chain activities. 6. Prioritise security: Prioritise security of supply chain operations by implementing additional security measures such as encryption, authentication, and access control. 7. Implement risk management strategies: Implement risk management strategies such as scenario planning, risk assessment, and contingency planning to minimise the impacts of war on supply chain operations. 8. Establish communication protocols: Establish communication protocols with suppliers to ensure timely and accurate information exchange. 9. Develop contingency plans: Develop contingency plans to protect supply chain operations in case of war. 10. Invest in training: Invest in training of personnel to ensure they are aware of the risks associated with war and how to respond to them.

Fig. 9. AI Chatbot responses to the chosen prompt

The SRA tool aims to identify and analyse events reported in the news, guide the user with ideas and technical options to anticipate those events, and, thus, be more resilient.

5 Conclusions

Since anticipation is one of the main constituents of ER, organisations need a tool to identify potential DEs from the news. It would add more value if the application could run SA and if it could guide towards solutions and preventive actions. The SRA tool has all those characteristics, and it is driven by AI techniques, meaning that users will not only be up to date with the latest news but will have access to robust, trained neural networks that can help them in the decision-making process.

Knowing that Google Scholar is a recognised search engine for scientific articles, SRA offers the option to start a specific search in Google Scholar, and this promotes that managers and all kinds of users get to know and use this search engine, leading to better-informed decisions and stronger links between academy and industry.

Regarding the users who have already tested SRA, besides the fact that they all consider it beneficial for enhancing ER, most of them provided new keywords and would like to customise it. This means that they find it helpful to anticipate DEs and to make better decisions. Following their feedback, now SRA has an option to include new words and phrases.

This work has theoretical and practical implications; combining constantly evolving powerful algorithms and techniques with a common goal generates synergy and opportunities to improve those techniques from an academic perspective. Moreover, from a practitioner's standpoint, this work contributes offering a user-friendly tool to take advantage of modern techniques, making it feasible for organisations to use them to enhance awareness and anticipation capabilities to build resilience. Future research should address improvements to this tool, training the chatbot with specific information through fine-tuning or retrieval augmented generation (RAG), defining effective prompts based on guidelines from Prompt Engineering, and developing a process to spread the findings along the organisations, so they can receive warnings for all the areas, enhancing ER.

Acknowledgments. This work was supported by the European Commission under the Erasmus+ Programme within the frame of CONTINUITY Project: Business Continuity Managers Training Platform with Reference No. 2021-1-IT01-KA220-VET-000033287 and the Regional Department of Innovation, Universities, Science and Digital Society of the Generalitat Valenciana within the frame of RESPECT Project: Resilient, Sustainable and People-Oriented Supply Chain 5.0 Optimization Using Hybrid Intelligence with Reference No. CIGE/2021/159.

Disclosure of Interests. The authors declared no potential conflicts of interest with respect to the research, authorship, and publication of this article.

References

1. Kent, D., Lee, J.: Supply Chain Stability Index. https://www.ascm.org/making-an-impact/research/supply-chain-stability-index/. Accessed 16 Oct 2023
2. Peerless Research Group, Logistics Management: Supply Chain Disruptions Survey. https://www.coursesidekick.com/management/430351. Accessed 16 Oct 2023
3. Taleb, N.N.N.: "Antifragility" as a mathematical idea. Nature **494**, 430 (2013). https://doi.org/10.1038/494430e
4. Sanchis, R., Canetta, L., Poler, R.: A conceptual reference framework for enterprise resilience enhancement. Sustainability. **12**, 1464 (2020). https://doi.org/10.3390/su12041464
5. Zhou, Z., Hosseini, S.: Evaluating supply chain resilience: a resilience capacity perspective. Int. J. Logistics Syst. Manag. **46**, 333–355 (2023). https://doi.org/10.1504/IJLSM.2023.134798
6. Sanchis, R., Marcucci, G., Alarcón, F., Poler, R.: Knowledge registration module design for enterprise resilience enhancement. In: IFAC-PapersOnLine, pp. 1029–1034. Elsevier B.V. (2021). https://doi.org/10.1016/j.ifacol.2021.08.122
7. Sanchis, R., Duran-Heras, A., Poler, R.: Optimising the preparedness capacity of enterprise resilience using mathematical programming. Mathematics. **8**, 1596 (2020). https://doi.org/10.3390/math8091596
8. Xiao, P.: Artificial Intelligence Programming with Python. John Wiley & Sons Inc., New Jersey (2022)
9. OpenAI: Introducing ChatGPT. https://openai.com/. Accessed 08 Apr 2024

10. Arias-Vargas, M., Sanchis, R., Poler, R.: Impact of predicting disruptive events in supply planning for enterprise resilience. IFAC-PapersOnLine **55**, 1864–1869 (2022). https://doi.org/10.1016/J.IFACOL.2022.09.670
11. Xie, M., Jiang, J., Shen, T., Wang, Y., Gerrard, L., Clarke, A.: A green pipeline for out-of-domain public sentiment analysis. In: Li, B., et al. ADMA 2022. LNCS, vol. 13087, pp. 190–202. Springer, Cham (2022). https://doi.org/10.1007/978-3-030-95405-5_14
12. Yu, B., et al.: Feature-guided multimodal sentiment analysis towards Industry 4.0. Comput. Electr. Eng. **100** (2022). https://doi.org/10.1016/j.compeleceng.2022.107961
13. Belhadi, A., Kamble, S., Fosso Wamba, S., Queiroz, M: Building supply-chain resilience: an artificial intelligence-based technique and decision-making framework. Int. J. Prod. Res. 1–21 (2021). https://doi.org/10.1080/00207543.2021.1950935
14. Naz, F., Kumar, A., Majumdar, A., Agrawal, R.: Is artificial intelligence an enabler of supply chain resiliency post COVID-19? An exploratory state-of-the-art review for future research. Oper. Manag. Res. (2021). https://doi.org/10.1007/s12063-021-00208-w
15. Deiva Ganesh, A., Kalpana, P.: Supply chain risk identification: a real-time data-mining approach. Ind. Manag. Data Syst. **122**, 1333–1354 (2022). https://doi.org/10.1108/IMDS-11-2021-0719
16. Nguyen, A., Pellerin, R., Lamouri, S., Lekens, B.: Managing demand volatility of pharmaceutical products in times of disruption through news sentiment analysis. Int. J. Prod. Res. (2022). https://doi.org/10.1109/MIS.2019.2918092
17. Xu, D., Tsang, I.W., Chew, E.K., Siclari, C., Kaul, V.: A data-analytics approach for enterprise resilience. IEEE Intell. Syst. **34**, 6–18 (2019). https://doi.org/10.1109/MIS.2019.2918092
18. Dzogbwu, T., Afrifa Jnr, S., Amoah, N., Fianko, S.K., Imdaadulah, A., De Beer, D.: Supply chain disruptions and resilience in manufacturing industry during covid-19: additive manufacturing intervention in perspective. J. Ind. Eng. Manag. **16**, 509 (2023). https://doi.org/10.3926/jiem.4526
19. Pham, T.H., Testorelli, R., Verbano, C.: Supply chain risk and its impact on performance: a structured literature review. J. Ind. Eng. Manag. **16**, 236–262 (2023). https://doi.org/10.3926/jiem.4719
20. Spieske, A., Birkel, H.: Improving supply chain resilience through industry 4.0: a systematic literature review under the impressions of the COVID-19 pandemic. Comput. Ind. Eng. **158** (2021). https://doi.org/10.1016/j.cie.2021.107452
21. Al-Banna, A., Rana, Z.A., Yaqot, M., Menezes, B.: Interconnectedness between Supply Chain Resilience, Industry 4.0, and Investment (2023). https://doi.org/10.3390/logistics7030050
22. Marinagi, C., Reklitis, P., Trivellas, P., Sakas, D.: The Impact of Industry 4.0 Technologies on Key Performance Indicators for a Resilient Supply Chain 4.0 (2023). https://doi.org/10.3390/su15065185
23. Bhoir, S.V., Patil, S.R., Mogul, I.Y.: Person-based automation with artificial intelligence Chatbots: a driving force of Industry 4.0. In: Artificial Intelligence and Industry 4.0, pp. 215–244. Elsevier (2022). https://doi.org/10.1016/B978-0-323-88468-6.00003-6
24. Phuong, M., Hutter, M.: Formal Algorithms for Transformers. DeepMind (2022)
25. Hutto, C.: VADER-Sentiment-Analysis. https://github.com/cjhutto/vaderSentiment#about-the-scoring. Accessed 16 Oct 2023
26. Crabtree, M.: What is Prompt Engineering? A Detailed Guide. https://www.datacamp.com/blog/what-is-prompt-engineering-the-future-of-ai-communication. Accessed 16 Oct 2023

A Novel Knowledge Formalization Framework to Support Decision-Making in Manufacturing Processes

Chiara Cimino[1(✉)], Mattia Torta[1], Paolo Albertelli[2], and Michele Monno[2]

[1] Consorzio MUSP, via Vittore Callegari 21/F, 29122 Piacenza, Italy
{chiara.cimino,mattia.torta}@musp.net
[2] Mechanical Engineering Department, Politecnico di Milano, via La Masa 1, 20156 Milan, Italy
{paolo.albertelli,michele.monno}@polimi.it

Abstract. Despite the recent efforts towards the introduction of advanced digital and automation technologies in the industry, most of the manufacturing processes are still governed and optimized by human expertise. This fundamental contribution is often in the form of explicit and implicit (or tacit) human knowledge, which is extremely difficult to document, structure and transfer to other people. In this scenario, manufacturing companies run the risk of losing this intrinsic asset, held by their personnel, worsening in turn the quality and efficiency of their products and processes. To cope with this knowledge loss problem, a novel approach to knowledge formalization is presented that lays the foundations for the realization of a digital platform equipped with interfacing capabilities as well as AI-based knowledge representation and reasoning tools. The paper focuses on evaluating the current scenario, generalizing the problem in the manufacturing context, identifying the main actors who will interact with the platform and proposing a possible procedure for knowledge formalization.

Keywords: manufacturing processes · knowledge representation · digital platform

1 Introduction and Motivation

1.1 The Implicit Knowledge in Manufacturing

Today's manufacturing landscape is characterized by an ever-increasing demand for high-performance, robust and resilient production processes, in order to keep up with the demand for low supply time and high-quality products. In this scenario, one of the key assets that companies have, and struggle to fully leverage, is the human component that leads, operates and oversees most of the manufacturing processes. The key role of skilled professionals, and the lack thereof, is indeed one of the three main sources of anxiety for businesses as recently pointed out by Deloitte in [1]; where skills and knowledge gaps are on the rise, as well as the average time to fill an open technical job [2]. While these gaps are still widening, a major turning point is unveiled in the

A. A. Juan et al. (Eds.): DSA ISC 2024, LNCS 14778, pp. 320–329, 2025.
https://doi.org/10.1007/978-3-031-78238-1_29

retirement rates of the active workforce. The recent pandemic coupled with the ever-increasing age of people involved in the manufacturing sector has drastically impacted the retirement trends, causing a wave of "excess retirements" in both the US and Europe [3]. This trend is visible in a variety of companies that must be properly equipped for the knowledge transfer of both explicit and implicit (or tacit) knowledge in different roles [4]. Indeed, focusing on the manufacturing companies, this trend could lead to a severe repercussion in terms of knowledge and expertise loss due to the departure of highly skilled technicians and operators. These skills range from practical know-how on manual operations, such as the assembly process of complex electro-mechanical components, to more conceptual knowledge about, for example, the configuration and coordination of intricate multi-phase production processes, i.e., the combination of different machining processes coupled with other complementary operations like heat treatments. In a machining process case, a series of tools are available to support the production planning and execution in every step; Computer-aided Design (CAD), Computer-aided Manufacturing (CAM), Manufacturing Execution System (MES) and Computer-Aided Process Planning (CAPP) systems help the operators to deal with almost every phase of the manufacturing journey [5]. Nevertheless, people involved in this journey still contribute with decisions and evaluations based on their experience by refining the production sequence derived from process planning software, optimizing cutting parameters and machining strategies for CAM programs, supervising the machining (execution) phase with in-process adaptation of cutting parameters and other technological evaluations that lead to better quality and productivity. More or less the same can be said in the manual assembly scenario, where experienced operators act on the process with a combination of explicit knowledge, derived from structured information such as drawings, textual instructions, or formalized best practices, and from implicit knowledge that evolves with the operators themselves during their careers. When the process is non-traditional and/or unconventional, such as the additive manufacturing processes, the procedures are not fully assessed and the contribution from experienced operators is extremely valuable for the development of the explicit knowledge since less a-priori knowledge is generally available and more difficult its creation process is. The case of additive manufacturing is indeed a clear example of a heavy experience-driven process, led by an extremely wide range of process parameters that can overwhelm inexperienced operators [6]. The same could hold for other kinds of unconventional processes such as demanufacturing and remanufacturing; in these cases the presence of knowledge-based tools is extremely important for all the procedures involved that could be implemented in a circular economy scenario where operators' training must be kept up-to-date and take less and less time [7].

1.2 Current Scenario of Implicit Knowledge Formalization

In recent decades, in coping with some of the aforementioned problems, several activities have been carried out to develop innovative approaches for knowledge management and knowledge representation in the manufacturing field. From the production planning and execution perspective, novel CAPP systems have been proposed taking human expertise into account to optimize the production sequence. In [8] a cloud platform has been developed to support the process planning phase by integrating sensor data and inputs from the skilled operators for real-time status estimation of shop-floor machines; this information, coupled with a process planning service, has led to better process coordination considering real-world dynamic manufacturing environment. A more direct integration of expert knowledge has been proposed in [9] as a distributed collaborative system for process planning. Moreover, the authors developed a cloud-based approach where expert heuristic knowledge of dislocated agents is stored in a knowledge repository and used to evaluate and optimize process plans. In this study, however, the knowledge repository is not well defined for the purpose. An interesting contribution is presented in [10] with a review of knowledge graph-based approaches for process planning in manufacturing. This technology has been used for modelling different aspects of a manufacturing process, such as resource allocation, CNC machining, and manufacturing assembly, enabling an automatic planning generation based on formalized knowledge. Regarding machining applications, the field is still not mature enough to guarantee useful results in terms of wide applicability and proper integration with the respective explicit knowledge of the field, and further research contributions need to be developed in this direction. The same can be stated for the additive manufacturing field, where human knowledge is even more crucial. Knowledge graph and ontology-based knowledge representation approaches have been proposed to structure domain experts' knowledge into an exploitable digital framework [11–13]; however, although some of them are quite promising in terms of process knowledge modelling, there seems to be a scenario specific bias as well as a lack of platform-level support for actual implementation. The same goes for highly manual process operations such as assembly procedures. In this case, AI and vision-based approaches have been developed to track human operations and assist them during the process. The tracking methodologies, such as the ones proposed in [14] and [15], are focused on modelling and recognising operators' movements and specific actions performed based on their know-how. These techniques, generally based on vision systems and AI-based image processing, can lead to a rather close description and supervision of the assembly sequence. The acquired knowledge can then be provided to support the assembly process with traditional means [16] as well as using more advanced AR technologies [17]. Other conceptual approaches have been developed for shop floor, they propose a method to support the operator with tools that include implicit knowledge but they are tailored for specific problems such as the process parameter choices [18] and fault analysis [19].

Overall, no matter if the explicit knowledge is well assessed, the implicit knowledge remains often embedded in the experts' subconscious and it is not used to increase the knowledge for a new operator that works on the same task or for the other workers at different process levels. Depending on the field of application, the implicit knowledge is generally in the form of custom implementations or modifications of a defined procedure with the aim of optimizing or simplifying it. From this point of view, the architecture here presented enables the formalization of both explicit and implicit knowledge by interacting with the main human actors, tracking their actions during a process, monitoring the effects of those actions and integrating knowledge reconciliation methodologies to merge the explicit knowledge with its implicit counterpart. Such a comprehensive tool for knowledge management in manufacturing processes could become a stepping stone to support decision-making at different hierarchical levels, with decisions based on the implicit and explicit expertise of human operators working in the same field, spanning its capabilities to different manufacturing scenarios. Despite all the efforts put forward to structure and further exploit the human experience in production processes, however, there seems to be a lack of a comprehensive tool capable of integrating this knowledge into a cohesive digital architecture, spanning multiple different production technologies and capable of integrating both the representation and the dispense of combined knowledge, explicit and implicit, for decision-making support and training purposes. This paper investigates the current manufacturing scenario, dealing with the mentioned processes and systems, identifying the gaps to be filled and proposing a novel approach to structure and model human knowledge dealing with manufacturing processes. The innovation proposed doesn't involve the platform-level tools, but rather it focuses on the implicit knowledge representation integrated with the explicit one.

2 Current Scenario and Platform Objective

In compliance with the knowledge formalization objective, we first consider the application context in which the research operates. Generally, any manufacturing process based on several subsequent transformations can be divided into several phases, where *qualified workers* act on the system through their expertise. Taking, for example, the single input of an agent on a part of a system that he/she governs with expertise, it is possible to state that – depending on the phase – he/she may make use of one or more software to make evaluations on the system, one or more HMI systems to visualize the results of a certain operation, a set of feedbacks that are completely subjective and derive entirely from his/her experience. In all cases considered, each worker - as depicted in Fig. 1 - will be able to interact with his/her reference system phase, visualize data/results pertaining to his/ her analysis and possibly repeat operations until, in his/her opinion, the system has arrived at an optimal result.

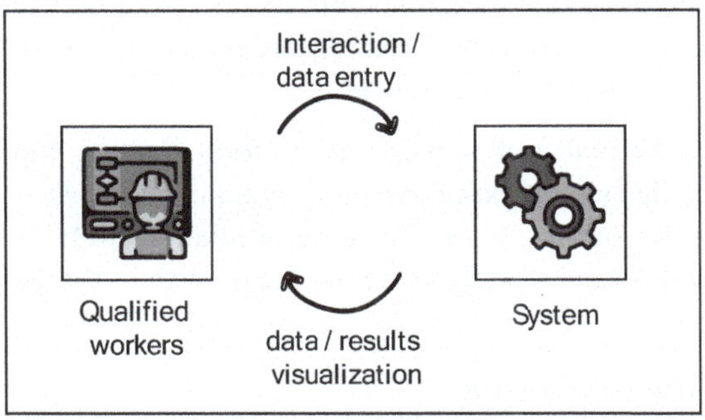

Fig. 1. Current scenario - agent abstraction at each step of a system.

Therefore, considering the entire system of transformation processes through time – each interacting with their system of reference – it will be based on the sequentiality of different decisions made by several workers. In particular, it is possible to describe the evolution of the system as per Fig. 2: at each stage of evolution, an agent will make decisions about the system at its stage that will be used as guidelines for the next one. It is possible to assume that, even without any formalization, the decisions made in the previous work phases will be translated into work constraints for the subsequent phases. In this context, if a decision made upstream is changed in one of the next phases it is possible to keep track of this change for the subsequent steps. Conversely, without a well-structured system, a modification in one of the last phases will never be visible to the workers who took some decision on the same aspects in an earlier step. This is the cause of fragmented information in the decision process that is strictly phase-specific: knowledge of the system resides in agents who look specifically at their phase without optimizing the decision on a system as a whole.

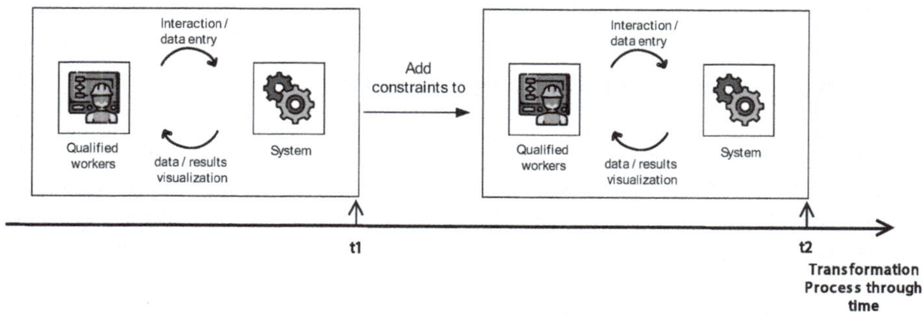

Fig. 2. Current scenario - agent abstraction in the sequential stages of a system.

Given the reference scenario for these types of systems, there are two gaps to be addressed:

- **Gap 1** - The reiteration of the analyses made by each worker in his or her work stage is purely implicit knowledge and is better the more experience he/she has.
- **Gap 2** - If several decisions over time concern the same domain of the system, it is still not possible to relate them across different qualified workers to improve the efficiency of the decision-making process.

The objective of the platform is, therefore, to formalize the implicit knowledge in order to improve the decision-making dynamics mentioned above. This article focuses on identifying the actor of this system, the type of relations to be instated for implicit knowledge formalization and describing a possible procedure for its formalization.

3 Knowledge Formalization

In the systems considered by this treatise (Fig. 2) - systems in which the operator's expertise is the main driver for optimization - different transformation processes follow one another, each bringing different constraints and modifications over time to the overall system. In this context, as illustrated in Fig. 3, the platform objective is to acquire knowledge from each stage of the transformation process. The main challenge relies on incorporating AI-based knowledge representation and reasoning (KR&R) tools to realize this knowledge relation for each worker. Such a formalization, once established, will allow the creation of a set of local/edge tools to interface process machines, and a series of enhanced human interaction tools, such as AR/VR, to acquire and provide know-how to human agents, facilitating operations and supporting decision making.

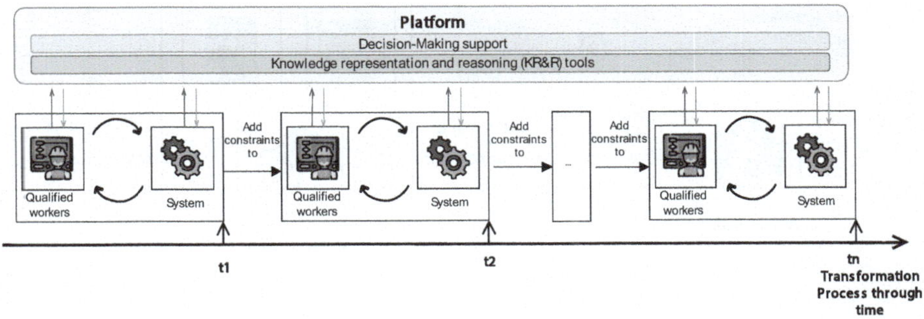

Fig. 3. Proposed platform scenario.

To overcome the described gaps, this section serves to identify the actors involved and the relations to be instated in order to define what *formalised implicit knowledge* means in the described context.

3.1 Knowledge Relations - The Actors

To formalise the knowledge from the workers and systems acting on the system over time, it is necessary to know (i) the types of systems with which the agents interact and (ii) to formalise the knowledge of the agents in a way that is compatible with the data acquired from the different systems involved in the transformation processes. With this perspective, workers and systems can be seen as *agents* that affect the system. In particular, it is possible to divide the agent as in the following, depending on their role within each of the transformation processes as can be visualised in Fig. 4:

- The *active agents* are the qualified workers, the knowledge-based agents that make decisions on the system based on their expertise on the system on which they are operating, which in turn helps them to speed up the transformation process. The inputs of this type of agent on the system are, as depicted in Fig. 4, the results from the system they interface with. Depending on the system behaviour, this type of agent makes decisions involving actions, i.e. changing parameters (also through physical actions on the system) and/or constraints on the system, which correspond to deductions regarding the state of the visualised system, just as occurs in the current scenario in Fig. 2;
- The *passive agents* represent the system with which the active agents interface through time, i.e. simulation systems, design systems but also processing machines. Passive agents always respond with the same output (usually, within minimal variations) to the same input under the same conditions of use. These systems are powered by the actions that an active agent can generate.

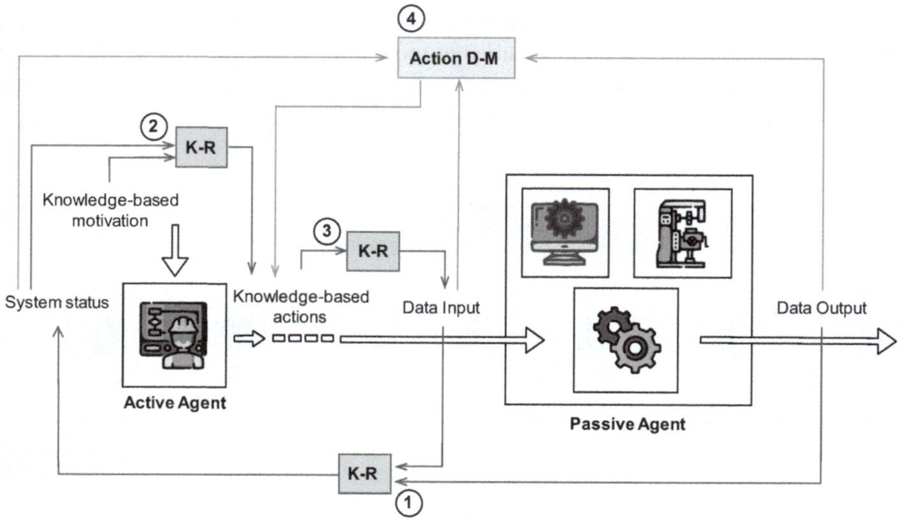

Fig. 4. Proposed platform scenario - actors, knowledge relations to be formalized (1, 2, 3) and the action decision-making tool (4) to be provided by the platform.

Each action performed by an active agent will affect the system, nevertheless, such an action may be objective and measurable, but also subjective and experience-based. It is evident that in the case of subjective actions and effects, these kinds of relations are not easy to monitor, therefore, information can be very easily lost in a standard process (i.e. the current scenario of Fig. 2). The platform's objective is to properly accumulate this information, analyse its quality and correctness – i.e. eliminate biased information – and make it usable for knowledge formalization.

3.2 Knowledge Formalization – Proposed Procedure

To fill the gap of knowledge loss for all the transformation processes of a system in which the operator's experience is the main driver for optimization, this paper proposes a possible knowledge formalization procedure by properly adding the missing relations.

Once described the actors involved in the generalization of this type of system, it is possible to understand how an active agent evaluates the state of the system of a passive agent (Data Output in Fig. 4) based on the modifiable parts of the passive agent it interfaces with (Data Input in Fig. 4), and its prior experience of working on that specific passive agent (knowledge-based motivations and actions). Therefore, a proposed formalization of the actions that an experienced active agent can implement on the system follows the steps below, as shown in Fig. 4:

1. Formalisation of the system state as seen by an active agent in terms of a pair of data that represent the input data given by an active agent to a passive agent and output data generated by this input;
2. Formalisation of the actions that can be taken by an active agent concerning the state of the system (formalised in step 1) and the knowledge-based motivation that depends on his/her experience in acting on a specific passive agent in that system state;

3. Formalisation of an active agent action and their interpretation in terms of input data to a specific passive agent;
4. Once all that information is formalised into the platform, the latter – knowing all the relevant information as illustrated in Fig. 4 – will be able to infer what type of actions to suggest to an active agent to perform an action on a specific observed system.

This is just a first novel attempt to apply knowledge formalization to the described systems. The platform must link all those different information and relations instated in order to be able to communicate to different active agents - who have different profiles depending on the type of activity and experience - the recommended actions.

4 Discussion and Conclusions

Starting from the implicit knowledge problem in manufacturing, this paper proposes an architecture for a platform that should be capable of formalizing such knowledge in order to use it to train properly other workers. In this respect, the paper proposes a formalization procedure based on a generalization of the system where the active agents and passive agents work together, and the system optimization relies on the expertise of the active agent. Such a formalization is, in this article, focused on transferring the knowledge in a predetermined way to the proposed platform for its fruition, answering to the identified Gap 1.

Future work will be focused on the formalization of the implicit knowledge in different case studies among the one treated in the introduction: a machining scenario, an additive manufacturing scenario and an assembly scenario. The following roadmap will be followed for each case study. Experienced active agents will be involved to identify their implicit knowledge and the type of passive agent (process/system) each of them interfaces with. Then, specific knowledge formalisation methodologies will be identified for the description of some specific activities that can (i) be constantly learnt in order to increase implicit knowledge and (ii) be integrated with information from the explicit knowledge coming from both active and passive agents. The ultimate goal is to develop a series of training methodologies to be implemented in the platform to speed up the training process with respect to standard procedures.

Contemporarily, dealing with different use cases, the research will explore the potential of the generalisation here stated to address the identified Gap 2. Indeed, it can be noted that the relationships between active agents interacting with different systems depend on the type of system they interact with, and actions are only correlated with each other if i.e. two systems are related and variations in one affect the other. Hence, the formalisation herein elaborated continues to be valid if knowledge relations are added to correlate all the passive agents that populate a system. Other future work will carry out the study of some use cases that can prove the feasibility of the approach proposed.

Acknowledgements. This work was supported by the Emilia-Romagna (RER) Regional Program PR FESR 2021–2027 (project KnowledgeX, grant number E37G22000450007) and by the Horizon Europe research and innovation program, grant agreement 101138627 – RemaNet.

Disclosure of Interests. The authors have no competing interests to declare that are relevant to the content of this article.

References

1. Deloitte Central Europe. 2023 Central Europe CFO Survey (2023)
2. Deloitte Central Europe. 2018 Deloitte skills gap and future of work in manufacturing study. A Deloitte series on the skills gap and future of work in manufacturing (2018)
3. Montes, J., Smith, C., Dajon, J. : The great retirement boom: the pandemic-era surge in retirements and implications for future labor force participation (2022). https://doi.org/10.17016/FEDS.2022.081
4. Igoa-Iraola, E., Díez, F.: Procedures for transferring organizational knowledge during generational change: a systematic review (2024). https://doi.org/10.1016/j.heliyon.2024.e27092
5. Xu, X., Wang, L., Newman, S.T.: Computer-aided process planning – a critical review of recent developments and future trends. Int. J. Comput. Integr. Manuf. **24**(1), 1–31 (2011). https://doi.org/10.1080/0951192X.2010.518632
6. Dey, A., Yodo, N.: A systematic survey of FDM process parameter optimization and their influence on part characteristics. J. Manuf. Mater. Process. **3**(3), 64 (2019). https://doi.org/10.3390/JMMP3030064
7. Tolio, T., et al.: Design, management and control of demanufacturing and remanufacturing systems. CIRP Ann. **66**(2), 585–609 (2017). https://doi.org/10.1016/J.CIRP.2017.05.001
8. Mourtzis, D., Vlachou, E., Xanthopoulos, N., Givehchi, M., Wang, L.: Cloud-based adaptive process planning considering availability and capabilities of machine tools. J. Manuf. Syst. **39**, 1–8 (2016). https://doi.org/10.1016/J.JMSY.2016.01.003
9. Milosevic, M., Lukic, D., Antic, A., Lalic, B., Ficko, M., Simunovic, G.: E-CAPP: a distributed collaborative system for internet-based process planning. J. Manuf. Syst. **42**, 210–223 (2017). https://doi.org/10.1016/J.JMSY.2016.12.010
10. Xiao, Y., Zheng, S., Shi, J., Du, X., Hong, J.: Knowledge graph-based manufacturing process planning: a state-of-the-art review. J. Manuf. Syst. **70**, 417–435 (2023). https://doi.org/10.1016/J.JMSY.2023.08.006
11. Sanfilippo, E.M., Belkadi, F., Bernard, A.: Ontology-based knowledge representation for additive manufacturing. Comput. Ind. **109**, 182–194 (2019). https://doi.org/10.1016/J.COMPIND.2019.03.006
12. Gouttebroze, S., Friis, J., Hovig, E.W., Boivie, K.: Toward semantic standard and process ontology for additive manufacturing. IOP Conf. Ser. Mater. Sci. Eng. **1281**(1), 012014 (2023). https://doi.org/10.1088/1757-899X/1281/1/012014
13. Dimassi, S., et al.: An ontology-based framework to formalize and represent 4D printing knowledge in design. Comput. Ind. **126**, 103374 (2021). https://doi.org/10.1016/J.COMPIND.2020.103374
14. Urgo, M., Tarabini, M., Tolio, T.: A human modelling and monitoring approach to support the execution of manufacturing operations. CIRP Ann. **68**(1), 5–8 (2019). https://doi.org/10.1016/J.CIRP.2019.04.052
15. Andrianakos, G., Dimitropoulos, N., Michalos, G., Makris, S.: An approach for monitoring the execution of human based assembly operations using machine learning. Procedia CIRP **86**, 198–203 (2019). https://doi.org/10.1016/J.PROCIR.2020.01.040
16. Keiser, D., Petzoldt, C., Walura, V., Leimbrink, S., Freitag, M.: Concept and integration of knowledge management in assembly assistance systems. Procedia CIRP **118**, 940–945 (2023). https://doi.org/10.1016/J.PROCIR.2023.06.162

17. Damgrave, R., Scheffer, S., Lutters, E.: Augmented reality support to employ tacit knowledge in non-conforming operations. Procedia CIRP **120**, 1475–1480 (2023). https://doi.org/10.1016/J.PROCIR.2023.09.196
18. Kernan Freire, S., Wang, C., Ruiz-Arenas, S., Niforatos, E. : Tacit knowledge elicitation for shop-floor workers with an intelligent assistant. In: Proceedings of the Conference on Human Factors in Computing Systems (2023). https://doi.org/10.1145/3544549.3585755
19. Hoerner, L., Schamberger, M., Bodendorf, F.: Using tacit expert knowledge to support shop-floor operators through a knowledge-based assistance system. Comput. Support. Coop. Work CSCW Int. J. **32**(1), 55–91 (2023). https://doi.org/10.1007/S10606-022-09445-4/TABLES/9

Predictive Maintenance for CNC Machines: Empowering Operators with Real-Time Insights

Miguel A. Mateo-Casalí(✉), Javier Mateos, Faustino Alarcón, and Francisco Fraile

Research Centre on Production Management and Engineering (CIGIP), Universitat Politècnica de València, Valencia, Spain
mmateo@cigip.upv.es
https://www.upv.es

Abstract. This paper describes a predictive model trained to predict failures of a computer numerical control machine. The model's objective is to support the shift from corrective maintenance operations to a preventive strategy where the model alerts operators of potential failures. The model is trained using computer numerical control data and alarm signals to learn situations that can lead to failure. When a possible issue is detected, the operators received a notification message while performing other routine tasks (hourly product quality checks). The aim is to improve process quality, reduce downtime and minimize the defective parts. At the same time, our approach empowers operators with real-time data and predictive insights, enabling them to autonomously perform maintenance tasks and make informed decisions to prevent equipment failures.

Keywords: Predictive maintenance · Dynamic control · CNC Machining · LSTM · RNN · Smart Manufacturing

1 Introduction

In the industrial sector, Computer Numerical Control (CNC) industrial equipment plays a crucial role in manufacturing high-precision metal parts. These automated systems enable machining tasks with precision and efficiency that are challenging to achieve through conventional methods. However, like any industrial equipment, CNC systems are susceptible to wear and failure, impacting the quality of the produced parts and the overall process efficiency. To address these challenges, companies often implement maintenance systems, such as corrective and preventive maintenance, to ensure optimal equipment performance. Nevertheless, these traditional approaches have limitations, often relying on periodic inspections or human intervention to detect and correct issues. There has been a growing interest in applying predictive models driven by artificial intelligence

A. A. Juan et al. (Eds.): DSA ISC 2024, LNCS 14778, pp. 330–341, 2025.
https://doi.org/10.1007/978-3-031-78238-1_30

techniques. These models allow for anticipating potential failures in CNC equipment by analysing performance data and operating conditions, facilitating the implementation of proactive maintenance strategies and reducing costs associated with unplanned downtime.

This case study aims to address this challenge by analysing machining data collected from the CNC machine (Nakamura Tome Turning Center model WT-100) to identify patterns of non-conformity in the process and establish predictive warning alarms. Successful implementation of this technology provides operators with advanced tools to avoid over-reliance on manual inspection equipment. By identifying potential failure patterns in the CNC production process before they occur, breakdowns and machine stoppages can be avoided, ensuring uninterrupted workflows and more efficient processes. In short, AI in smart manufacturing improves the final product's quality and optimises resources and overall productivity.

Considering the paper's objective, this article is divided into 5 sections. A first introductory section, followed by an analysis of the state of the art on the prevention problem in numerical control manufacturing. Then we will discuss how the data has been acquired and prepared and continue with the model building in Sect. 4. Finally, we will present the conclusions and next steps for the future.

2 Literature Review

In the industrial realm of numerical control manufacturing, predictive techniques utilising multiple sensors have been employed to enhance the precision of predictions concerning potential tool or machine failures during the CNC process. By combining sensor readings from CNC equipment, the process can be continuously monitored, aiding in understanding the process behaviour during production [1]. The scalability of this approach primarily lies in acquiring, storing, and generating deep learning models based on these sensors. This has led to the successful development and implementation of artificial neural networks such as Back-Propagation Neural Networks (BPNN) [2,3] or Localised Fuzzy Neural Network (LFNN) [4,5] for tool control and machine-tool relations compared to traditional regression and classification methods.

A specific focus has been on demonstrating the suitability of Long Short-Term Memory (LSTM) Recurrent Neural Networks (RNN) for handling multi-sensor data in milling processes. LSTMs are adept at retaining information over long sequences, capturing long-term dependencies effectively, and offering great model flexibility when dealing with multi-sensor data [6,7]. These models have proven effective in predictive maintenance methods, making them ideal for real-time integration into dynamic adaptive control methods within the milling industry [8].

Several research studies have highlighted the application of LSTM in milling processes [9]: A LSTM approach was used to forecast surface roughness during milling of an S45C steel dataset, showcasing its effectiveness in predicting surface roughness. A data-driven model was proposed for predicting the Remaining Useful Life (RUL) of milling tools based on convolutional neural networks (CNN) and stacked LSTM networks [10]. A study proposed a deep learning model based on LSTM architecture for a recurrent neural network to estimate cutting tool wear amounts, demonstrating the effectiveness of LSTM in capturing long-term dependencies for modelling sequential data [11]. Similarly, another study presented a novel approach using LSTM neural networks to reconstruct process forces in a five-axis milling center compared to model-based methods [12,13]. Long Short-Term Memory for Remaining Useful Life (RUL) prediction has shown promise in monitoring RUL based on milling process cycles [14]. These findings collectively enhance the potential of LSTM-based models to improve predictive maintenance and process monitoring within the milling industry.

As extracted from this literature review, Long and Short Term Memory (LSTM) Recurrent Neural Networks (RNNs) offer significant advantages for process monitoring and control in the milling industry. They stand out for their ability to capture complex patterns and long-term dependencies in data from multiple sensors during milling operations. This enables them to accurately predict parameters such as surface roughness, tool wear and remaining tool life [1]. Real-time integration of LSTM models into dynamic adaptive control methods has improved production efficiency, minimizing downtime and reducing maintenance costs by predicting potential tool or machine failures. Recent studies have demonstrated their effectiveness in predicting surface roughness [10] and remaining life [12,15], as well as in reconstructing process forces on milling centers. The adoption of LSTM models for predictive maintenance [16] and process monitoring promises to improve product quality and productivity in the milling industry through advanced AI techniques.

3 Dataset Acquisition and Preparation

The data acquisition process for LSTM predictive model training is based on collecting the twelve (12) signals provided by the Nakamure-Tome WT-100 CNC machine during CNC part production. These signals are distributed so that six (6) correspond to the top machining turret and the remaining six (6) to the bottom machining turret. Durind CNC machine operation, signal parameters are recorded every second, a time stamp is applied to all and each data signal and later stored in a local database. In addition to collecting CNC data in local databases, an alarm log is recorded throughout the production process and on different production days. This alarm log also contains a timestamp that allows the filtering of the signals from the CNC machine to obtain only the data that has caused the appearance of alarms that are of interest to the case study. By correlating the time stamps of the alarm log with the collected signal data, the training dataset is built for further training and LSTM model fitting.

3.1 CNC Signal Data Acquisition

This section defines the twelve signals received by the CNC machine and briefly describes the multi-sensor signal, data frequency and the measurement units in which they are presented. In the use-case, CNC signal data provides information regarding parameters such as speed, feed rate, spindle motor load, servo load, cutting time and tool temperature. Data collection frequency is one second, enabling the study of different time-bucket signal patterns. Multi-sensor data provided for the model build and train is presented in Table 1.

Table 1. Multi-sensor CNC data

Sensor	Description	Unit
Speed	Spindle motor speed	RPM
Feed Rate	Cutter velocity against the workpiece	mm/min
Spindle Motor Load	Spindle motor working load	%
Servo Load	Rated current consumption	%
Cutting Time	Cutter time against the workpiece	s
Temperature	Spindle motor temperature	°C

Each of the sensor data gathered from CNC equipment complies with the data structure specification on Table 2, which provides a brief signal description and data type proposed for all the sensor variables monitored.

Table 2. CNC signal data structure

Signal Parameter	Description	Data Type
'start'	Date and time of signal start time	TIMESTAMP
'end'	Date and time of signal end-time	TIMESTAMP
'value'	Value of the measured signal	INT

In additionally to the data signal acquisition for the machining towers, an identifier is used to differentiate and locate the signals whether they are related to the upper turret (P1) or to the lower one (P2).

3.2 CNC Alarm Data Acquisition

CNC alarm data provides information regarding the malfunctioning of the Nakamura CNC equipment and attends to the data structure defined in Table 3. The parameter alarm number code is the parameter that needs to be predicted according to pattern behaviour in the CNC signal data. Both data structures are related thanks to the coordination and synchronization of the timestamps that help to relate signal data event with potential upcoming alarms thanks to the use of LSTM RNN models.

Table 3. Alarm Data Structure

Header	Description
DateAndTimeOfOcurrence	Date and time of alarm in "YYYY-MM-DD HH:MM:SS"
TimeSpanOfOccurrence (minute)	Duration of the alarm in minutes
EquipmentName	Name of the equipment related to the alarm
MachineName	Name of the CNC machine
AlarmKind	Type of alarm (e.g., OT for 'Other', SV for 'Servo'...)
AlarmNumber	Alarm code number
AlarmMessage	Alarm Message

3.3 Data Extraction and Dataset Preparation

In a deep learning model, preparing a quality dataset is an essential task that directly affects neural network training and behaviour. For this purpose, the data pipeline that follows the data collected from the CNC machine sensor and the alarm data is defined. All data is stored in an on-prem relational database (PostgreSQL) that will allow the insertion of newly generated data and the query from local devices.

To better understand the approach taken, it is important to think about the architecture used within the data pipeline. Firstly, the CNC signal and alarm data are read, filtered and inserted into the database. This process is facilitated by a Python data importer script that aggregates all the pre-processed data and creates a hypertable within the database. This hypertable stores the readings per alarm event, providing a structured repository for subsequent analysis.

Once the data has been collected and stored, a data preparation script is used to improve the consistency and coherence of the data. Specifically, the data preparation script acts as a time synchroniser, ensuring that timestamps across different datasets are accurately aligned. By imputing missing data, normalising values and downsampling where necessary, the data hypertable is optimised for further analysis.

The prepared data is then transformed into a materialised view, which is input to our Long Short-Term Memory (LSTM) architecture. This architecture, built using tools such as Keras Tuner [17] in Python, is designed to analyse sequential data patterns effectively. By harnessing the power of LSTM networks, we aim to uncover insights and predictive patterns within CNC signal and alarm data, ultimately improving operational efficiency and decision-making.

4 Model Building and Training

This section proposes a model build for multisensor sequence learning that covers the LSTM model architecture design, hyperparameter optimization and training process.

4.1 Model Building

When multi-sensor data comes from different sensor sources it is common to find different data acquisition rates, irregular intervals or missing data. By using LSTM models these irregularities can be easily handled without affecting the inference performance (Fig. 1).

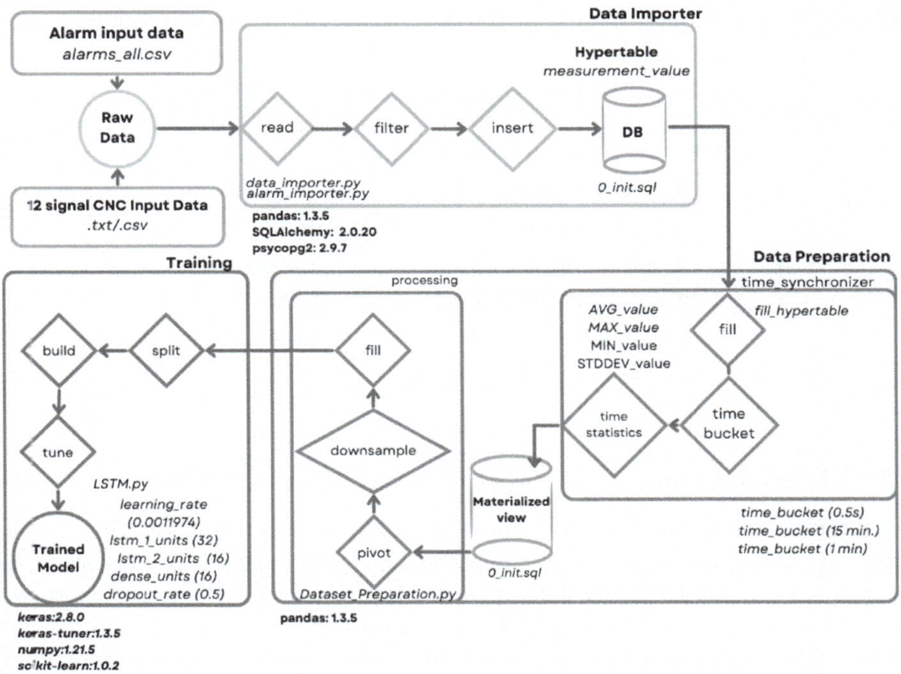

Fig. 1. Data Pipeline

LSTM-based models have demonstrated state-of-the-art performance in various multi-sensor applications [18] making them the most suitable model for the proposed use-case. Our LSTM model architecture is based on Keras native Python library which contains two (2) LSTM layers two (2) dense layers and one (1) dropout layer.

The first LSTM layers consists of 32 neurons with a Rectified Linear Unit (ReLu) activation function, configured to returns sequences. In contrast, the second LSTM layer comprises 16 neurons, also ReLU activation but returns only the output from the last step. Following both LSTM layers, a fully connected dense layer is added. This one connects every neuron from the previous LSTM layer containing 16 dense unit matching the 16 neurons from the previous LSTM layer.

The fourth layer incorporated is a dropout layer that prevents models from overfitting as it has been validated in several deep learning models [19,20] in various milling multi-sensor use cases. Thanks to setting to zero a fraction of the input units during each training iteration, which is known as dropout rate. Lastly, a one-neuron dense layer provides a binary output that determines the occurrence potential of an alarm activation. By utilizing the sigmoid activation function, this layer produces a binary output normalized to the range [0,1] that represents the model's confidence in the classification outcome.

All the hyperparameters used to build the LSTM model architecture have been optimized through the use of Keras tuner as can be seen in Table 4.

Table 4. Optimal Hyperparameters for LSTM Layers

Layer ID	Layer Parameter	Keras Optimal	Activation Method
1	'LSTM 1 Neurons'	'32'	'ReLU'
2	'LSTM 2 Neurons'	'16'	'ReLU'
3	'Dense 1 Units'	'16'	'Default'
4	'Dropout Rate'	'0.5'	N/A
5	'Dense 2 Units'	'1'	'Sigmoid'

4.2 Model Training

Once the model architecture is defined and its hyperparameter selection is validated the LSTM training process can start. The materialized view data stored is used as our dataset for the training process. Before starting the training process the dataset is subdivided with the help of train-test-split function into two (2) subsets. Training subset is composed of 80% of the dataset while the remaining 20% is used for validation and testing purposes.

Commonly, when the dataset is small training parameters are optimized by deploying a Random Search hyperparameter tuning process with Keras Tuner. For the proposed model architecture, the hyperparameter tuning objective is to maximize the validation metric that measures the area under ROC curve (AUC). ROC AUC metric validation is specially important in binary classification tasks [21,22]. Key hyperparameters to be optimized include the learning rate, which controls the step size during the training process. Batch size determines the number of samples processed on each epoch iteration affecting the training duration. Epochs specify the number of model iterations through the dataset before overfitting happens and after model convergence is reached. Lastly, Max Trials and

Executions define the extent to which the optimization process is carried out. The optimal hyperparameters used for the training process are presented in the Table 5.

Table 5. Optimal Model Training Hyperparameters

Hyperparameter	Best Result
Learning Rate	0.0003
Batch Size	32
Epochs	100
Max Trials	10
Executions	1

4.3 Model Integration Deployment

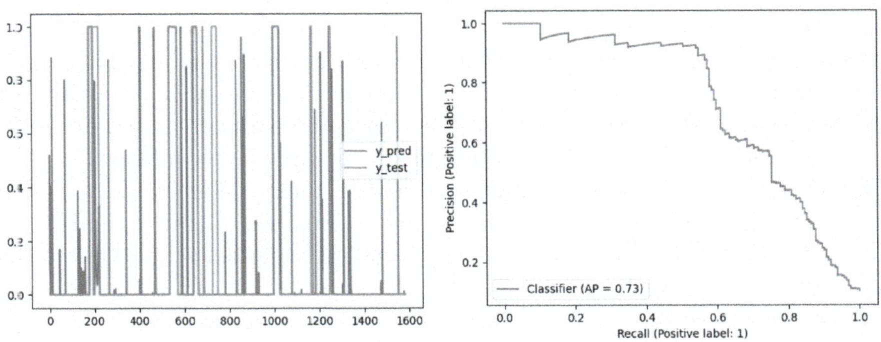

Fig. 2. Model Prediction vs. True Labels (left) and Precision-Recall Curve (right)

Once our algorithm is launched, it produces two different outputs that represent different aspects of the performance of the trained machine-learning model (Fig. 2):

1. **Graph of Predictions vs. True Labels:** This chart shows the model predictions (y_pred) overlaid with the true labels (y_test). This allows you to visualise how the model predictions compare to the actual values in the validation dataset. Ideally, the model predictions should closely follow the true labels, so a good fit is shown as a close overlap between the two lines.

2. **Precision-recall curve (precision-recall curve):** This chart shows how the precision and recall of the model vary as the classification threshold is adjusted. Precision measures the proportion of correct optimistic predictions among all positive predictions, while recall (also known as sensitivity) measures the proportion of positive cases the model correctly identified. The idea is to maximise both accuracy and recovery, so this chart helps to assess the balance between accuracy and recovery and select the optimal threshold for each specific application.

The two charts generated by this algorithm represent different aspects of the performance of the trained machine-learning model. First, the Accuracy and Recall Graph (accuracy-recall curve) illustrates how the model's accuracy and recall vary as the classification threshold is adjusted. Precision indicates the proportion of correct optimistic predictions among all positive predictions, while recall (also known as sensitivity) measures the proportion of positive cases the model correctly identified. The ideal is to maximise both accuracy and recall. This graph is essential for assessing the trade-off between accuracy and recall and selecting the optimal threshold for a specific application.

4.4 Model Results in Factor

The i4Q Intelligent Manufacturing Line Reconfiguration Toolkit (i4QLRT), developed in the i4Q project, is a set of optimization microservices that employ simulation to enhance manufacturing line configuration and achieve quality targets. Leveraging AI algorithms, it calibrates machine parameters and optimizes line configuration to increase productivity and simplify line reconfiguration. The i4QLRT solution is interoperable, deployable on Linux or Windows servers, and adaptable to various industries. It offers pre-trained and abstract models and integrates seamlessly with other services through standards such as OAuth and APIs.

In alignment with these capabilities, the i4QLRT solution has been implemented alongside the model developed within FACTOR company. FACTOR Ingeniería y Decoletaje S.L, based in Valencia, Spain, specialises in metal machining and precision turning, serving industries such as aeronautics, automotive and medical. Its priority objective is to guarantee product quality to the customer, avoiding defects that impact efficiency, cost, and customer satisfaction. FACTOR's primary focus lies in enhancing factory efficiency by implementing i4Q solutions. Their main objective is to improve Overall Equipment Effectiveness (OEE) by eliminating defective parts, reducing machine downtime, and optimizing overall equipment efficiency. i4QLRT contributes to this goal by predicting defective parts and optimizing machine performance, thus complementing FACTOR's efforts to enhance factory efficiency (Fig. 3).

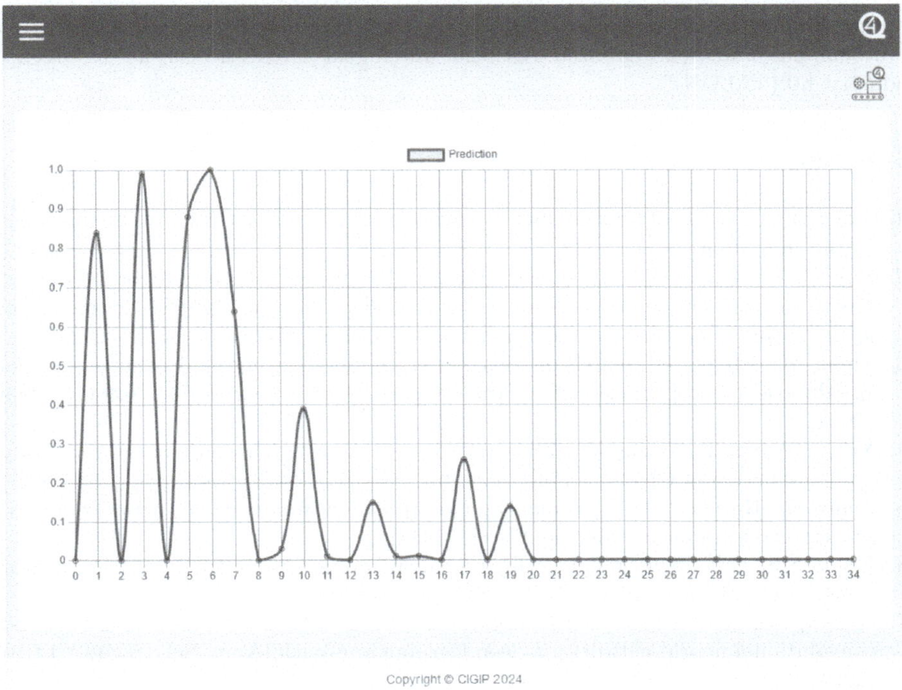

Fig. 3. Result Line Reconfiguration Toolkit (LRT) plot

5 Conclusions and Future Steps

In conclusion, the LSTM-based predictive model developed for CNC machine failure prediction shows significant potential for improving manufacturing processes. Thanks to the extensive data acquisition, preparation, model building and training processes, the model provides valuable information about potential equipment failures, enabling proactive maintenance strategies and optimising operational efficiency.

Future work will require further refinement of the model, incorporating new data sources and refining the hyperparameters to improve prediction accuracy. On the other hand, predicting what type of alarm or failure will occur should be possible. In addition, continuous monitoring and evaluation of model performance will be critical for ongoing optimisation and adaptation to changing manufacturing environments. All of this will be realized in the TALON project with the collaboration of the i4Q project's LRT solution with the goal of providing a better optimization solution that facilitates improved factory efficiency, driving continuous improvement and innovation in manufacturing processes.

Acknowledgment. The research leading to these results received funding from the European Union H2020 programs with grant agreement No. 958205 "Industrial Data Services for Quality Control in Smart Manufacturing (i4Q)", and the Horizon

Europe Framework Programme (HORIZON) with grant agreement No. 101070181 "Autonomous and Self-organized Artificial Intelligent Orchestra-tor for a Greener Industry 4.0 (TALON)".

References

1. Jiang, Y., et al.: Tool wear prediction upgrade kit for legacy CNC milling machines in the shop floor, p. 10 (2022)
2. Lin, C.-Y., Wu, S.-C., Hu, C.-H., Jian, B.-L.: Spindle thermal error prediction based on back propagation neural network and hybrid taguchi genetic algorithm for a computer numerical control machine tool. Sens. Mater. **35**, 4397 (2023)
3. Zhou, M., Yin, L.: Research on thermal error modeling method of machine tool spindle based on optimized BP neural network. J. Phys.: Conf. Ser. **2694**, 012069 (2024)
4. Yu, J., Wang, T., Wang, W., Zhao, G., Yao, Y.: Fuzzy reasoning based reliability fault prediction of CNC machine tools. Int. J. Adv. Comput. Sci. Appl. **14** (2023)
5. Zhou, E., Harrison, D.K., Link, D., Healey, R.: A fuzzy neural network for in-process error compensation, pp. 63–66 (1994)
6. Dimla Snr, D.E.: Sensor signals for tool-wear monitoring in metal cutting operations - a review of methods. Technical report 8 (2000)
7. Arslan, M., et al.: Tool health monitoring using airborne acoustic emission and convolutional neural networks: a deep learning approach. Appl. Sci. (Switz.) **11**(6) (2021)
8. Mbilong, P.M., Aarab, Z., Belouadha, F.Z., Kabbaj, M.I.: Enhancing fault detection in CNC machinery: a deep learning and genetic algorithm approach. Ingenierie Syst. d'Inform. **28**(5), 1361–1375 (2023)
9. Jun, W., Kui, H., Cheng, Y., Zhu, H., Shao, X., Wang, Y.: Data-driven remaining useful life prediction via multiple sensor signals and deep long short-term memory neural network. ISA Trans. **97**, 241–250 (2020)
10. Manjunath, K., Tewary, S., Khatri, N.: Surface roughness prediction in milling using long-short term memory modelling. Mater. Today: Proc. **64**, 1300–1304 (2022)
11. An, Q., Tao, Z., Xu, X., El Mansori, M., Chen, M.: A data-driven model for milling tool remaining useful life prediction with convolutional and stacked LSTM network. Meas.: J. Int. Meas. Confederation **154**, 107461 (2020)
12. Zhang, B.S.J.: Recurrent neural networks with long term temporal dependencies in machine tool wear diagnosis and prognosis. Eng. Appl. Artif. Intell. (2017)
13. Denkena, B., Bergmann, B., Stoppel, D.: Reconstruction of Process Forces in a Five-Axis Milling Center with a LSTM Neural Network in Comparison to a Model-Based Approach. J. Manuf. Mater. Process. **4**(3) (2020)
14. Huang, Y.C., Chen, Y.H.: Use of long short-term memory for remaining useful life and degradation assessment prediction of dental air turbine handpiece in milling process. Sensors **21**(15) (2021)
15. Feng, Y.: A holistic approach for improving milling machine cutting tool wear prediction. Appl. Intell. **53**(24), 30329–30342 (2023)
16. Lee, H., Kang, D.H., Jeong, S.C.: A study on the predictive maintenance system for CNC facilities based on intelligent cutting tool load prediction. ICIC Express Lett. **16**(3), 265–273 (2022)
17. O'Malley, T., et al.: Keras Tuner (2019). https://github.com/keras-team/keras-tuner

18. Dimla, D.E., Lister, P.M., Leighton, N.J.: Multi-sensor integration method of signals in a metal cutting operation via application of multi-layer perceptron neural networks. Technical report 440 (1997)
19. Liu, H., Liu, Z., Jia, W., Lin, X., Zhang, S.: A novel transformer-based neural network model for tool wear estimation. Meas. Sci. Technol. **31**(6) (2020)
20. Chen, J., Chen, D., Liu, G.: Using temporal convolution network for remaining useful lifetime prediction. Eng. Rep. **3**(3), 1–17 (2021)
21. Pedregosa, F., et al.: Scikit-learn: machine learning in Python. J. Mach. Learn. Res. **12**, 2825–2830 (2011)
22. Wu, S., Flach, P., Ferri, C.: An Improved model selection heuristic for AUC, pp. 478–489. Springer, Heidelberg (2007)

Deep Learning Applied to Automated Visual Defect Detection for Wind Towers Painting Process

Joan Lario[1](✉), Natalia Pérez García-de-la-Puente[2], Javier Mateos[1], and Salih Aksu[3]

[1] Research Centre on Production Management and Engineering (CIGIP), Universitat Politècnica de València, Camino de Vera S/N, 46022 València, Spain
joalafe@upv.es
[2] Instituto Universitario de Investigación en Tecnología Centrada en el Ser Humano, Universitat Politècnica de València, Camino de Vera S/N, 46022 València, Spain
[3] GRI Renewable Industries, Madrid, Spain

Abstract. Zero Defect Manufacturing (ZDM) strategies focus on promptly and precisely detecting defects to reduce material and energy consumption, and avoid product failure while the product is in use. Integrating a computer vision defect detection system is an important approach to improving the quality of inspection policies and the performance of the manufacturing process. Last year's high attention has been paid to surface image defect detection based on deep learning algorithms. The deep learning methods on automated vision systems require large quantities of annotated data. Acquiring defects is a costly and time-consuming task in industrial contexts since defects only occur in small percentages, data is biased through the predomination of non-defective samples, and there is a lack of publicly available datasets to train the algorithms. This study aims to evaluate the performance of a deep learning algorithm based on the dataset employed and the training parameters selected.

Keywords: Defect detection · artificial intelligence · computer vision · quality inspection · deep learning · surface defects

1 Introduction

Non-destructive inspection technologies (NDIT) are evolving quickly due to the introduction of new sensors, machine learning algorithms, big data, and the Internet of Things (IoT) [1–3]. The zero defects manufacturing strategy is a data and technology-driven paradigm that reduces waste and cost and increases product quality and process productivity [4]. To deploy real-time NDIT on the manufacturing shop floor, the solutions should integrate new artificial intelligence technologies to improve the accuracy and automation needs for production implementation [5]. The NDITs are focused on the execution of repetitive inspection tasks that conventionally are performed out of the line manually or even with destructive methods.

A. A. Juan et al. (Eds.): DSA ISC 2024, LNCS 14778, pp. 342–352, 2025.
https://doi.org/10.1007/978-3-031-78238-1_31

In the painting and coating industrial processes, defects such as pits, holes, underfill, and scratches appear due to failures related to equipment, raw materials or production environment conditions [6]. The painting of wind towers acts as a surface treatment to increase the corrosion resistance of the steel; in the long term, this corrosion mechanism can result in equipment failure due to the high demand conditions of the offshore environment. The painting defects increase the costs incurred by the manufacturers, including reprocessing, warranty, and reputation costs, and reduce the product's service life.

Manual visual inspection is the conventional technique for detecting defects in painting or coating treatments [7]. The painting process quality can be improved if the visual inspection is performed by an automated vision system rather than manually by operators. The performance of automatic vision systems is influenced by the robustness, accuracy and time efficiency in the defect detection process; this article covers these topics and provides an early use case for painting inspection solutions.

Integrating non-destructive inspection solutions based on deep learning algorithms will improve production quality, which can be considered a core competency that industrial companies should possess to improve their performance [8]. Automatic vision inspection systems significantly improve the accuracy and reliability of defect detection since the inspection does not rely on aspects such as operator training, mental or physical fatigue [9, 10].

In recent years, industrial production in various fields (electronic production, steel manufacturing, etc.) has benefited from applying deep learning artificial intelligence algorithms [11]. One of the main advantages that position the deep learning algorithm as a standard in supervised computer vision is that it automatically learns and extracts features with strong predictive value, reducing the preprocessing and fine-tuning required. Deep learning can be considered a subfield of machine learning and artificial intelligence, which, in the last decade, has been demonstrating outstanding performance in defect detection in computer vision [12–14].

The most common methodology to train deep learning algorithms relies on supervised machine learning [6]. The approach employs the human annotation of the images for classification and object detection for training. Frequently, in industrial environments the main defect samples found in painting or surface defects are assigned to a few types of defects or category, and in the general dataset the vast majority of the data accounts for non-defective samples [5, 15]. The lack of balanced data on deep learning algorithms can produce inaccurate results due to prioritisation performed by the standard loss functions [16]. Therefore, to deploy automatic deep learning visual inspection solutions in industrial product quality detection, a vast amount of data should be acquired, labelled and trained, increasing the algorithm's human labour and computational burden. The main challenge to applying this technology is that defect rates are low in manufacturing environments, increasing the efforts to obtain a representative dataset since the companies do not offer public dataset [17, 18].

The current author's research objectives are constructing a novel dataset of painting defects and performing several training scenarios to investigate the evolution of the average precision based on the deep learning model employed and the size of the dataset employed.

2 Research Methodology

The deployment of an automatic vision system involves several steps, image acquisition (hardware), image processing, and machine learning, to close the quality assurance loop (Fig. 1). These new Zero Defects Zero Waste (ZDZW) solutions for quality assurance involve various sensors or technologies to create the inspection platform for the final integration with the manufacturing control system.

The accuracy, efficiency and performance of the machine vision-based visual detection equipment not only rely on the artificial intelligence algorithms, but another important parameter is the image acquisition hardware. The conventional computer vision inspection system comprises three modules: image acquisition, image processing, and machine learning (Fig. 1).

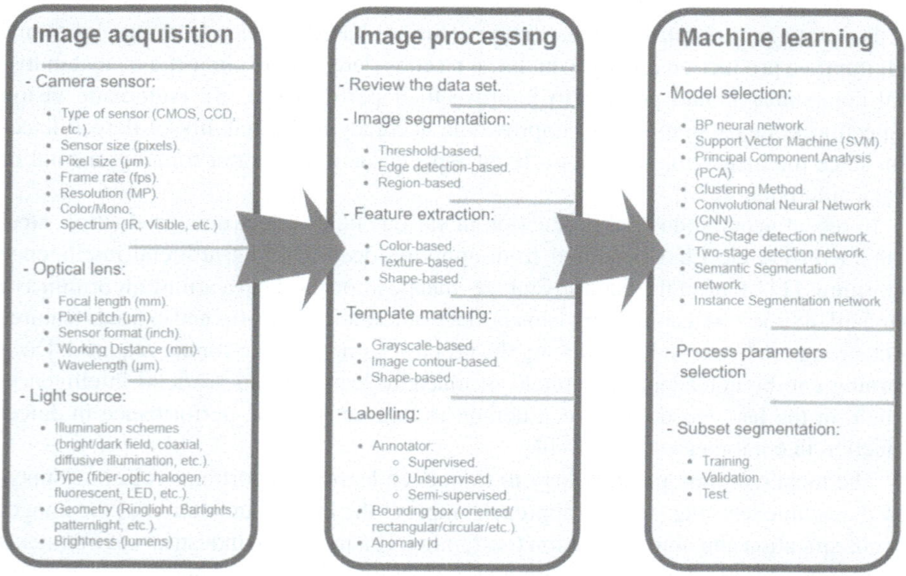

Fig. 1. Automatic vision system diagram for painting inspection.

2.1 Image Acquisition Hardware

The sensor size and resolution, speed and image stabilization are factors that influence the machine-vision inspection process to detect small defects that occur in the painting process. The camera Basler acA5472-5gc has been selected for the image acquisition system, which uses the Sony IMX183 color CMOS sensor of 20 Megapixels, pixel size 2.4×2.4 μm and sensor size of 13.13×8.76 mm. Another basic hardware element for image acquisition is the optical lens, which influences the contrast, aberration, working distance, focal length and field of view. A focal length lens Fujinon Lens CF25ZA-1S F1.8 f25 mm 4/3, which provides for a field of view (FOV) of around 30 cm in width and 20 cm in height (around 18 px/mm of spatial resolution), is employed. The selected camera and lens allow the acquisition of images from the wind tower for defect detection with a working distance of around 50–55 cm. Strategically mounting up to six cameras on an autonomous robotic platform enhances the efficiency and speed of the quality inspection process.

Due to the high inspection surface of the wind towers, the image acquisition is performed in an open workshop with poor lighting to capture the defects with a correct illumination scheme and source. The correct design lighting system improves visual contrast, reduces environmental interference and highlights defects. Current research employs barlight model MBJ SBL-0150 LED light source, with incidence angle of 40°, due to their ease of integration, low consumption, and high lifespan.

2.2 Image Processing and Training Dataset

Deep learning model training relies on using high-quality datasets and amounts of annotated defects to enhance the precision and dependability of Artificial Vision systems. The datasets employed in the current research cover a wide range of defects that occur during the painting process of the wind towers. Critical defect classes include pinholes, blistering, inclusions, scratches, delamination, and crumples, each of which has significant implications for the quality of the painted surface (Fig. 2). After completing the data acquisition for the wind energy use case, all images undergo annotation using squared Bounding Boxes (BBs), facilitating the identification of anomalies within the image.

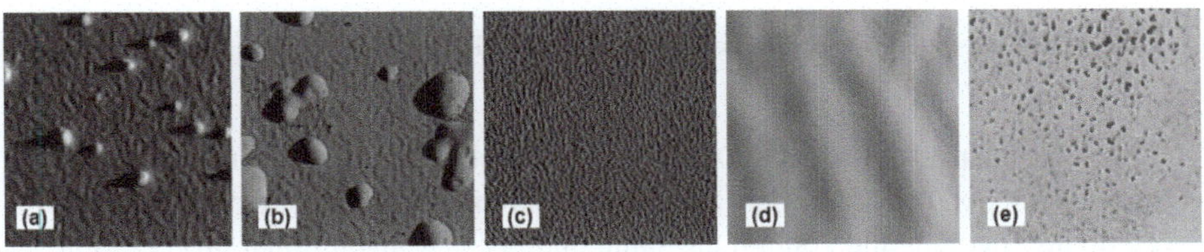

Fig. 2. Painting defects on wind tower section: (a) inclusion, (b) pinhole, (c) crumple, (d) delamination and (e) scratch.

The deep learning training carried out by supervised learning techniques learns from the annotated subset to extract meaningful patterns and features that will be employed to predict class labels of unseen datasets. Current research is composed of three separate training datasets. The datasets for each training iteration are described in Table 1. The dataset acquired has been divided into three subsets terms with different weight each one, which are train (70%), test (20%) and validation (10%). The different YOLO v8 models have been trained on the train in a subset of, and the evaluated on the validation subset.

2.3 Machine Learning

The current authors' research employs Convolution Neural Network (CNN) architecture for feature extraction and image defect detection of the painting defects. The algorithm

Table 1. Training dataset per iteration.

Dataset	Images	Anomalies	Inclusions	Pinhole	Scratch	Delamination	Crumple
1st Iteration	299	342	313	11	15	2	1
2nd Iteration	401	450	403	20	24	2	1
3rd Iteration	443	554	441	63	47	2	1

employed in the current research is YOLO v8, which employs a deep neural network and fixed-grid regression to carry out its functions. One of the main advantages of the deep learning defect detection algorithm is that perform the feature extraction, selection, and classification of painting defects.

Table 2. Deep learning models and specifications.

Model	Layers (units)	Neurons (M)	FLOPs[a](B)
YOLOv8n	365	3.2	8.7
YOLOv8s	365	11.2	28.6
YOLOv8m	365	25.9	78.9
YOLOv8l	365	43.7	165.2
YOLOv8x	365	68.2	257.8

[a]Number of floating point operations needed (in CPU or GPU) to run a single model inference

The performance of the YOLOv8 algorithm architecture is directly related to the selected model. For YOLOv8, there are six (6) different models that are based on the same algorithm architecture and contain the same number of convolutional layers (Table 2). However, the organization the neurons within the layers provides different performances to each model. Models with fewer neurons tend to perform better in time-constraint environments due to simple layer configuration. On the other hand, models with a vast number of neurons tend to provide significantly better mean Average Precision (mAP) results at the expense of computational resources when big datasets are employed for training purposes. Models with a higher number of neurons require a greater number of floating point operations, which translates into higher quality hardware required to achieve real-time inspection inference.

The defect detection model is configured for the training process according to the hyperparameter settings described in Table 3. These parameters define the training process and are, among others, Learning rate initial (lr0) and final (lrf), provide information on how quickly the model updates each iteration. Epochs define the training time and represent the model passing through the dataset. It is needed for the model to converge, and too many epochs may result in overfitting models. Batch Size provides the number of images passed to the model in each epoch iteration. Image Size provides the resolution conversion to the training dataset. The YOLOv8 backbone architecture is based on a

Cross Stage Partial Network (CSP-Net). During the training process, the NVIDIA GPU RTX model is used for accelerated computation.

Table 3. Deep learning training parameters.

Model hyperparameter	Description	Value
Learning rate	Affects model training time and iteration update time (lr0 and lrf)	0.01
Epochs	Number of times the model iterates through the whole dataset	150
Batch Size	Number of images passed simultaneously to the model	[4–16]
Image Size	Image resolution conversion	[640–2088]

3 Results and Discussion

Three parameters should be considered to evaluate the performance of a defect detection algorithm. The first parameter is the acquisition time, which refers to the time consumed to acquire and label the dataset needed for training, verification and testing. The second is the computational time required to train the artificial intelligence algorithm according to the parameters selected. The last parameter is the response time, which refers to the time employed from the instant that a task detection is created until the algorithm provides the detection result. The response time should be at least three times lower than the production time to avoid stoppage due to abnormal behaviour of inspection equipment. Figure 3 shows the classification results after execution the prediction model on the test dataset.

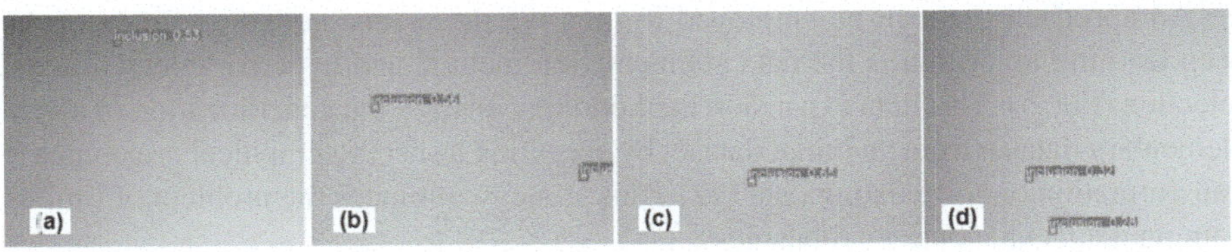

Fig. 3. Classification results of painting images after prediction model execution.

To evaluate the performance of the prediction model training by deep learning, the mean Average Precision (mAP) has been employed to evaluate the accuracy and Intersection over Union (IoU) to measure the overlap between two Bounding Boxes (BBs). The first performance metric, mAP50, determines a prediction as a truth of an IoU limit over 50%. The second performance metric is the mAP50-95, which determines a prediction as a truth of an IoU limit over 95%. At the first training scenario the deep learning algorithm was trained with YOLOv8n model employing an image resolution of 640 pixels and three different datasets (Table 1). The performance metrics used are summarized in Table 4. The results show that an increase of 212 anomalies in the third training set improved the mAP50 by 133% and the mAP50-95 by 83% concerning the first dataset trained.

Table 4. Performance metrics for scenario 1 of Deep learning training.

Iteration	Images	Anomalies	Resolution (pixels)	Model	mAP50	mAP50-95
1st Iteration	299	342	640	YOLOv8n	0.27	0.12
2nd Iteration	401	450	640	YOLOv8n	0.37	0.16
3rd Iteration	443	554	640	YOLOv8n	0.63	0.22

Creating comprehensive datasets is a laborious and time-consuming process; since defective samples are naturally rare in the industrial process, the data collected should be annotated by skilled operators to identify the defects and consume a considerable amount of data for the training model phase. All these factors inhibit the further spread of deep learning algorithms for automatic industrial inspection. Machine learning training presents several strategies for dealing with small datasets, such as data augmentation, transfer learning, pre-trained models, or synthetic training data [6, 19–21]. The second training scenario trained the deep learning algorithm by modifying the YOLOv8 hyperparameters, image resolution and employing the two datasets with more labelled anomalies.

As shows in Fig. 4 the YOLOv8s employing 1080 pixels image resolution achieve better parameter-accuracy than the other configurations studied, surpassing other models that employs same image size. This improvement in accuracy has been observed in both data, and the best results improve in 107% the mAP50 and 100% the mAP50-95. Also, the increase of annotated defects in 3rd iteration has increase the accuracy the performance of YOLOv8n (Fig. 4).

An approach that can be employed to improve the precision and accuracy of the deep learning algorithm is the data augmentation method and hyperparameters model selection (Fig. 5). This data expansion methodology enables the generalization of a supplementary dataset from an initial dataset by executing a specific graphical or geometric transformation on the existing data [22]. This strategy mitigates the problem of limited training data, enhancing its robustness.

The vision inspection system for painting defects requires high-resolution images to be processed with high accuracy, which requires a longer response time that may affect the shop floor deployment. One of the main challenges of automatic inspection based on the deep learning algorithm is the real-time performance of online quality assurance. Even though the deep learning model has been trained on-premise with the datasets, it takes a long time to train and calculate the inspection results if the task is performed Central Processing Unit (CPU) (Table 5). The computation time can affect the real-time response and execution efficiency of the automatic inspection solutions; this issue can be improved by integrating high-performance processing units to complete defect recognition such as Graphics Processing Units (GPUs).

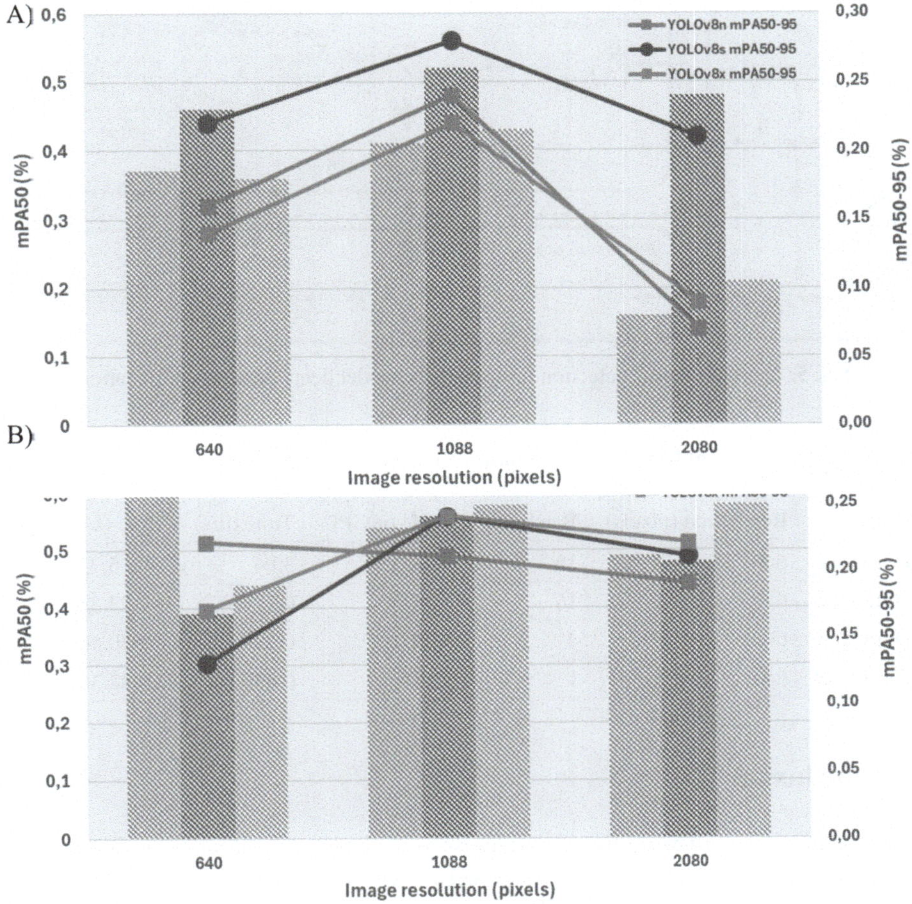

Fig. 4. Parameter accuracy based in dataset and deep learning model employed. A) 2nd interaction dataset. B) 3rd iteration dataset.

The capability of autonomous machine vision systems to detect painting defects during wind tower production is crucial for quality assurance of the final product quality. In the wind energy sector, to improve manufacturing performance, the NDIT solutions must send feedback to the manufacturing process to repair, reprocess or adjust process parameters to decrease energy and materials consumption, as well as productive resources. Finally, to close the quality loop for defect detection, the information of the deep learning algorithm could be integrated with an Autonomous Guided Vehicle (AGV) and Augmented Reality Platform (ARPs). Thanks to AGV and ARP Solutions, the painting inspector can locate the defect in the real world and repair it before sending it to the next working station. The interoperability between different ZDZW solutions will increase the manufacturing resource utilization rate and quality output, decrease manufacturing costs, and improve sustainable manufacturing.

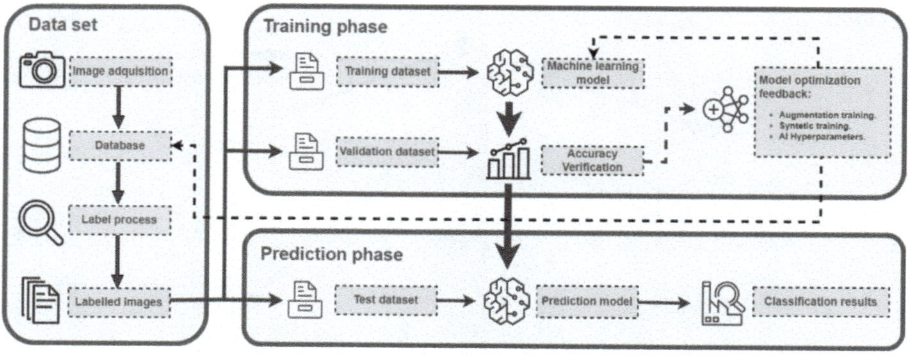

Fig. 5. Painting defect detection flow chart for model deep learning optimization.

Table 5. Time consuming for Deep learning training.

Model	Resolution (pixels)	Batch Size	CPU or GPU	Time (it/s)	Time (s/epoch)
YOLOv8n	640	16	GPU	3.84	5.73
YOLOv8n	640	16	CPU	0.17	127.38
YOLOv8x	640	4	GPU	0.32	1070.55
YOLOv8x	640	1	CPU	0.004	72432.36

4 Conclusion

This article has summarised the research efforts to deploy an automatic vision system based on deep learning algorithms for detecting surface defects in the industrial wind tower painting process using a ZDZW strategy.

Deploying industrial vision technology can improve defect detection capabilities in early process stages, reducing material and energy consumption, equipment utilization and improving productivity. The deep learning algorithms applied to painting defect detection have led to a model that enables the identification of small defects (>40 mm^2) in huge inspection surfaces (>500 m^2), which are challenging tasks to be performed manually by operators.

Regular defect data collection is crucial for model training to adapt to new defect types and ensure balanced datasets. Adding more annotated anomalies enhances the model's ability to detect and classify anomalies in various conditions, improving solution reliability and effectiveness. To effectively implement automatic computer vision inspection equipment for painting or coating defect detection, methods such as data augmentation should be employed to increase the training data set for deep learning algorithms.

Acknowledgments. The ZDZW project has received funding from the European Union's Horizon Europe program under grant agreement No 101057404. Views and opinions expressed are, however, those of the author(s) only and do not necessarily reflect those of the European Union. Neither the European Union nor the granting authority can be held responsible for them.

Disclosure of Interests. The authors have no competing interests to declare that are relevant to the content of this article.

References

1. Morgan, J., Halton, M., Qiao, Y., Breslin, J.G.: Industry 4.0 smart reconfigurable manufacturing machines. J. Manuf. Syst. **59**, 481–506 (2021). https://doi.org/10.1016/j.jmsy.2021.03.001
2. Yao, F., Ding, Y., Hong, S., Yang, S.-H.: A Survey on evolved LoRa-based communication technologies for emerging Internet of Things applications. Int. J. Netw. Dyn. Intell. 4–19 (2022). https://doi.org/10.53941/ijndi0101002
3. Penumuru, D.P., Muthuswamy, S., Karumbu, P.: Identification and classification of materials using machine vision and machine learning in the context of industry 4.0. J. Intell. Manuf. **31**(5), 1229–1241 (2020). https://doi.org/10.1007/s10845-019-01508-6
4. Psarommatis, F., Sousa, J., Mendonça, J.P., Kiritsis, D.: Zero-defect manufacturing the approach for higher manufacturing sustainability in the era of industry 4.0: a position paper. Int. J. Prod. Res. **60**(1), 73–91 (2022). https://doi.org/10.1080/00207543.2021.1987551
5. Azamfirei, V., Psarommatis, F., Lagrosen, Y.: Application of automation for in-line quality inspection, a zero-defect manufacturing approach. J. Manuf. Syst. **67**, 1–22 (2023). https://doi.org/10.1016/j.jmsy.2022.12.010
6. Zhou, Y., Yuan, M., Zhang, J., Ding, G., Qin, S.: Review of vision-based defect detection research and its perspectives for printed circuit board. J. Manuf. Syst. **70**, 557–578 (2023). https://doi.org/10.1016/j.jmsy.2023.08.019
7. Agnisarman, S., Lopes, S., Chalil Madathil, K., Piratla, K., Gramopadhye, A.: A survey of automation-enabled human-in-the-loop systems for infrastructure visual inspection. Autom. Constr. **97**, 52–76 (2019). https://doi.org/10.1016/j.autcon.2018.10.019
8. Huang, S.H., Pan, Y.C.: Automated visual inspection in the semiconductor industry: a survey. Comput. Ind. **66**, 1–10 (2015). https://doi.org/10.1016/j.compind.2014.10.006
9. Taylor, W.A.: Statistical Procedures for the Medical Device Industry. Taylor Enterprises, Incorporated (2017). https://books.google.es/books?id=F50etAEACAAJ
10. Chrysler Corporation, Ford Motor Company, and General Motors Corporation. Advanced product quality planning (APQP) and control plan. Retrieved October, no. June 1994, p. 128 (1994)
11. Adibhatla, V.A., Chih, H.-C., Hsu, C.-C., Cheng, J., Abbod, M.F., Shieh, J.-S.: Defect detection in printed circuit boards using you-only-look-once convolutional neural networks. Electronics **9**, 1–16 (2020)
12. Krizhevsky, A., Sutskever, I., Hinton, G.E.: ImageNet classification with deep convolutional neural networks. Commun. ACM **60**(6), 84–90 (2017). https://doi.org/10.1145/3065386
13. Sarker, I.H.: Deep learning: a comprehensive overview on techniques, taxonomy, applications and research directions. SN Comput. Sci. **2**(6), 1–20 (2021). https://doi.org/10.1007/s42979-021-00815-1
14. Wang, J., Ma, Y., Zhang, L., Gao, R.X., Wu, D.: Deep learning for smart manufacturing: methods and applications. J. Manuf. Syst. **48**, 144–156 (2018). https://doi.org/10.1016/j.jmsy.2018.01.003

15. Gao, Y., Li, X., Wang, X.V., Wang, L., Gao, L.: A review on recent advances in vision-based defect recognition towards industrial intelligence. J. Manuf. Syst. **62**, 753–766 (2022). https://doi.org/10.1016/j.jmsy.2021.05.008
16. Yang, L., Jiang, H., Song, Q., Guo, J.: A survey on long-tailed visual recognition **130**(7) (2022). https://doi.org/10.1007/s11263-022-01622-8
17. Niu, S., Li, B., Wang, X., Lin, H.: Defect image sample generation with GAN for improving defect recognition. IEEE Trans. Autom. Sci. Eng. **17**(3), 1611–1622 (2020). https://doi.org/10.1109/TASE.2020.2967415
18. Zhang, G., Cui, K., Hung, T.Y., Lu, S.: Defect-GAN: high-fidelity defect synthesis for automated defect inspection. In: Proceedings - 2021 IEEE Winter Conference on Applications of Computer Vision, WACV 2021, pp. 2523–2533 (2021). https://doi.org/10.1109/WACV48630.2021.00257
19. Prunella, M., et al.: Deep learning for automatic vision-based recognition of industrial surface defects: a survey. IEEE Access **11**, 43370–43423 (2023). https://doi.org/10.1109/ACCESS.2023.3271748
20. Schmedemann, O., Baaß, M., Schoepflin, D., Schüppstuhl, T.: Procedural synthetic training data generation for AI-based defect detection in industrial surface inspection. Procedia CIRP **107**, 1101–1106 (2022). https://doi.org/10.1016/j.procir.2022.05.115
21. Mezher, A.M., Marble, A.E.: Computer vision defect detection on unseen backgrounds for manufacturing inspection. Expert Syst. Appl. **243**, 122749 (2024). https://doi.org/10.1016/j.eswa.2023.122749
22. Lyu, C., et al.: RTMDet: An Empirical Study of Designing Real-Time Object Detectors, pp. 63–65 (2022). http://arxiv.org/abs/2212.0778

A Model for the Multidepot Multiple Travelling Salesman Problem and Advanced Resolution Strategies

Juan Pablo Fiesco, Beatriz Andres(✉), and Raul Poler

Research Centre on Production Management and Engineering (CIGIP), Universitat Politècnica de València (UPV), Camino de Vera S/N, 46022 Valencia, Spain
{jfiesco,bandres,rpoler}@cigip.upv.es

Abstract. This paper delves into the Multidepot Multiple Travelling Salesman Problem (MDMTSP), an extension of the TSP that presents additional complexities. A review of the state of the art is performed, from which a MDMTSP mixed integer linear programming model is extracted. For this model, a new objective function and new variables are added to cover the need to balance the number of cities to be visited by each salesman. Various strategies are also investigated to reduce the computation times associated with the NP-Hard MDMTSP. The proposed mathematical model is not only an abstract theory, but will also be integrated into the AI-Delivery Optimiser solution of the AIDEAS project. This integration represents a step towards the manufacturing industry of the future, where artificial intelligence and process optimisation will work together to drive competitiveness and business success.

Keywords: Multiple Travelling Salesmen · Problem Solution Approaches · Modelling

1 Introduction

The Multiple Travelling Salesman Problem (MTSP) expands upon the Travelling Salesman Problem (TSP) by introducing multiple salesmen tasked with visiting a set of cities exactly once before returning to their starting point, typically a depot, at the minimum travel cost. The MTSP has applications in optimisation problems like vehicle routing problems (VRPs), task assignments, machine and operator scheduling in manufacturing, inventory management and warehouse replenishment, or order-picking problems in warehouses [1]. The MTSP prohibits multiple visits and subtours by distinguishing it from the task assignment problem. In the scenario with a single depot, all the salesmen initiate and conclude their tours at a sole location, but when multiple depots are present (MDMTSP), each hosts a specific number of salesmen with flexibility to return to either their original or different depots [2]. As the MDMTSP is a NP-hard problem, the objective of this paper is to provide various advanced resolution strategies to reduce computation times. This paper has been done within the scope of the AI-Driven Industrial Equipment

A. A. Juan et al. (Eds.): DSA ISC 2024, LNCS 14778, pp. 353–360, 2025.
https://doi.org/10.1007/978-3-031-78238-1_32

Product Life Cycle Boosting Agility, Sustainability and Resilience (AIDEAS) Horizon Europe project. AIDEAS proposes a set of intelligent solutions to support the manufacturing phase in the machinery industry. Solutions include the procurement optimiser (AI-PO), the fabrication optimiser (AI-FO) and the delivery optimiser (AI-DO). The current paper focuses on the AI-DO, which is an optimisation solution applied to support the product transportation problem. Accordingly, the paper pays attention to the development of advanced techniques to address the MDMTSP. So this paper is organised as follows: Sect. 2 summarises the state of the art. Section 3 formulates the MDMTSP and validates the model with short experiments. Section 4 describes a set of advanced strategies to reduce the MDMTSP resolution times. Finally, conclusions are discussed in Sect. 5.

2 State of the Art

In the MTSP context, the authors discussing the problem coincide with the fact that more efficient heuristic algorithms are required, especially in large-scale datasets, which highlights the need to develop innovative resolution strategies in this domain. Although exact methods promise optimal solutions, they prove impractical for larger instances due to the problem's complexity. Along these lines, [2] emphasises the efficacy of branch-and-bound methods, especially when paired with well-selected initial bounds. A more recent study by [3] underscores the use of both exact and heuristic techniques. Heuristic approaches, such as genetic algorithms (GA) or ant colony (AC), have gained favour for their efficiency in tackling the MTSP. Furthermore, hybrid algorithms that blend metaheuristics with local searches or clustering methods [4] and matheuristics that combine exact methods with heuristics or metaheuristics [5] have surfaced to mitigate computational complexity and convergence time. These diverse solution methodologies reflect a concerted endeavour to efficiently address the MTSP by striving to strike a balance between optimality and practicality.

For its applicability, the TSP has traditionally been used in transport, product distribution, production planning and logistics. Moreover, recent advances in drone technology have led to academy and industry interest being shown in investigating the application of unmanned aerial vehicles (UAVs) [6]. Some studies in the literature focus on the cooperation of trucks and drones for last-mile delivery [7, 8]. In light of this, [3] describe eight more possible application areas, including multirobot task allocation and scheduling, disaster management, and monitoring and surveillance.

This paper focuses on the MDMTSP, a TSP extension challenge that has been addressed in various ways. The work of [9] is particularly noteworthy, which delves into the application of multiple depots and multiple peddlers to the facility location problem. Their study incorporates crucial factors, such as vehicle autonomy and service frequency requirements, by providing valuable insight into real-world logistics scenarios. Another interesting contribution is that by [10] who introduces the multidepot electric vehicle location routing problem with time windows by extending the classic VRP to include electric vehicles and energy supply options.

3 Problem Formulation

The MTSP involves optimising routes for multiple salesmen, each starting and ending their tour at a designated depot while visiting a set of cities exactly once. In the variant MDMTSP, the problem is compounded by considering multiple potential depots from which salesmen can initiate their routes. The objective is to divide the set of cities into subsets, each corresponding to a tour for a specific salesman, to ensure that every city is visited exactly once in each subset. Additionally, each tour must commence and conclude at its respective depot, with a minimum requirement of two nodes visited during each tour. The goal may include minimising the total distance travelled by all the salesmen or balancing their workload. Figure 1 depicts a two-route graph by illustrating the paths that each sales person will take to visit cities and to return to their respective depots.

From the analysis of the state of the art, the mixed integer linear programming (MILP) outlined in [11] is taken as a reference to propose advanced techniques and to improve resolution times. The MILP uses three decision variables:

- $X_{kij} = 1$, if salesperson k visits node j immediately after node i; 0, otherwise.
- $z_i = 1$, if node i is the starting (origin) node during a tour of any salesperson; 0, otherwise.
- u_i = visiting rank of node i

Objective Function (1) seeks to minimise the total cost, represented by the combined length of m tours.

$$Minimize \sum\nolimits_{i=1}^{n} \sum\nolimits_{j=1:i\neq j}^{n} (d_{ij} \sum\nolimits_{k=1}^{m} X_{kij}) \tag{1}$$

Subject to:

$$\sum\nolimits_{i=1:i\neq j}^{n} \sum\nolimits_{k=1}^{m} X_{kij} = 1 \forall j = 1, 2, \ldots, n \tag{2}$$

$$\sum\nolimits_{i=1:i\neq p}^{n} X_{kip} - \sum\nolimits_{j=1:j\neq p}^{n} X_{kpj} = 0 \forall p = 1, \ldots, n, k = 1, 2, \ldots, m \tag{3}$$

$$\sum\nolimits_{i=1:i\neq j}^{n} \sum\nolimits_{k=1}^{m} X_{kij} \geq 1 \forall k = 1, 2, \ldots, m \tag{4}$$

$$u_i - u_j + L \cdot \sum\nolimits_{k=1}^{m} X_{kij} \leq L - 1 + L \cdot z_j \forall i, j = 1, 2, \ldots, n : i \neq j \tag{5}$$

$$1 \leq u_i \leq L \forall i = 1, 2, \ldots, n \tag{6}$$

$$\sum\nolimits_{i=1}^{n} z_i = m \tag{7}$$

$$X_{kij} \in \{0, 1\} \forall i, j = 1, \ldots, n : i \neq j, k = 1, 2, \ldots, m \tag{8}$$

$$z_i \in \{0, 1\} \forall i = 1, \ldots, n \tag{9}$$

Constraint (2) ensures that each node is visited exactly once during tours. Flow conservation Constraint Set (3) mandates that if a salesperson visits a node, (s)he must also depart from it. Constraint Set (4) ensures that each salesperson is assigned to a tour. Constraints (5) and (6) prevent undesirable subtours, with L representing the maximum number of nodes that a salesperson can visit, calculated as $L = n - 2m + 2$. Furthermore, constraint (5) allows that in the case where index i is the last node visited before arriving its depot node (j), and where u_i takes the highest according to its visit order (L or approximately L), the variable X_{kij} can also take the value one respecting the limitation on the right-hand side of the constraint because z_j is equal to one.

Constraint (7) specifies the m starting nodes corresponding to m tours. Lastly, constraints (8) and (9) define the decision variables utilised in the model.

With this formulation of the problem, what is minimised is the total distance travelled by all the salespersons. However, another aspect to consider is the load of cities assigned by each agent; this amount should be equitable to not saturate a salesperson. Thus [11] propose a new MILP model for the MDTSP with a min-max objective by adding the variable *Smax*. Inspired by this work, we propose a new MILP that minimises both the total route, and the deviation between the maximum and minimum number of locations to visit, by penalising the solution with an imbalance in route allocation.

STMAX = number of locations visited on the longest route.

STMIN = number of locations visited on the shortest route.

The new Objective Function (10) minimises the total cost of m tours and the bias of the maximum number of locations visited and a minimum number. Constraints (2)– (9) remain the same, while constraints (11) and (12) calculate the maximum and minimum number of locations visited by all the salespersons.

$$Minimise \sum\nolimits_{i=1}^{n} \sum\nolimits_{j=1:i\neq j}^{n} (d_{ij} \sum\nolimits_{k=1}^{m} X_{kij}) + (STMAX - STMIN) \cdot c \tag{10}$$

$$\sum\nolimits_{i=1}^{n} \sum\nolimits_{j=1:i\neq j}^{n} X_{kij} \geq STMIN \forall k = 1, 2, \ldots, m \tag{11}$$

$$\sum\nolimits_{i=1}^{n} \sum\nolimits_{j=1:i\neq j}^{n} X_{kij} \leq STMAX \forall k = 1, 2, \ldots, m \tag{12}$$

Next a set of experiments validating the proposed model is presented by considering small, medium and large data size instances. A table with the experiment results is offered (Table 1). Each instance is represented by the number of customers, the number of trucks and depots, which are the same, the balance of the number of customers visited by each truck (100% indicates that trucks visit the same number of customers), the total cost, the GAP and the resolution time. The solver software used in this research is GUROBI, which is generally employed to solve MILP problems. The proposed MILP has been implemented as a Python application running on a CPU with 12th Gen Intel(R) Core(TM) i7-12700 2.10 GHz processor (6 GB RAM). The GAP is calculated as the absolute difference between the target value of the best known solution and the current best solution, divided by the absolute value of the best known solution.

By analysing the results in Table 1, it can be concluded that the exact method used can achieve optimal solutions in seconds for small instances. However, for large instances, the solution is far from optimal and limits the time to 15 min. This shows the need to explore new strategies to solve the problem.

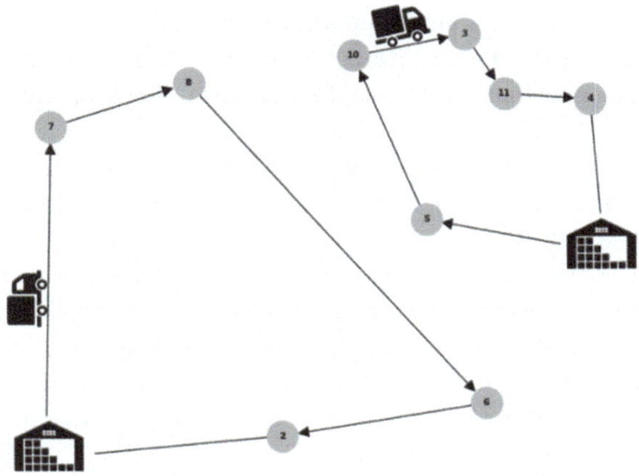

Fig. 1. Example of an MDMTSP and a solution.

Table 1. Instance experiments results.

Instance	No. Customers	No. Trucks	No. Depots	Balance	Total cost	GAP	Time (sec.)
2_10_S1	10	2	2	100.00%	428	0.00%	2.24
2_54_M1	54	2	2	100.00%	780	13.04%	900
2_89_M2	89	2	2	97.82%	1016	11.27%	900
2_96_M3	96	2	2	100.00%	1068	16.59%	900
2_100_M4	100	2	2	100.00%	1064	19.15%	900
4_30_L1	30	4	4	88.88%	564	26.96%	900
4_73_L2	73	4	4	95.00%	1118	37.31%	900
4_91_L3	91	4	4	95.83%	1324	56.48%	900
4_95_L4	95	4	4	96.00%	1350	55.91%	900

4 Advanced Resolution Strategies for the MDMTSP

Solving the MDMTSP entails devising efficient algorithms capable of handling the combinatorial complexity inherent in determining optimal routes while respecting the constraints imposed by multiple depots and the requirement of visiting each city exactly once. The MDMTSP is an NP-hard problem, which means that finding an optimal solution can take an exponential amount of time depending on the size of the problem. This section proposes a set of advanced techniques to improve the MDMTSP resolution times without compromising the solution quality, but by achieving the optimal bounds.

(i) Evolutionary approaches: metaheuristics, such as GA, AC or simulated annealing (SA), which experiment with different parameter settings, such as population sizes, mutation rates and stopping criteria, to find the combination that best works for a specific dataset [11].
(ii) Hybrid algorithms: they employ the combination of GA, AC or SA metaheuristics to make the most of the strengths of the different methods and to improve convergence speed [4].
(iii) Improved construction heuristics: it investigates more sophisticated construction heuristics for generating initial routes, such as the modified Nearest Neighbour Algorithm, the Nearest Insertion Algorithm or the Best Insertion Heuristic [12].
(iv) Relaxation approaches: they are employed to transform the MILP into a continuous problem to obtain initial solutions. They use gradual improvement techniques to arrive at a more optimal solution [13].
(v) Neighbourhood search techniques: they implement neighbourhood search techniques, such as local search or larger neighbourhood search, to explore nearby solutions and iteratively improve the current solution.
(vi) Parallelisation techniques: they involve the simultaneous execution of multiple computations to more efficiently solve the problem. Parallel GA use parallelism in the evaluation and evolution of candidate solutions, multiple populations of solutions can be evolved concurrently, and crossover and mutation operations are performed independently on different subsets of the population [14].
(vii) Decomposition Techniques and Clustering: they decompose the problem into several smaller ones, define "clusters" around Depots. They divide the MTSP into a series of subproblems, each representing a tour for an individual salesman. They sequentially optimise each tour, possibly using a TSP solver, and combine solutions to form the overall solution. They divide cities into clusters and assign each cluster to a different salesman. Finally, the routes connecting clusters are optimised [15].
(viii) Matheuristics: this combine exact methods with heuristics or metaheuristics to provide presolutions to the MILP [5].

It can be concluded that there is not an unique resolution strategy for improving the MDMTSP, and that the selected strategies to solve the MDMTSP depend on the problem characteristics and the specific requirements of the application. Experimenting with various techniques and adjusting parameters can help making significant improvements in the speed and quality of solutions.

5 Conclusions and Future Research Lines

In conclusion, this paper presents a brief state of the art of the MDMTSP, analyses an MILP proposed by Karabulut et al. (2021) and proposes a new formulation by adding the minimisation of load balance of visited locations to the objective function. Additionally, various strategies are explored to reduce the computation times associated with this combinatorial problem. These strategies include evolutionary approaches, hybrid algorithms, improved construction heuristics, relaxation approaches, neighbourhood search tools, parallelisation techniques, decomposing approaches, clustering and matheuristics.

For a real-world application of this problem, the potential of combining the MTSP with depot management and depot location design problems is highlighted for offering a comprehensive perspective to address logistics and distribution challenges.

Computational experiments were carried out on small, medium and large datasets, of which the last ones are capable of simulating real-world problems. Small instances approximate optimal solutions in an average of two seconds with 0% of GAP. Medium instances do not achieve the optimal solution, although longer computation times imply a reduction in the GAP. Finally, the resolution of large datasets exponentially increase computational times without achieving an optimal solution, which demonstrates that the MDMTSP is an NP-hard problem that needs advanced techniques to obtain good solutions in shorter computational times. To all this, a set of nine advanced resolution strategies is proposed.

It is noteworthy that this research aligns with the AIDEAS project and its solution, the AIDEAS Delivery Optimiser, an AI-based tool capable of optimising product storage and delivery. This solution not only focuses on optimising storage space, storage conditions and product transportation, but also provides optimisation for logistics planning and scheduling.

As future research directions, further exploring the application of the proposed advanced strategies to improve the computational efficiency of the MDMTSP is suggested, as well as integrating these solutions into practical systems like the AIDEAS Delivery Optimiser to enhance logistics management in real business environments. Therefore as a next step to this research, the decomposition of the problem into smaller ones by defining clusters around depots and utilising matheuristics to provide presolutions to the MILP, are proposed to be implemented.

Acknowledgements. The research that led to these findings received funding from the Horizon Europe Ref.101057294 "AI-Driven Industrial Equipment Product Life Cycle Boosting Agility, Sustainability, and Resilience (AIDEAS)"; the Regional Department of Innovation, Universities, Science, and Digital Society of the Generalitat Valenciana "Programa Investigo" Ref. INVEST/2022/330, which the European Union supported - NextGenerationEU with Plan de Recuperación, Transformación y Resiliencia; the Regional Department of Innovation, Universities, Science, and Digital Society of the Generalitat Valenciana Ref. CIAEST/2022/39 "Doctoral research staff to carry out research stays in companies in the Valencia Region"; and the Vice-Rectorate for Research of the Universitat Politècnica de València Ref. PAID-06-23 "Intelligent optimisation algorithms and models for production planning and replenishment (ATENNEA)".

References

1. Matai, R., Singh, S., Lal, M.: Traveling salesman problem: an overview of applications, formulations, and solution approaches. In: Traveling Salesman Problem, Theory and Applications. InTech (2010). https://doi.org/10.5772/12909
2. Bektas, T.: The multiple traveling salesman problem: an overview of formulations and solution procedures. Omega (Westport) **34**(3), 209–219 (2006). https://doi.org/10.1016/j.omega.2004.10.004
3. Cheikhrouhou, O., Khoufi, I.: A comprehensive survey on the multiple traveling salesman problem: applications, approaches and taxonomy. Comput. Sci. Rev. **40**. (2021). https://doi.org/10.1016/j.cosrev.2021.100369

4. Jiang, C., Wan, Z., Peng, Z.: A new efficient hybrid algorithm for large scale multiple traveling salesman problems. Expert Syst. Appl. **139** (2020). https://doi.org/10.1016/j.eswa.2019.112867
5. Guzman, E., Andres, B., Poler, R.: Models and algorithms for production planning, scheduling and sequencing problems: a holistic framework and a systematic review. J. Ind. Inf. Integr. **27** (2022). https://doi.org/10.1016/j.jii.2021.100287
6. Khoufi, I., Laouiti, A., Adjih, C.: A survey of recent extended variants of the traveling salesman and vehicle routing problems for unmanned aerial vehicles. Drones **3**(3), 1–30 (2019). https://doi.org/10.3390/drones3030066
7. Zhang, R., Dou, L., Xin, B., Chen, C., Deng, F., Chen, J.: A review on the truck and drone cooperative delivery problem Unmanned Syst. (2023). https://doi.org/10.1142/S2301385024300014
8. Rinaldi, M., Primatesta, S., Bugaj, M., Rostáš, J., Guglieri, G.: Development of heuristic approaches for last-mile delivery TSP with a truck and multiple drones. Drones **7**(7) (2023). https://doi.org/10.3390/drones7070407
9. Chan, Y., Baker, S.F., Chan, Y., Dij, D.: The Multiple Depot, Multiple Traveling Salesmen Facility-Location Problem: Vehicle Range, Service Frequency, and Heuristic Implementations (2005). www.elsevier.corn/locate/mcm
10. Paz, J.C., Granada-Echeverri, M., Escobar, J.W.: The multi-depot electric vehicle location routing problem with time windows. Int. J. Ind. Eng. Comput. **9**(1), 123–136 (2018). https://doi.org/10.5267/j.ijiec.2017.4.001
11. Karabulut, K., Öztop, H., Kandiller, L., Tasgetiren, M.F.: Modeling and optimization of multiple traveling salesmen problems: An evolution strategy approach. Comput. Oper. Res. **129** (2021). https://doi.org/10.1016/j.cor.2020.105192
12. Ursani, Z., Ursani, A.A.: Augmented tour construction heuristics for the travelling salesman problem. Int. J. Ind. Optim. **4**(2), 131–144 (2023). https://doi.org/10.12928/ijio.v4i2.7875
13. Ricciardi, L.A., Vasile, M.: Solving multi-objective dynamic travelling salesman problems by relaxation. In: GECCO 2019 Companion - Proceedings of the 2019 Genetic and Evolutionary Computation Conference Companion, pp. 1999–2007. Association for Computing Machinery, Inc., (2019). https://doi.org/10.1145/3319619.3326837
14. Dhanalakshmi, R., Parthiban, P., Anbuchezhian, N.: Optimisation of multiple travelling salesman problem using metaheuristic methods. Int. J. Enterp. Netw. Manag. **13**(3), 199–215 (2022). https://doi.org/10.1504/IJENM.2022.125803
15. Romanuke, V.V.: Deep clustering of the traveling salesman problem to parallelize its solution. Comput. Oper. Res. **165** (2024). https://doi.org/10.1016/j.cor.2024.106548

Smart Meter Reference Load Profiles and Peak Demand Models

Paula Carroll[1], Fábio Silva[2(✉)], Farah Tahir[2], Brian O'Regan[2], and Pádraig Lyons[2]

[1] School of Business, University College Dublin (UCD), Dublin, Ireland
paula.carroll@ucd.ie

[2] International Energy Research Centre (IERC), Tyndall National Institute, University College Cork (UCC), Cork, Ireland
{fabio.silva,padraig.lyons}@ierc.ie

Abstract. Smart grids are components of future clean energy systems. They create new challenges and opportunities for the efficient and reliable management of the electricity system. The aggregate behaviour of groups of electricity users, such as the magnitude and timing of peak demand are important factors in determining the required capacity of the Low Voltage (LV) distribution network in the context of a transition to low carbon heat and transport. In this paper we use smart meter data from consumer behaviour trials in Ireland to create Reference Load Profiles (RLPs), and to model peak residential electricity demand. We find differences in the peak demand and profile by day type, and that Burr, and Weibull models are useful to estimate peak demand.

Keywords: Reference Load Profile · Smart Grid · Peak Demand Model

1 Introduction: Clean Energy Targets

The EU aims to achieve climate neutrality by 2050 and has set targets for its member states [7]. Ireland, an EU member, has set out its clean energy and climate plans in [9,14]. The transport and residential sectors account for approximately two thirds of overall energy consumption representing, respectively, 17.1% and 9.0% of the total amount of greenhouse gas emissions in Ireland in 2022 as per the Climate Action Plan 2023 (CAP23) progress reports [12]. Much of this energy is currently sourced through fossil fuels. The electrification of heat and transport will change the energy balance model and will significantly increase the requirements on the electrical distribution network infrastructure.

Electricity distribution systems are typically planned and dimensioned to facilitate the expected maximum demand of groups of users. The After Diversity Maximum Demand (ADMD) measure is used by Distribution System Operators (DSO) in the design of distribution networks where coincident demand is aggregated over a sample of customers. Traditional statistical models that underpin the design rules and approaches are based on historical consumption data

A. A. Juan et al. (Eds.): DSA ISC 2024, LNCS 14778, pp. 361–375, 2025.
https://doi.org/10.1007/978-3-031-78238-1_33

from small samples of customers. Such models were appropriate when consumption patterns were relatively homogeneous. However, the adoption of low carbon technologies such as heat pumps, electric vehicles, photovoltaic generation and energy storage will change these patterns. DSOs and other agents must reassess residential load models, especially peak load, to facilitate network dimensioning and investment planning. The availability of smart meter data will allow more reliable models of residential energy consumption to be created.

In this paper, we review residential Reference Load Profile (RLP) creation methods and peak demand estimation techniques. We then propose and evaluate statistical models using smart meter data from consumer behaviour trials conducted in Ireland. We focus on answering the following research question:

1. Are there differences in RLPs by day of the week?
2. Which statistical models are a good fit to residential peak load distributions?

We find that there are significant differences in RLPs by day and that Burr and Weibull models offer useful approaches to model peak residential demand.

2 Residential Electricity Load Modelling

Challenges from the uptake of heat pumps are outlined in [15]. National scale problems include peak demand and ramp rate increase. At the local scale the Low Voltage (LV) distribution network may experience excessive voltage drop and insufficient thermal capacity of the LV feeder and transformer. The impact of electric vehicles on the grid are assessed in [5]. They suggest that a coordinated charging approach is needed to allow the mass adoption of electric vehicles so that power losses can be minimised and the load factor can be maximised.

Before anticipating future load issues, we first need to understand current residential RLPs and peak demand patterns. Paatero et al. [20] describe statistical modelling and simulation of two Finnish hourly appliances and lighting data sets from 1082 households/buildings to create residential RLPs. The focus in [22] is on load patterns for consumer groups using frequency domain analysis. They note the smoothed load patterns when taking the average over the group and rely on the customer class group defined by the utility company suitable for the customers' needs and lifestyle.

In contrast, Irish smart meter data [8] is used in [11] to clustering consumers with similar usage patterns. An analysis of the clusters finds that how strongly householders feel can convince other occupants of the building to reduce their energy usage is a significant factor in cluster membership. A machine learning approach to creating RLPs is described in [19]. They compare Symbolic Aggregate ApproXimation with other data reduction techniques for electricity load profile clustering using a sample of high resolution load data for 234 consumers at 15-minute intervals. Multiple linear regression is used by [18] to explore bottom-up approaches to load modelling: a dwelling and occupant characteristics model to describe consumption patterns of various households; and an electrical appliances model describing how electricity is consumed. The results show that occupant characteristics (head of household age, and household composition) impact peak electricity consumption times.

Next we consider peak demand models. Early standards to estimate residential peak demand are described in [1]. Mean demand is estimated from field measurements. The model allows for variability due to customer behaviour due to temperature and other causes. A simple regression model is used: $\overline{D} = P + Q/\sqrt{N}$, where $\overline{D}$ is the average demand, P and Q are regression coefficients, and N is the portfolio size (number of customers). They fit models per customer type and per half hour intervals and focus on winter peak demand, historically highest in temperate climates like Ireland and the UK. Meeting demand with a 90% probability was deemed an appropriate risk level.

Next we consider the digitalisation of the electricity system. Both statistical and machine learning approaches offer insights from the digitised grid but face limitations, including challenges in reconciling the discrepancies between the empirical evidence from metered data and outputs from simulations based on statistical models as outlined in [21]. Smart Meters (SMs) are crucial for EU greenhouse gas reduction and renewable energy goals [6]. Ireland's National Smart Metering Programme was announced in 2017 [4]. The rollout is to be completion by 2024 using the high level design enabling daily data transfer between consumers and suppliers [3]. SM infrastructure will significantly increase the amounts of data collected by participants in electricity markets [10]. This will aid accurate residential energy consumption models and policy design. SMs also enable customer engagement via tariffs and Demand Side Response (DSR), vital for economic transition to low carbon economy, integrating Renewable Energy Source (RES) and technologies like heat pumps and Electric Vehicles (EVs). Smart meter data availability may address concerns raised in [21] and [20]. Bottom-up models in [20] are used to test several DSR strategies, assuming household loads can be managed separately to enable DSR measures on an individual appliance level simultaneously in several households. They do not discuss household appliances management technology, their main concern is to understand how to adjust load profiles to reduce peak demand by shifting some appliance demand to off-peak times. New rules and protocols concerning data collection and communication between the different participants in the market are being developed. However, concerns about privacy arise with smart meters [17]. For example, third parties may intercept signals for illegal purposes, allow intrusive surveillance of the individuals' usage for purposes other than billing and may lead to unfair tariff design practices. Such concerns must be addressed in strategic energy technology implementation plans.

3 Methodology

We focus on developing useful models to answer our two research questions using smart meter Customer Behaviour Trial (CBT) data available from [8]. The Irish regulator, Commission for Energy Regulation (CER), conducted a large scale smart meter CBT in 2009–2010 to assess the performance of smart meters and their impact on consumer behaviour. The CBTs aimed to assess customer response to price incentives. The anonymised trial data are available for research purposes. We note that, as the Smart Meter Data (SMD) are anonymised, we cannot say where on the network these customers reside.

The CER used a stratified random sampling framework to ensure a representative sample in terms of household size, geographic location, and socio-economic indicators. Over 5,000 participants were recruited, each representing a household, i.e. a number of persons sharing a single residential unit. Pre- and post-trial surveys focused on electricity usage attitudes, trial expectations, dwelling and electrical appliances. The CBT recorded electricity usage at half-hourly intervals, producing 269 MB of time-series data per household. Participants' half-hourly KW usage data from July to December 2009 served as benchmark. After this, the CBT participants were divided into test and control groups to test tried different tariffs and Demand Side Management (DSM) stimuli.

The trial ran from January to December 2010, coinciding with severe weather events. We focus on the benchmark period. We removed meters of participants who dropped out or where technical issues that caused significant data loss. We created several views of the data in order to answer our research questions. We aggregated the kW half hour data by day of the week to calculate average daily half hourly load profiles. Peak demand values (kWhh) and their occurrence times were extracted per customer daily to create distributions of peak demand patterns. Peak demand time series were smoothed by averaging for each meter per day type during the benchmark period. The tagged data by day (i.e., Monday, Tuesday, ...) allows us to compare weekend to weekday profiles. A sample of average peak half hour kW demand by day is shown in Table 1.

Table 1. Sample Peak Daily Average Demand (kWhh)

Meter Id	Number Occupants	Monday	Tuesday	Wednesday	Thursday	Friday	Saturday	Sunday
1002	1	0.7100	0.6754	0.8407	0.7582	0.7390	0.7685	0.6724
1003	3	2.3253	2.3231	2.4708	2.3898	2.3441	2.4913	2.5336
1004	4	4.7401	4.6716	4.6140	4.0250	4.3486	4.4139	4.7045

Note that we use half-hourly kW data resolution to align with Ireland's single electricity market practices. Occupancy per household is available in the CBT survey data, which is crucial for load models as noted in [18]. However, this data will not be available in Ireland's smart meter rollout [3].

We also created another view of time-series load data, similar to the approach taken in [2]. We reduce the data in size by calculating explanatory variable/features of the data such as the moments (mean, variance, skewness) and other descriptive statistics for each household such as the time and values of the peak usage during the morning periods. We then built and tested several statistical models under the following headings:

1. RLPs - Exploratory Data Analysis;
2. Peak Demand - Probability Density;
3. Peak Demand - General Linear Models.

Firstly, we evaluate the distance between the weekday and weekend average RLPs, and identify the time when peak demand occurs. Next we conduct univariate analysis of the peak demand and evaluate the fit of several probability density functions from the exponential family of distributions. Metrics such as the log likelihood, Akaike Information Criterion (AIC), Bayesian Information Criterion (BIC), Kolmogorov-Smirnov D (KS), Anderson-Darling (AD) and Cramer-von Mises W-Sq (CvM), are used to assess the Goodness of Fit (GOF) of the statistical Probability Density Function (PDF) to the Empirical Density Function (EDF). These metrics quantify the distance between the Probability Density Function (PDF) and Empirical Density Function (EDF) in different ways and seek the most parsimonious model by penalising models with more parameters. The lower the distance scores the better the model fit.

Finally, we evaluate General Linear Models (GLM) for peak demand and occupancy time. These extend the simple linear regression model in [1]. Data is divided 60:20:20 for train, validate, and test subsets. We employ elastic net approach to regularisation to mitigate over-fitting, yielding balanced regression models between number of coefficients and their size. Various metrics determine the best-fitting GLM regression model and parameter values for empirical data.

4 Results and Analysis

Recall our primary research questions relate to RLP and peak demand models. Question 1 asks if there differences in RLPs by day of the week. Analysis of the daily load profiles (averaged per customer per day), show similarities among weekdays, and stronger differences between weekdays and weekend days. This is consistent with the findings in [18,20]. Figure 1 shows the average load profiles per customer by day type over the benchmark period.

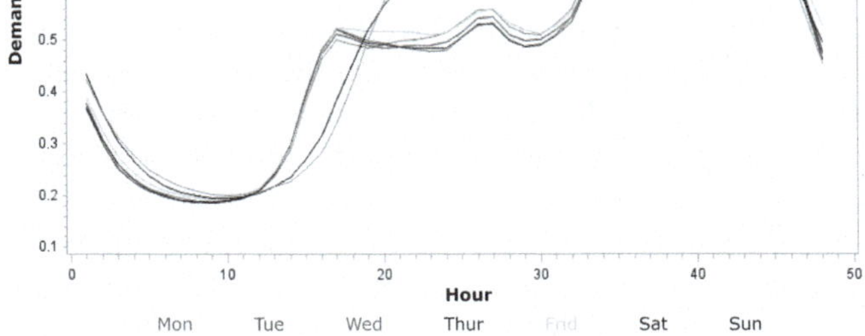

Fig. 1. Reference Load Profile (RLP) by day

Figure 2 shows a dendrogram for the clustered daily RLPs. Midweek days show the strongest similarities. We can see more clearly that weekdays form one cluster and that weekend days form a second cluster. Within those clusters we can see that there are some differences between the weekdays. Public holidays in Ireland generally occur on Mondays, the dendrogram shows that for example Mondays are somewhat dissimilar to other weekdays. The clustering shows that Fridays are also a bit different to the other weekdays.

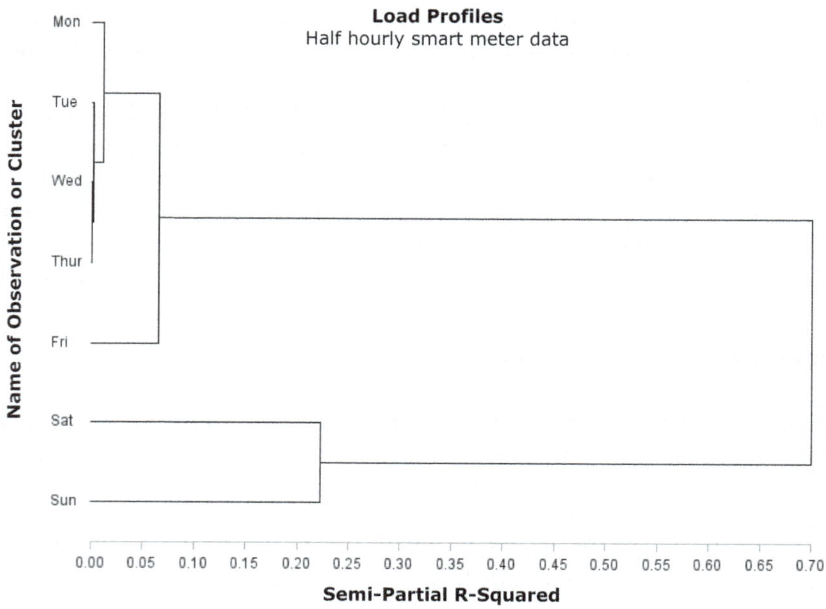

Fig. 2. Reference Load Profile (RLP) Cluster Dendrogram

We also see some differences between Saturdays and Sundays. Figure 3 shows the distances between sample weekday and weekend day types. We can again see that the biggest difference is between weekday and weekend patterns of behaviour. In particular we note the difference in Fig. 3b in terms of the height of the peaks, since the LV network must be dimensioned to cater for the higher of the two. We next address our second research question and report on the statistical models that offer a good fit to residential peak load distributions. Recall we processed the data to create a view of the peak demand for each customer each day. Figure 4 shows the distributions of peak demand per consumer in kW half hour for weekdays and weekends over the benchmark period. We see relatively high kW values reported for the 90, 95 and 99 percentiles and note that on all days the pattern of peak demand is right skewed. This tells us a small number of consumers generate high demand.

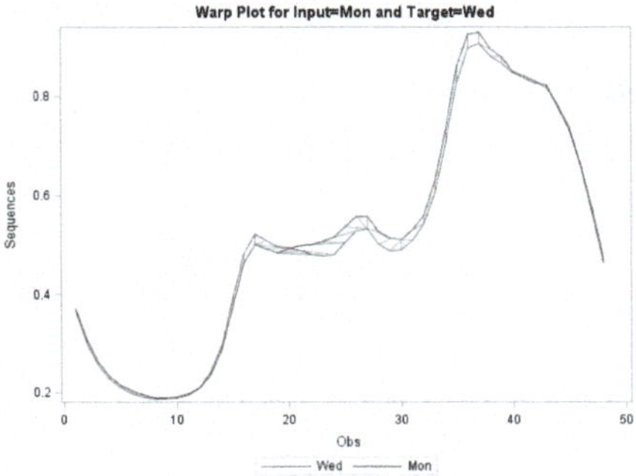

(a) Distance Mondays - Wednesdays

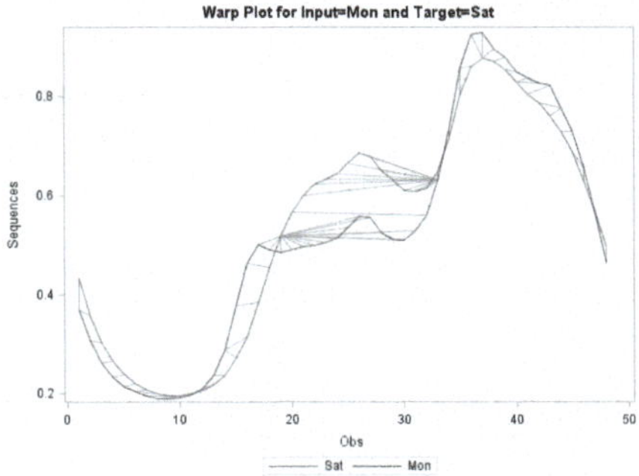

(b) Distance Mondays - Saturdays

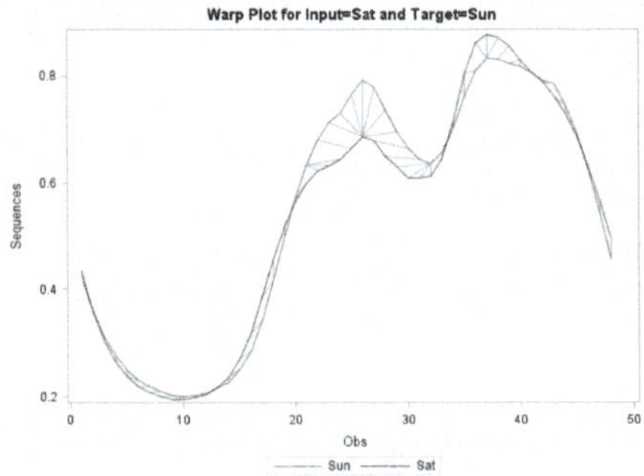

(c) Distance Saturdays - Sundays

Fig. 3. Comparison of Reference Load Profile (RLP) by day type

Comparing the empirical distribution of average peak demand on Mondays (a weekday) with Saturdays (a weekend day), we see evidence in Fig. 5 that the distributions are different, and that the average peak residential demand on Mondays is higher than on Saturdays. Statistical tests confirm the hypothesis, $p = 0.0035$ from a Kolmogorov-Smirnov Two-Sample Test.

The right tailed empirical patterns we observed suggest a probability density from the exponential family of distributions could be used to model the peak demand. We test the fit of a number of models from the exponential family to the average peak demand. Preliminary analysis shows that Burr and Weibull models are a good fit for the distributions of average Monday peak demand. Table 2 shows sample model fit metrics for average peak Monday demand. The log likelihood measure suggests the Burr mode as the best fit, all the other metrics suggest the Weibull model. The Weibull is a simpler model (i.e., has less parameters), and may be more familiar to the electricity community as it is used to fit wind power models, e.g., [16].

Figure 6 shows the Burr and Weibull PDFss fitted to the empirical data, and both offer reasonable solutions. Similarly, looking at Saturdays (typical weekend days), Burr and Weibull models offer a good model fit of average peak demand, see Fig. 7.

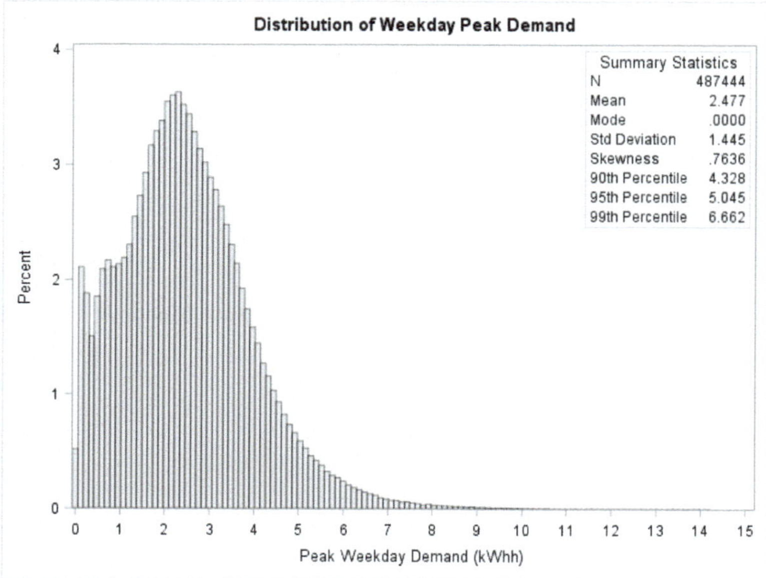

(a) Weekday Peak Demand Distribution

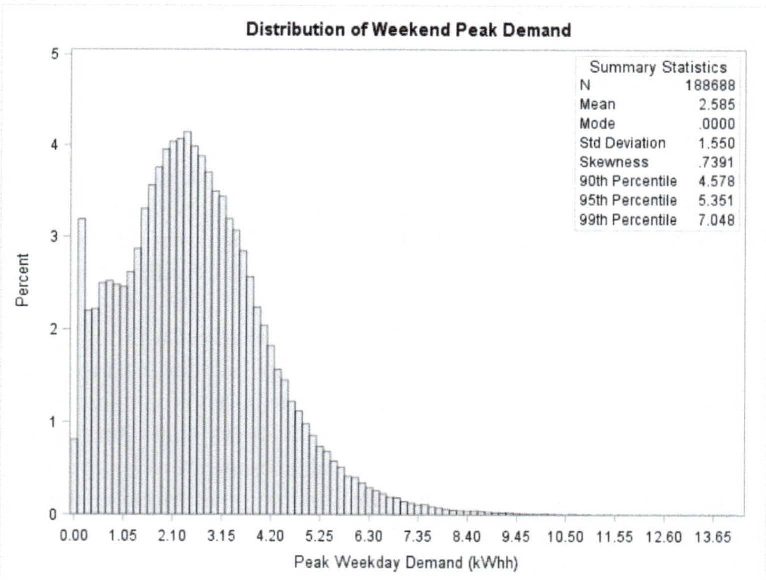

(b) Weekend Peak Demand Distribution

Fig. 4. Comparison of weekday and weekend Peak Demand

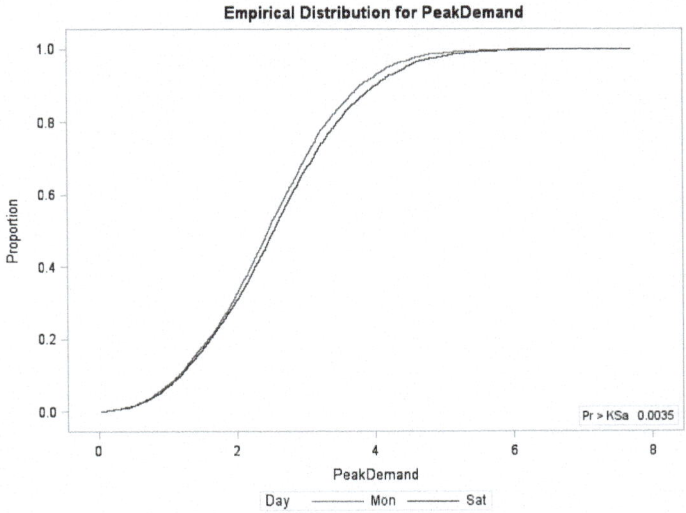

Fig. 5. Distance between the average peak demand empirical distributions on Mondays versus Saturdays

Table 2. Sample PDF model fit results (Mondays)

Distribution	−2 Log Likelihood	AIC	BIC	KS	AD	CvM
Burr	11293	11299	11318	1.50	3.00	0.51
Exp	15015	15017	15024	17.43	601.71	118.99
Gamma	11629	11633	11646	4.07	30.23	5.28
Igauss	13064	13068	13080	8.93	146.75	28.63
Logn	12378	12382	12395	6.17	79.33	13.58
Pareto	15016	15020	15032	17.44	601.76	119.00
Gpd	15015	15019	15032	17.43	601.71	118.99
Weibull	11294	11298	11311	1.49	2.94	0.51

In summary we suggest two-parameter Weibull models for average peak demand. The two-parameter Weibull model PDF is given by:

$$f(t) = \frac{\beta}{\eta}\left(\frac{t}{\eta}\right)^{\beta-1} e^{-\left(\frac{t}{\eta}\right)^{\beta}} \tag{1}$$

where β is a shape parameter and η is a scale parameter.

In further analyses, we are interested in when peak demand of the individual consumers occurs, as the LV network design is concerned with coincident peak load. In all the RLPs, we see that aggregate peak demand occurs in the evening hours. Taking a different view of the data, we examine the peak demand per

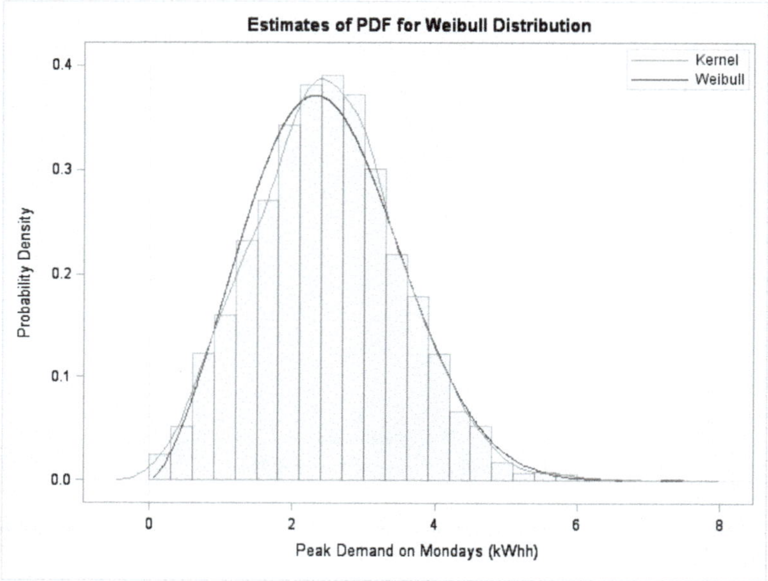

(a) Average Monday Peak Demand: Weibull

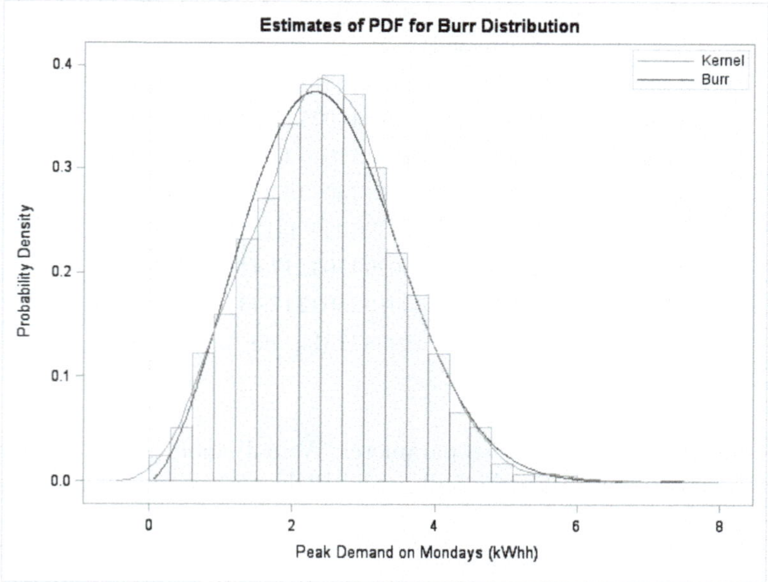

(b) Average Monday Peak Demand: Burr

Fig. 6. Comparison of Weibull and Burr models for average Monday Peak Demand

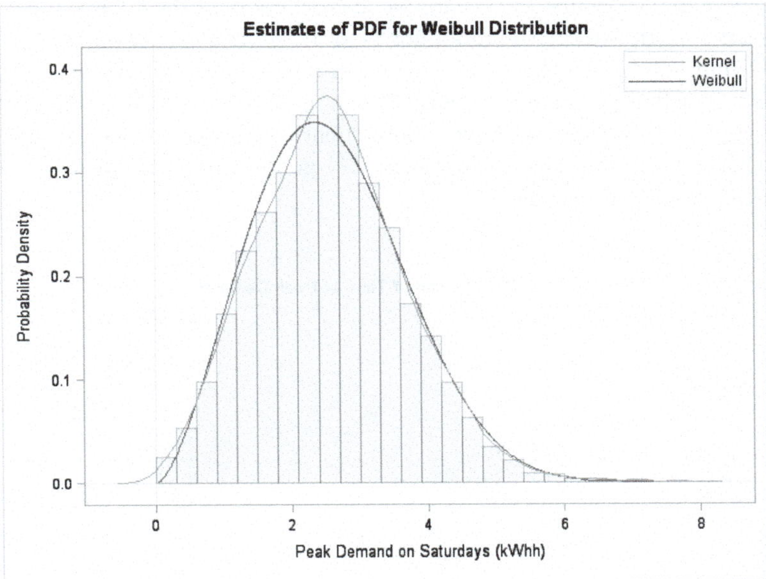

(a) Average Saturday Peak Demand: Weibull

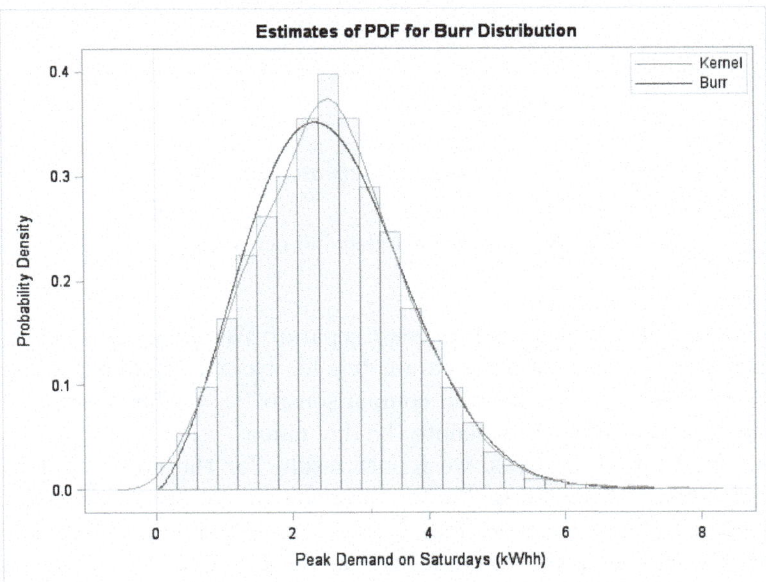

(b) Average Saturday Peak Demand: Burr

Fig. 7. Comparison of Weibull and Burr models for average Saturday Peak Demand

consumer and explore when that occurs. Figure 8 shows the half hour time slot when peak demand occurs for the sample of consumers in the CER CBT over the benchmark period. We see the mode for peak usage is at 18.00 on the 24 h clock (the 36th half hour, with the average at approximately 16.00. We see a degree of variability in when consumers generate their peak demand. It is interesting to observe that some customers have complementary timing of peak usage to the norm, with a small percentage (approx. 5%) generating their peak demand in the hours just after midnight.

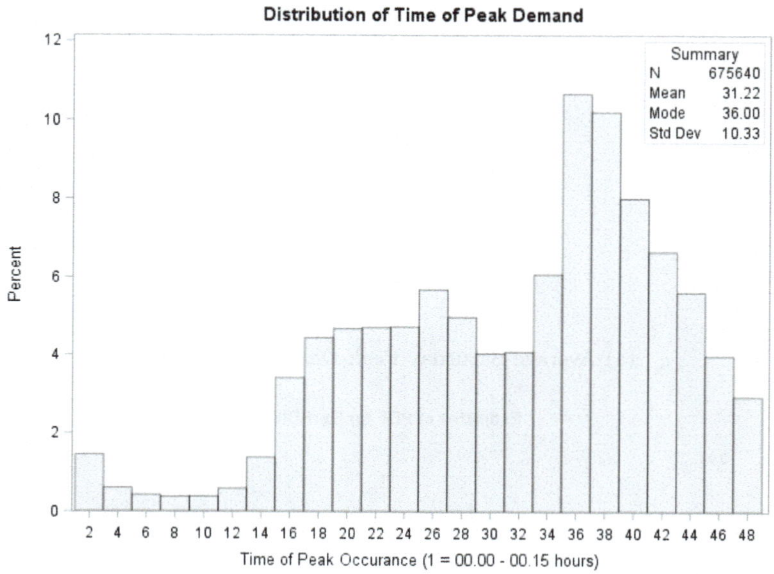

Fig. 8. Time of Peak demand occurrence

We note that it may be useful to create separate Weibull models per day type or per time step, e.g., we estimate parameters for average Monday peak demand as $\beta_M = 2.582$ and $\eta_M = 2.792$, in comparison to $\beta_S = 2.500$ and $\eta_S = 2.897$ where the M and S subscripts denote the day name.

In our final set of analyses, we report results for the GLM models which are an extension to the simple linear regression model in [1]. Several models were tested using the time series reduced data described and were successful at predicting the average peak demand, but the GLM models did not prove useful to also predict the time of peak demand occurrence.

For brevity we report one of the useful models for average peak demand which returns an R^2 value of 0.92 indicating the model can explain 92% of the variation in average peak demand. The residual errors are approximately normally distributed suggesting the linear model is a good fit. The estimated coeficients provide the following multiple linear regression model:

$$APD = 2.386 * SD + 0.091 * max - 0.949 * MORNMAX + 1.099 * mean + 0.822 * MORNPEAK + 0.283 * MORNRANGE \quad (2)$$

where APD is the Average Peak Demand, SD is the Standard Deviation, max is the maximum of the consumer's daily peak demands in the sample, MORNMAX is the Maximum Energy Usage During Morning Periods (kWhh summed over the morning time slots), mean is the average usage, MORNPEAK is the value of the Peak Usage During the Morning, and finally, MORNRANGE is the Range of Usage in the Morning Hours (06.00–10.00). These explanatory variables were all reported as significant ($p < 0.0001$), the intercept coefficient was not statistically significant.

Again, the regression models could be further developed by parameterising per time slot or day type. The addition of other explanatory variables and interaction terms created models with higher R^2 values. However this came at the risk of over-fitting and an increase in the complexity of the model.

5 Discussion and Conclusion

This paper explores the methods used to create RLPs and develop models for peak residential electricity demand, also making a novel contribution in providing insights into the PDFs and regression models suitable to model peak demand. These models will be useful to those concerned with evaluating the capacity of Photovoltaic (PV) distribution networks and provide a rigorous starting point from which to evaluate scenarios for new end-use applications in the smart grid. This models will present opportunities to de-risk investment decisions with use of these appropriate models.

We acknowledge the limitations of our work, as with all statistical model fitting, there is a risk of over-fitting. We have used machine learning data reduction approaches rather than traditional time-series approaches. In future research we may compare the two approaches. In addition, the CBT took place in 2009. Improvements in electrical appliances and building energy ratings can reduce consumption, although the rate of change is relatively slow since the life cycle of homes and appliances is relatively long, and the average annual usage has not changed significantly to date. Applying our analysis to more recent data would be a useful next step. However, General Data Protection Regulation (GDPR) and commercial sensitivity concerns have impeded the availability of open source smart meter data for research. Data sharing agreements and new business models which allow consumers extract value from their data are needed. As noted in [13], interest in local supply markets mainly comes from new actors such as local energy community groups, social enterprises and municipalities. Such groups may be interested in developing innovative business models to share and extract value from smart meter data. Meanwhile, the Irish CBT remains one of the most cited smart meter studies. The large sample size and the quality of the study design which ensured the quality and reliability of the data.

We have presented several useful reproducible models. As the grid becomes further digitised, data from consumers and their end-use applications will become more widely available subject to the limitations we have noted. The study methods could be implemented by electricity network operator using their own data and could also be used as a prototype for electricity market participants to derive value from new smart meter data sources. We argue that our models have their uses in facing the challenges of resolving conflict between fast paced technological innovation and slower/longer term investment decisions when the need for transformation to achieve climate change targets is so urgent.

We conclude that the differences in weekend and weekday usage patterns suggest RLPs per day type should be considered. In response to our research question on which statistical models offer a good fit to residential peak load distributions we conclude that Weibull models describe the average peak demand PDFs well, and that multiple linear regression models are useful to provide estimates of average peak demand for a portfolio of consumers.

We see that there is potential for further development of the models recommended by our study. In particular we intend to tailor the Weibull peak demand models per time slot t by parameterising the model: $\beta \rightarrow \beta_t$ and $\eta \rightarrow \eta_t$.

Finally, in this study we have not addressed the challenges of ensuring data privacy and fairness to all consumers. We acknowledge in the context of the Big Data revolution that this is a major concern that must be addressed in smart energy transition implementation plans.

Acknowledgments. The authors acknowledge the support of the International Energy Research Centre (IERC), the Tyndall National Institute, the University College Cork (UCC), and the University College Dublin (UCD), Ireland.

References

1. Britain, E.C.G., Britain, C.E.C.G.: Report on statistical method for calculating demands and voltage regulation on L.V. radial distribution systems. ACE report, Electricity Council (1981). https://books.google.ie/books?id=_ynHtgAACAAJ
2. Carroll, P., Murphy, T., Hanley, M., Dempsey, D., Dunne, J.: Household classification using smart meter data. J. Off. Stat. **34**(1), 1–26 (2018). https://doi.org/10.1515/jos-2018-0001
3. CER: Commission for energy regulation national smart metering programme smart metering high level design. Decision Paper 046, CER (2014). Accessed June 2016
4. CER: Review of typical domestic consumption values for electricity and gas customers. Decision Paper CER/17042, Commission for Energy (2017)
5. Clement-Nyns, K., Haesen, E., Driesen, J.: The impact of charging plug-in hybrid electric vehicles on a residential distribution grid. IEEE Trans. Power Syst. **25**(1), 371–380 (2010). https://doi.org/10.1109/TPWRS.2009.2036481
6. COM: A policy framework for climate and energy in the period from 2020 to 2030. Technical report 15, European Union (2014). https://bit.ly/3TX7dSu

7. European Comission: 2030 climate targets - European Commission. https://bit.ly/3xK6IE4
8. Commission for Energy Regulation (CER): Cer smart metering project - electricity customer behaviour trial, 2009-2010 [dataset]. Irish Social Science Data Archive. SN: 0012-00, 1st edn. (2012). www.ucd.ie/issda/CER-electricity. Accessed January 2018
9. DECC: National Mitigation Plan. Technical report 2017, Department of the Environment, Climate and Communications (DECC), Ireland (2017)
10. Figueiredo, V., Rodrigues, F., Vale, Z., Gouveia, J.B.: An electric energy consumer characterization framework based on data mining techniques. IEEE Trans. Power Syst. **20**(2), 596–602 (2005). https://doi.org/10.1109/TPWRS.2005.846234
11. Fitzpatrick, J., Carroll, P., Ajwani, D.: Creating and characterising electricity load profiles of residential buildings. In: Lemaire, V., Malinowski, S., Bagnall, A., Guyet, T., Tavenard, R., Ifrim, G. (eds.) AALTD 2020. LNCS (LNAI), vol. 12588, pp. 182–203. Springer, Cham (2020). https://doi.org/10.1007/978-3-030-65742-0_13
12. GoI: Climate Action Plan 2023 - Progress Report - Q3 2023. Technical report Q3 2023, Government of Ireland (GoI), Ireland (2023)
13. Hall, S., Roelich, K.: Business model innovation in electricity supply markets: the role of complex value in the united kingdom. Energy Policy **92**, 286–298 (2016). https://doi.org/10.1016/j.enpol.2016.02.019, https://www.sciencedirect.com/science/article/pii/S030142151630060X
14. Ireland, D.: Ireland's transition to a low carbon energy future 2015–2030. White paper, Department of Communications, Energy & Natural Resources (2015). https://bit.ly/3vL5k3q. Accessed January 2018
15. Love, J., et al.: The addition of heat pump electricity load profiles to GB electricity demand: evidence from a heat pump field trial. Appl. Energy **204**, 332–342 (2017). https://doi.org/10.1016/j.apenergy.2017.07.026
16. Lowery, C., O'Malley, M.: Impact of wind forecast error statistics upon unit commitment. IEEE Trans. Sustain. Energy **3**(4), 760–768 (2012). https://doi.org/10.1109/TSTE.2012.2210150
17. McKenna, E., Richardson, I., Thomson, M.: Smart meter data: balancing consumer privacy concerns with legitimate applications. Energy Policy **41**, 807–814 (2012). https://doi.org/10.1016/j.enpol.2011.11.049, https://bit.ly/3U5Vkto. Modeling Transport (Energy) Demand and Policies
18. McLoughlin, F., Duffy, A., Conlon, M.: Characterising domestic electricity consumption patterns by dwelling and occupant socio-economic variables: an Irish case study. Energy Build. **48**, 240–248 (2012). https://doi.org/10.1016/j.enbuild.2012.01.037
19. Notaristefano, A., Chicco, G., Piglione, F.: Data size reduction with symbolic aggregate approximation for electrical load pattern grouping. IET Gener. Transm. Distrib. **7**(2), 108–117 (2013). https://doi.org/10.1049/iet-gtd.2012.0383
20. Paatero, J.V., Lund, P.D.: A model for generating household electricity load profiles. Int. J. Energy Res. **30**(5), 273–290 (2006). https://doi.org/10.1002/er.1136
21. Ramírez-Mendiola, J.L., Grünewald, P., Eyre, N.: The diversity of residential electricity demand – a comparative analysis of metered and simulated data. Energy Build. **151**, 121–131 (2017). https://doi.org/10.1016/j.enbuild.2017.06.006
22. Zhong, S., Tam, K.S.: A frequency domain approach to characterize and analyze load profiles. IEEE Trans. Power Syst. **27**(2), 857–865 (2012). https://doi.org/10.1109/TPWRS.2011.2170592

Intraday Multireservoir Hydropower Optimization with Alternative Deep Reinforcement Learning Configurations

Rodrigo Castro Freibott[1], Álvaro García Sánchez[1(✉)], Francisco Espiga-Fernández[1], and Guillermo González-Santander de la Cruz[2]

[1] Industrial Engineering, Business Administration and Statistics Department, ETSII, Universidad Politécnica de Madrid, José Gutiérez Abascal 2, 28006 Madrid, Spain
{rodrigo.castro.freibott,francisco.espiga.fernandez}@alumnos.upm.es, alvaro.garcia@upm.es

[2] baobab soluciones, José Abascal 55, 28003 Madrid, Spain
guillermo.gonzalez@baobabsoluciones.es

Abstract. Managing multireservoir hydropower systems in an intraday context poses unique challenges due to the need for frequent decisions in response to fluctuating energy prices. While Reinforcement Learning (RL) methods have been applied to long-term management, this paper addresses the gap in short-term planning within a single day. We use an alternative RL algorithm and investigate various modeling approaches for the intraday multireservoir optimization problem. Through extensive experiments using real hydropower system data, we analyze the performance of different RL agents and benchmark them against random and greedy policies. Results demonstrate that optimal modeling choices, including reward adjustment, sufficient forecast information, and grouping of actions, significantly impact performance. Moreover, our findings suggest that Soft Actor-Critic, a RL algorithm that has not been applied before in this domain, is a viable alternative to methods such as Q-learning. Overall, this study contributes to the understanding of RL techniques in hydropower optimization and provides valuable insights for practical implementation in real-world scenarios.

Keywords: Daily optimization · Hydropower generation · Multireservoir · Reinforcement Learning

This work was supported both by the project Project IA4TES (Advanced Intelligent Technologies for Sustainable Energy Transition) with file number TSI-100408-2021 from the 2021 AI R&D Missions Program, within the framework of the Spain Digital Agenda 2025 and the National Artificial Intelligence Strategy, funded by the Recovery, Transformation, and Resilience Plan and co-financed with European funds from the Recovery and Resilience Facility (RRF), Next Generation EU and the Spanish Agencia Estatal de Investigación for the support provided by the Ministerio de Ciencia e Innovación of Spain (Grant Ref. PID2022-137748OB-C31 funded by MCIN/AEI/10.13039/501100011033) and "ERDF A way of making Europe".

A. A. Juan et al. (Eds.): DSA ISC 2024, LNCS 14778, pp. 376–392, 2025.
https://doi.org/10.1007/978-3-031-78238-1_34

1 Introduction

The operation of a multireservoir hydropower system is a challenging task, involving many factors such as energy price, hydrological information, and the non-linear dynamics of power systems. Despite this complexity, researchers have developed Reinforcement Learning (RL) methods that find optimal long-term planning strategies. With these strategies, decisions are made periodically over the course of a year. Our paper explores a different problem, that of finding short-term planning strategies which involve many decisions every day, and uses an alternative RL algorithm than those tested in the literature.

The subsequent sections of this paper are structured as follows. We commence with a review of existing literature concerning this subject (Sect. 2), providing a foundation for defining the objectives of the paper (Sect. 3). We then explain the short-term multireservoir problem in detail (Sect. 4). Following this, we present various ways to model the problem within the RL framework (Sect. 5) and how we implemented them (Sect. 6). Subsequently, we explain the methods used to benchmark RL: the greedy and random policies (Sect. 7). Finally, using data from an actual hydropower system, the RL algorithm is applied under several modeling options (Sect. 8), yielding results (Sect. 9) that inform our conclusions (Sect. 10).

2 Literature Review

This literature review summarizes the studies, problem statements and methodologies undertaken in this field before we carried out our own work.

In the field of hydropower optimization, Reinforcement Learning (RL) has been applied in numerous works of the literature [2–5,7–9]. The following paragraphs summarize the problem statement and methodology of these studies.

Regarding the problem statement, the literature focuses on long-term management strategies spanning months or years. Their objectives include minimizing spillage or meeting demand requirements [3], maximizing energy generation [5,8,9], or a combination of both [2,4]. Additionally, one study aims to maximize income by considering electricity prices [7].

As for the methodology, the most studied method in the literature is Q-learning [2,5,8], along with some variations [3,9]. However, one work uses an evolutionary direct policy method [4], while another uses a supervised learning technique [7].

The reviewed literature reveals a gap in the problem statement. None of the studies addresses the short-term intraday management problem, which is the central focus of our research. In this problem, the goal is to adapt the hydropower station to changes in energy price within the same day.

This problem is significantly different from the long-term management documented in existing research. Intraday optimization requires a much higher temporal resolution, with each period spanning minutes instead of days. With this finer time discretization, one cannot neglect the delays between water release and energy production, as well as the difference between the reservoir and turbine outflows.

Another gap of the reviewed literature concerns the methodology. To our knowledge, the only value-based RL method applied in existing literature is Q-learning. This method requires discrete inputs and outputs, so the continuous inputs and outputs in hydropower optimization are discretized to allow the use of Q-learning. In contrast, this paper pioneers the study of a discretization-free method, specifically Soft Actor-Critic [6], which has not been previously explored in this context.

3 Objectives

In this paper, we set out to tackle these three open research questions:

1. How can we effectively model the *intraday* multireservoir management problem using Reinforcement Learning (RL)?
2. Among the various modeling choices available, which one proves to be the most effective in addressing this problem?
3. Beyond Q-learning, how do alternative RL algorithms fare when applied to the multireservoir management problem?

These questions form the core objectives of our study, aiming to fill the gaps in the understanding of RL techniques applied to multireservoir management.

4 Intraday Multireservoir Optimization Problem

4.1 Introduction

A multi-reservoir hydropower system can be divided into several parts, each consisting of a reservoir, a dam, a channel, and a power group, as shown in Fig. 1.

The dam enables water storage in the reservoir, which is released by opening the channel gate. The extent to which the gate is opened determines the outflow of the reservoir. Over time, this water reaches the power group, where it generates energy, and then proceeds to the next reservoir, if there is one.

The revenue generated from this energy production depends on the energy price at the corresponding hour. Adapting production to these energy prices may require changing the outflow multiple times in the same day, hence the term "*intraday* multireservoir optimization problem".

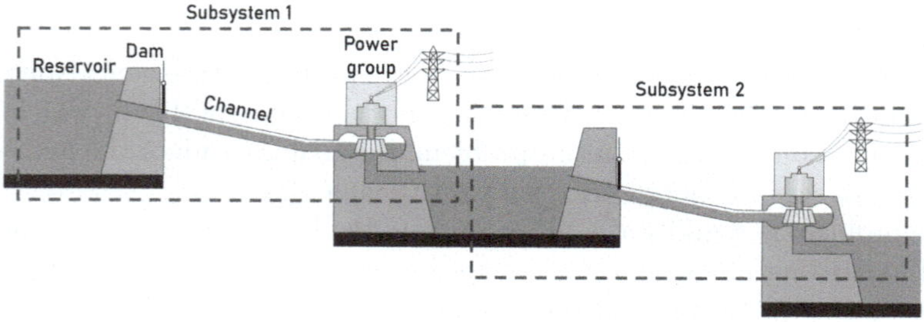

Fig. 1. Elements of a multireservoir hydropower system.

4.2 Details

Decision and Objective. The problem consists of determining the outflow from each reservoir at each period throughout the day. The goal is to maximize the revenue earned with the generated energy, subtracting operational costs.

Input Data. We assume a forecast of the energy price for each hour of the day is known. We also assume we have an estimate of the natural inflow to each reservoir at each period. The natural inflow is the incoming water from the river or precipitation, which is notable in the first reservoir. However, subsequent reservoirs mainly rely on water from the previous subsystems rather than from the river or rain.

Time Discretization. Time is divided into fixed intervals or periods. Within each interval, the outflow from each reservoir is determined, and it remains constant throughout that interval. To evaluate *intraday* strategies, a fine time discretization is needed, requiring periods that span only minutes.

When using such small periods, the delay between water release and power generation cannot be ignored. Consequently, the outflow from one period does not generate energy in the same period but rather in a subsequent one. This delay depends on the length of the channel.

Revenue and Operational Costs. Revenue for each period is calculated by multiplying the energy price for the corresponding hour by the generated power within that period. The power generated depends non-linearly on the water that is passing through the power group (turbine flow). In turn, the turbine flow depends on the past outflows of the previous reservoir.

Each power group has multiple turbines installed. If the turbine flow is greater than a certain threshold, an additional turbine in the power group needs to be activated, triggering a startup cost.

4.3 Motivation

The intraday variant of the hydropower optimization problem is of great interest in markets where prices shift every 15 min. Quick adaptability is essential in this context.

Conventional optimization techniques like linear programming require a lot of computational time for complex problems such as hydropower optimization. For this reason, they are too slow to handle the high-frequency changes present in this context. In contrast, Reinforcement Learning (RL) methods, once trained, can make decisions in seconds. Furthermore, unlike traditional approaches, RL methods can be extended to stochastic contexts, as shown in [7]. These are the reasons RL was chosen in this work.

4.4 Conclusion

This detailed case of the hydropower optimization problem, which considers time delays, revenue and operational costs, is the main focus of our work. The remaining pages of this paper describe how Reinforcement Learning (RL) was used to address this problem.

5 Deep Reinforcement Learning

5.1 Introduction

The main components of the Reinforcement Learning (RL) framework are the *environment* and *agent*. Their respective functions are illustrated in Fig. 2.

The environment is a model that encodes the relevant variables of the hydropower system (volume, power, etc.) into a single *state vector*, $\mathbf{s}_t$, for every period t. The agent is a function that maps the state vector into an *action vector*, $\mathbf{a}_t$.

The environment can interpret this action vector, implementing the corresponding outflows within a simulated hydropower system, and consequently updating the state variables. Subsequently, the environment computes a *reward*, r_t, and the next state vector, $\mathbf{s}_{t+1}$. The new state vector can be given to the agent once again, repeating the process. The cycle finishes when the agent has chosen an action for each period of the day.

The agent is implemented as one or more Artificial Neural Networks that map $\mathbf{s}_t$ to $\mathbf{a}_t$. Using a training algorithm built upon Machine Learning, the networks can be trained to maximize the expected *return*. The return is defined as the cumulative sum of rewards within a day, $R = \sum_t r_t$. Maximizing the return (R) instead of the immediate reward (r_t) avoids short-sighted behavior.

In this paper, we used the RL algorithm known as Soft Actor-Critic [6], in which the agent is divided into two networks: the *actor* and the *critic*. The actor processes a state vector ($\mathbf{s}_t$) to produce an action vector ($\mathbf{a}_t$), while the critic evaluates the state-action pair (($\mathbf{s}_t, \mathbf{a}_t$)) to predict the expected return associated with that action in that state. The critic is used only to train the actor: once the agent has been trained, the critic is not used anymore. In our experiments, we implemented the critic and the actor with the Artificial Neural Network architectures explained in Sect. 8.2.

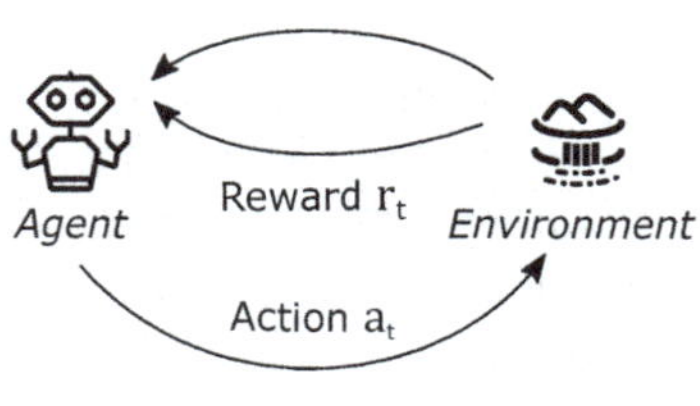

Fig. 2. Components of the Reinforcement Learning (RL) framework.

The following sections describe the implementation details of the environment. Firstly, Sect. 5.2 explains how the hydropower system is simulated. Then, Sect. 5.3 indicates the computation of the reward (r_t) derived from the generated income, followed by Sect. 5.4, which explains the definition of the state vector ($\mathbf{s}_t$) derived from the hydropower system's state. Finally, Sect. 5.5 describes the translation of the action vector ($\mathbf{a}_t$) into outflows. We will cover different ways to model these variables, with the goal of fulfilling Objective 1 of this paper as discussed in Sect. 3.

5.2 Environment

The process of training the RL agent entails numerous iterations wherein the agent suggests an action and obtains the reward and subsequent state. The computation of these rewards and states requires either direct access to the actual system or a simulated environment. Due to the extensive number of training iterations and the duration of each iteration, which mirrors several minutes in the real system, we opted to use a simulated environment.

The simulation of the hydropower system follows the description of the problem outlined in Sect. 4. The accuracy of the simulator was confirmed using data from the actual system, which is detailed in Sect. 8.1.

In essence, the simulator operates by taking outflows as input and giving income as output. Specifically, in period t, the simulation model takes as input the decided outflow ($\hat{q}^{\text{out}}_{d,t}$) for every dam d and updates its turbine flow ($q^{\text{turb}}_{d,t}$), actual outflow ($q^{\text{out}}_{d,t}$), volume ($v_{d,t}$), power ($w_{d,t}$), and income ($i_{d,t}$).

Firstly, the turbine flow of every power group d is updated with Eq. 1. Given the time delay between releasing water and reaching the power group, the turbine flow is a function of the past outflows instead of the current outflow. To be more specific, the turbine flow is a moving average of the past outflows, taking the periods between the maximum lag ($L^{\max}_d$) and the minimum lag ($L^{\min}_d$) corresponding to the channel's length.

$$q^{\text{turb}}_{d,t} = \frac{q^{\text{out}}_{d,t-L^{\min}_d} + \cdots + q^{\text{out}}_{d,t-L^{\max}_d}}{L^{\max}_d - L^{\min}_d + 1} \quad \forall d = 1, \ldots, D \tag{1}$$

Secondly, the actual outflow from each dam d is updated using Eq. 2. In essence, this equation ensures that the actual outflow ($q^{\text{out}}_{d,t}$) remains equal to the decided value ($\hat{q}^{\text{out}}_{d,t}$) unless the volume, updated with this discharge, would fall below the minimum threshold. In such cases, the outflow is adjusted to maintain the volume at the minimum level. Note that the second argument of the minimum function is derived by solving Eq. 3 for $q^{\text{out}}_{d,t}$ when $v_{d,t} = V^{\min}_d$.

$$q^{\text{out}}_{d,t} = \min\{\hat{q}^{\text{out}}_{d,t}, Q^{\text{in}}_{d,t} + q^{\text{turb}}_{d-1,t} + \frac{v_{d,t-1} - V^{\min}_d}{\Delta t}\} \quad \forall d = 1, \ldots, D \tag{2}$$

Thirdly, the volume of every reservoir d is updated with the volume balance equation shown in Eq. 3. Note the volume of reservoir d increases with the natural inflow ($Q^{\text{in}}_{d,t}$), increases with the turbine flow in the preceding dam ($q^{\text{turb}}_{d-1,t}$), and decreases with the actual outflow ($q^{\text{out}}_{d,t}$). The natural inflows are treated as known parameters.

$$v_{d,t} = v_{d,t-1} + \Delta t \left(Q^{\text{in}}_{d,t} + q^{\text{turb}}_{d-1,t} - q^{\text{out}}_{d,t} \right) \quad \forall d = 1, \ldots, D \tag{3}$$

Fourthly, the power and number of active turbines of every power group d is updated with Eq. 4. This equation calculates the power produced with the current turbine flow, as well as the number of active turbines required to process this flow. Both values are given by a non-linear function, denoted as f.

$$w_{d,t}, n_{d,t} = f(q^{\text{turb}}_{d,t}) \quad \forall d = 1, \ldots, D \tag{4}$$

Finally, the income of every subsystem d is updated with Eq. 5. Here, the generated power ($w_{d,t}$) is multiplied by the length of the period (Δt) to get the produced energy, which is in turn multiplied by the electricity price (P_t) to get the generated revenue. The final income is calculated by subtracting the operational costs from this revenue. These costs are computed by multiplying the new turbine startups ($\max\{n_{d,t} - n_{d,t-1}, 0\}$) with the startup cost (S). The energy prices and startup cost are treated as known parameters.

$$i_{d,t} = P_t w_{d,t} \Delta t - S \max\{n_{d,t} - n_{d,t-1}, 0\} \quad \forall d = 1, \ldots, D \tag{5}$$

Note the volume and outflow must adhere to specified minimum and maximum thresholds, as shown in Eq. 6 and Eq. 7. Any values beyond these limits are clipped within the simulator.

$$V^{\min}_d \leq v_{d,t} \leq V^{\max}_d \quad \forall d = 1, \ldots, D \tag{6}$$

$$0 \leq q^{\text{out}}_{d,t} \leq Q^{\max}_d \quad \forall d = 1, \ldots, D \tag{7}$$

To utilize the simulator in RL, it is necessary to map the income to the reward, the simulator's variables to the state vector, and the agent's action vector to the outflows. These mappings will be explained in Sects. 5.3, 5.4 and 5.5, respectively.

5.3 Reward Definition

Reward Formula. We compute the reward r_t given to the agent using Eq. 8. Here, $i_{d,t}$ is the income generated by dam d, $\max_t P_t$ is the peak price of the day, $h_{d,t}$ is a variable indicating if the water hammer constraint (explained below) has been violated for reservoir d, and H is the corresponding penalty.

Note that income will consistently be greater on days when prices are higher, regardless of the agent's actions. Therefore, it is not advisable to directly use the income as the reward, as it would reinforce the agent's behavior on days with high prices. This is why we divide the income by the peak price in Eq. 8. Similarly, the water hammer penalty is also divided by the peak price to maintain values on the same scale.

$$r_t = \frac{\sum_d (i_{d,t} - H \cdot h_{d,t})}{\max_t P_t} \tag{8}$$

Water Hammer Constraint. The "water hammer constraint" prevents sudden changes in the direction of outflow variation; that is, it forbids the outflow to start increasing if it has just been decreasing, and vice versa. In particular, it prevents such changes of direction during the last K periods. Consequently, the variable $h_{d,t}$ of Eq. 8 is computed with Eq. 9, where $\delta_{d,t} = q^{\text{out}}_{d,t} - q^{\text{out}}_{d,t-1}$ is the outflow variation at period t.

$$h_{d,t} = \begin{cases} 0 & \text{if } \delta_{d,t} \cdot \delta_{d,t-1} \cdot \ldots \cdot \delta_{d,t-K} > 0 \\ 1 & \text{otherwise} \end{cases} \tag{9}$$

Violating the water hammer constraint does not impact the income directly, but it could lead to the occurrence of a water hammer in the channel, posing risks such as noise, vibration, or even structural collapse. To mitigate this risk, we incorporate penalties for violating this constraint into the reward, with the anticipation that the agent will learn to fulfill this constraint and prevent water hammer incidents.

Adjusted Reward. Here, we propose a modification to the reward computation that adjusts it based on the performance of the greedy policy outlined in Sect. 7.

The rationale behind this adjustment is as follows. In Eq. 8, we normalize the income by the maximum price to mitigate day-dependent biases in the reward. Despite this adjustment, our experiments indicated that this method falls short in fully mitigating bias, as agents consistently achieve higher rewards on days with greater inflows.

To address this issue, we decided to shift and scale the reward by the average reward of the greedy policy. Under this approach, the average reward achieved by the greedy policy throughout the day (r^{greedy}) is computed at the beginning. Then, this value is used to transform the rewards given to the agent at each period of the day, as shown in Eq. 10. This way, the agent receives the unbiased adjusted reward (r'_t) instead of the original reward (r_t).

Furthermore, note this formula does not scale the reward when $r^{\text{greedy}} < 1$ and does not shift it when $r^{\text{greedy}} < 0$. This precaution is taken to prevent artificially inflating the agent's reward in days where the performance of the greedy policy is poor.

$$r'_t = \frac{r_t - \max\{r^{\text{greedy}}, 0\}}{\max\{r^{\text{greedy}}, 1\}} \tag{10}$$

Section 9 will compare the results of using the basic reward of Eq. 8 and the adjusted reward of Eq. 10.

5.4 State Definition

Global State Vector. The state vector $\mathbf{s}_t$ is given by Eq. 11. In principle, this vector contains all the information the agent needs to make its decision.

The prices of future periods were included to help the agent decide whether to release water or save it for later. These prices are gathered in a price vector,

$\mathbf{P}_{t,t+N} = (P_t, P_{t+1}, \ldots, P_{t+N})$, where N is a parameter indicating how far the agent looks ahead, hereinafter called "lookahead" parameter. Besides the prices, the state vector contains the current state of each dam, represented by a dam state vector that will be defined later, $\mathbf{s}_{d,t}$.

$$\mathbf{s}_t = (\mathbf{P}_{t,t+N}, \mathbf{s}_{1,t}, \ldots, \mathbf{s}_{D,t}) \tag{11}$$

Dam State Vector. Each dam state vector $\mathbf{s}_{d,t}$ is defined by Eq. 12. The vector is formed by the past state of the dam, the past outflows, the past variations, and the future inflows, as detailed in the following paragraphs.

The first component of the vector represents the state of the dam before the agent applies its action ($a_{d,t}$). As shown in Eq. 13, it includes the previous volume ($v_{d,t-1}$), turbine water ($q^{\text{turb}}_{d,t-1}$), power ($w_{d,t-1}$), and number of active turbines ($n_{d,t-1}$). Moreover, to help the agent understand how its previous action resulted in this state, the vector includes the difference between the previously assigned outflow and the actual one, $\kappa_{d,t-1} = \hat{q}^{\text{out}}_{d,t-1} - q^{\text{out}}_{d,t-1}$, which we call "outflow cut".

In addition, the dam state vector includes the past outflows, $\mathbf{q}^{\text{out}}_{d,t-L_d^{\max}-1,t-1} = (q^{\text{out}}_{d,t-L_d^{\max}-1}, \ldots, q^{\text{out}}_{d,t-1})$. Giving the past outflows serves two functions. Firstly, it may allow the agent to understand how the previous turbine water ($q^{\text{turb}}_{d,t-1}$) is calculated with the outflows from from $t - L_d^{\max} - 1$ to $t - L_d^{\min} - 1$ as shown in Eq. 1. Secondly, it may also allow the agent to understand how the reward is inderictly given by the outflows from from $t - L_d^{\max}$ to $t - L_d^{\min}$. These outflows will determine the current turbine flow ($q^{\text{turb}}_{d,t}$) and thus the reward (r_t).

Since the reward also has a penalty for not fulfilling the water hammer constraint, the outflow variations in the last K periods are also given to the agent. These can be computed as the difference between two lagged outflow vectors, $\mathbf{\Delta}_{d,t-K,t-1} = \mathbf{q}^{\text{out}}_{d,t-K,t-1} - \mathbf{q}^{\text{out}}_{d,t-K-1,t-2}$.

Finally, to allow the agent to plan ahead, the dam state vector also contains the future natural inflows, $\mathbf{Q}^{\text{in}}_{d,t,t+N} = (Q^{\text{in}}_t, Q^{\text{in}}_{t+1}, \ldots, Q^{\text{in}}_{t+N})$. This expression uses the same value for the lookahead parameter as the future prices, N.

$$\mathbf{s}_{d,t} = (\boldsymbol{\sigma}_{d,t-1}, \mathbf{q}^{\text{out}}_{d,t-L_d^{\max}-1,t-1}, \mathbf{\Delta}_{d,t-K,t-1}, \mathbf{Q}^{\text{in}}_{d,t,t+N}) \tag{12}$$

$$\boldsymbol{\sigma}_{d,t-1} = (v_{d,t-1}, q^{\text{turb}}_{d,t-1}, w_{d,t-1}, n_{d,t-1}, \kappa_{d,t-1}) \tag{13}$$

These vectors represent all the data provided to the agent during training and testing. Yet, the original values of these variables were not directly used. Instead, they underwent one or more transformations beforehand, as explained in the subsequent paragraphs.

Normalization. Normalization ensures the input values remain within a fixed range, typically between 0 and 1. This adjustment is crucial to prevent the agent from disproportionately emphasizing variables with larger scales, such as the volume in our problem.

Each subsystem's volume, outflow, turbine flow, power, and turbine count were normalized using their physical minimum and maximum values. Inflow was normalized using its highest historical value, whereas price, like in the reward (see Eq. 8), was normalized using the peak value of the day.

Quantile-Based Pseudo-Discretization (QPD). "Quantile-based Pseudo-Discretization" (QPD) is a novel technique used to space out the values of each state variable more evenly.

This transformation was motivated by the examination of normalized values shown in Fig. 4. These histograms reveal that certain variables consistently maintain proximity to a fixed value. This pattern is observed in the natural incoming flows of all dams except the first ($\mathbf{Q}^{\text{in}}_{d,t,t+N}; d > 1$), the outflow cut ($\kappa_{d,t-1}$), and the past variations ($\mathbf{\Delta}_{d,t-K,t-1}$).

To achieve a more uniform distribution of values across all state variables, including the specific ones mentioned, we apply the QPD transformation. To do this, a sample set of state vectors ($\{\mathbf{s}_t\}$) must be available to calculate the percentiles of each component. Using these percentiles, the transformed state vector is computed with Eq. 14, where p is a function that maps each value to the percentile it belongs to. The percentile is divided by 100 to keep the values within the range $[0, 1]$.

$$\mathbf{s}'_t = \frac{p(\mathbf{s}_t)}{100} \tag{14}$$

Principal Component Analysis (PCA). Principal Component Analysis (PCA) is dimensionality reduction technique [1] that we applied to reduce the number of components of the state vectors $\mathbf{s}_t$ while retaining most of the information. The desired level of retained information, or *variance*, is tipically expressed as a percentage of the original variance.

To apply PCA, a sample set of state vectors ($\{\mathbf{s}_t\}$) is used to compute the transformation that reduces the number of components while preserving the desired variance. The learned transformation can then be applied to any state vector ($\mathbf{s}_t$), resulting in a transformed state vector with fewer components ($\mathbf{s}'_t$).

Section 9 details the outcome of using the normalized state vectors, the QPD-transformed vectors, the PCA-transformed vectors, or vectors transformed by both QPD and PCA.

5.5 Action Definition

Single-Step Action. The action vector $\mathbf{a}_t$ is transformed into outflows using Eqs. 15, 16. Note Eq. 16 was defined such that each component of the action vector belongs to the interval $[-1, 1]$. This was a requirement of the Soft Actor-Critic implementation used in this paper (see Sect. 6) in which actions are occasionally sampled from a normal distribution of mean 0.

$$\mathbf{a}_t = (a_{1,t}, \ldots, a_{D,t}) \tag{15}$$

$$\hat{q}_{d,t}^{\text{out}} = \frac{a_{d,t}+1}{2} Q_d^{\max} \quad \forall d = 1, \ldots, D \tag{16}$$

Multi-step Action. In addition to the single-step action introduced earlier, this paper experimented with a "multi-step action", in which the agent decides the outflow of multiple future periods at the same time. To be more specific, the agent gives the outflow of the next A periods instead of the next single period. This essentially replaces Eq. 15 with Eq. 17.

The adoption of this multi-step action was implemented to mitigate the delay between action and reward inherent in the original single-step action. With the original approach, the agent selects the outflow for a given period, while the reward for that period is calculated based on turbine flow, which is influenced by past outflows (refer to Eq. 8 and Eq. 5). Consequently, the agent does not receive immediate feedback on its action but rather for past actions. In contrast, by grouping multiple actions together, the reward becomes dependent on the initial actions within the group rather than past actions.

$$\mathbf{a}_t = (a_{1,t}, \ldots, a_{D,t}, \ldots, a_{1,t+A-1}, \ldots, a_{D,t+A-1}) \tag{17}$$

When using this type of action, the reward and state vector must be adapted accordingly.

Firstly, the reward must be replaced by the cumulative reward over the periods affected by the action, as shown in Eq. 18.

$$r'_t = r_t + r_{t+1} + \ldots + r_{t+A-1} \tag{18}$$

Secondly, the state vector must be expanded so the agent has i) enough information about the past to understand the effect of its last action, and ii) enough information about the future to be capable of planning ahead. With this goal in mind, Eqs. 11, 12 are replaced by Eqs. 19, 20, 21. Note that both sets of equations are equivalent when $A = 1$.

$$\mathbf{s}'_t = (\mathbf{P}_{t,t+N+A-1}, \mathbf{s}'_{1,t}, \ldots, \mathbf{s}'_{D,t}) \tag{19}$$

$$\mathbf{s}'_{d,t} = (\boldsymbol{\Sigma}_{d,t-1,t-A}, \mathbf{q}^{\text{out}}_{d,t-L_d^{\max}-A,t-1}, \boldsymbol{\Delta}_{d,t-K-A+1,t-1}, \mathbf{Q}^{\text{in}}_{d,t,t+N+A-1}) \tag{20}$$

$$\boldsymbol{\Sigma}_{d,t-1,t-A} = (\boldsymbol{\sigma}_{d,t-1}, \ldots, \boldsymbol{\sigma}_{d,t-A}) \tag{21}$$

Section 9 will showcase the outcomes obtained using both the single-step and multi-step actions across various values of A.

6 Implementation

The Reinforcement Learning (RL) environment was created in Python using the Gym/Gymnasium framework [11], while the RL agent's training was conducted utilizing the open-source Soft Actor-Critic implementation within Stable-Baselines3 [10].

7 Benchmark Methods

The developed Reinforcement Learning (RL) algorithms will be benchmarked against two policies: a random policy and a greedy policy.

The random policy samples each outflow from a uniform distribution, $\hat{q}_{d,t}^{\text{out}} \sim \mathcal{U}(0, Q_d^{\max})$. In contrast, the greedy policy attempts to assign the maximum flow at each time step, $\hat{q}_{d,t}^{\text{out}} = Q_d^{\max}$.

In a real hydropower system, the greedy policy corresponds to the decision of keeping the gates fully open at all times. In essence, this makes the system behave like a run-of-the-river plant. Although this method aligns with the current practice of the hydropower plant that inspired this study, it neglects leveraging the reservoirs to save water for periods of peak pricing.

8 Experimental Setting

8.1 Available Data

For our experiments, we configured the environment described in Sect. 5.2 with the data of a real hydropower system comprising two reservoirs, named "dam1" and "dam2". The specific characteristics of these reservoirs are featured in Table 1. Furthermore, the power groups of the system are depicted in Fig. 3, illustrating the function f present in Eq. 4, which maps the turbine flow to the generated power and number of turbines.

To train and test the agents, three years' worth of operational data from the hydropower system was at hand. This dataset included various variables such as natural inflow, volume, actual outflow, turbine flow, and power for each subsystem, recorded every 15 min.

Among these variables, only the inflows and the initial volume and outflows of each day were used to initialize the environment; the environment computed the remaining variables instead of relying on the dataset. Additionally, this dataset was complemented with energy price information from OMIE [12], enabling the calculation of income.

Table 1. Characteristics of each subsystem.

Variable	Dam 1	Dam 2
Minimum volume	$V_1^{\min} = 34045\,\text{m}^3$	$V_2^{\min} = 17117\,\text{m}^3$
Maximum volume	$V_1^{\max} = 70882\,\text{m}^3$	$V_2^{\max} = 58343\,\text{m}^3$
Maximum outflow	$Q_1^{\max} = 14.15\,\text{m}^3\text{s}^{-1}$	$Q_2^{\max} = 11.27\,\text{m}^3\text{s}^{-1}$
Minimum lag	$L_1^{\min} = 1$	$L_2^{\min} = 3$
Maximum lag	$L_2^{\max} = 1$	$L_2^{\max} = 5$
Maximum power	4.6 MW	8.48 MW
Turbines installed	2	3

8.2 Parameter Values

We discretized time into intervals or periods of 15 min to match the temporal resolution of the available data. Moreover, we assumed there is risk of water hammer when sudden changes in outflow variation happen within half an hour, equivalent to $K = 2$ periods.

Regarding the operational costs, two scenarios were considered. In the first scenario, startups and water hammer risks have a penalty of $S = 50$ and $H = 25$ euro, respectively. Conversely, in the second scenario, startup costs and water hammer risks are considered insignificant, resulting in no penalties: $S = 0$ and $H = 0$. All agents were trained in both scenarios, and their final performance in each is detailed in Sect. 9.

Finally, the implementation specifics of Soft Actor-Critic are outlined. Both the actor and the critic were implemented as Multilayer Perceptrons with two hidden layers, each consisting of 256 neurons and employing ReLU activation. The discount factor was set to $\gamma = 0.99$ and the soft update coefficient to $\tau = 0.005$. Two learning rates were tested, $\alpha = 3 \times 10^{-4}$ and $\alpha = 1 \times 10^{-3}$, but no significant difference was found between them. Therefore, only the results for the former value are presented in Sect. 9. In applying Principal Component Analysis (PCA) to the states, the explained variance was set to be 98% of the original variance.

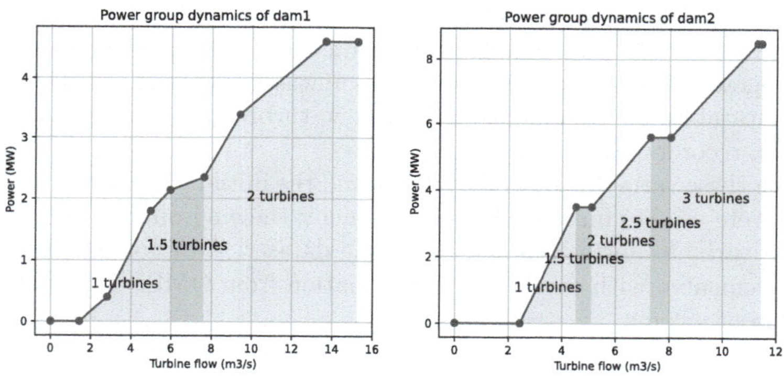

Fig. 3. The power-turbine characteristic flow curve for each power group. It shows the power generated with a specific turbine flow and indicates the number of turbines needed to handle that flow. When the required number of turbines is not a whole number, indicating uncertainty, the agent is given a startup penalty.

8.3 State Variables Analysis

The Reinforcement Learning (RL) environment was run with the data and parameters specified in Sects. 8.1 and 8.2. The environment was first used to simulate 1000 days of operation with a random policy. The normalized state vectors experienced by this agent were analyzed to produce the histogram shown in Fig. 4. This histogram prompted us to train some agents with states transformed via Quantile-based Pseudo-Discretization (QPD), as explained in Sect. 5.4.

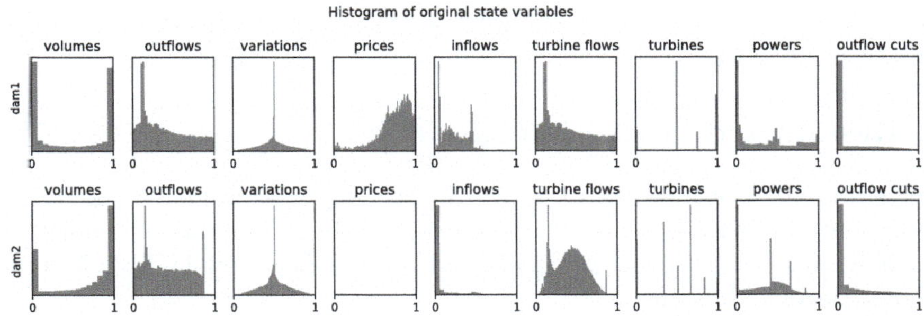

Fig. 4. Histogram of the normalized state variables experienced by a random policy throughout 1000 simulated days. Prices are not present in the second dam to avoid redundancy.

8.4 Training and Testing

Each Reinforcement Learning (RL) agent was trained in 99,000 steps using the inflows and prices of the available data. This number of training steps is equivalent to 1000 days with the single-step action, but to $1000 \cdot A$ days with the multi-step action.

In order to fulfill the Objective 2 outlined in Sect. 3, many different modeling choices were tested. To be specific, the following types of agents were trained:

- Agents in different scenarios. Some agents received startup and water hammer penalties in the reward ($S = 50, H = 25$), while others had no operational costs ($S = 0, H = 0$).
- Agents with different levels of information about the future. The most short-sighted agent was given the price and inflow of the next 16 periods, while the most far-sighted agent was given the values of the next 96 periods. The tested values are $N = 16$, $N = 32$, $N = 64$, and $N = 96$.
- Agents that saw the original or transformed state vectors. The transformations tested were Quantile-based Pseudo-Discretization (QPD), Principal Component Analysis (PCA), and QPD followed by PCA.
- Agents with the single-step and multi-step action. The multi-step action was tested considering $A = 9$, $A = 20$, $A = 33$, and $A = 99$ periods. These values were chosen such that no action is executed incompletely at the end of the day of 99 periods.
- Agents which received the normal reward as feedback, and those which were given the adjusted reward (refer back to Sect. 5.3).

Training was done on a computer with a 3 GHz Intel Core i7-9700 CPU and 16 GB of RAM. With this system, training a single agent for 99,000 steps took between 37 min and 7 h, depending on the configuration (multi-step actions and adjusted rewards noticeably increase training time).

Following the training phase, all agents underwent testing on 11 carefully selected days from the dataset at hand. These days include the driest day (lowest inflows) and the rainiest day (highest inflows), as well as 9 intermediate cases. The average income throughout these 11 days is taken as the performance measure of each agent.

9 Results and Discussion

The performance of all agents outlined in Sect. 8.4 was evaluated against the greedy and random policies outlined in Sect. 7. Table 2 illustrates the difference in average income between the agents and these policies, while Table 3 details the daily income differences.

Table 2 shows that PCA and QPD state transformations typically result in worsened outcomes. Concerning the lookahead periods for the agent, the optimal values are generally $N = 32$ and $N = 64$, which correspond to about half a day of forecasts. In terms of action steps, single-step actions ($A = 1$) yield the best performance when operational costs are considered, whereas multi-step actions with $A = 99$ are preferable in the absence of such costs. Multi-step actions with lower values for A diminish agent performance. Lastly, adjusting the reward tends to increase income.

The agents with the highest average income are highlighted in blue in Table 2 and their daily performance is given by Table 3. This table shows the best agents

Table 2. Performance of each trained agent over the greedy and random policies, with and without operational costs (S, H). Each value shows the difference in average income as a percentage (%). Highlighted in blue are the performances of the best agent in each scenario.

Trained agent				$S = 50, H = 25$		$S = 0, H = 0$	
Transform	**Lookahead**	**Action**	**Reward**	**Greedy**	**Random**	**Greedy**	**Random**
None	$N = 16$	$A = 1$	Unadjusted	+2.21%	+81.71%	-6.10%	+28.76%
			Adjusted	+0.88%	+79.35%	+0.01%	+37.13%
		$A = 9$	Unadjusted	-16.27%	+48.87%	-6.66%	+27.99%
		$A = 20$		-24.58%	+34.09%	-14.48%	+17.26%
		$A = 33$		-21.61%	+39.38%	-5.46%	+29.63%
		$A = 99$	Unadjusted	-1.62%	+74.91%	-0.17%	+36.88%
			Adjusted	-12.04%	+56.38%	+1.25%	+38.83%
	$N = 32$	$A = 1$	Unadjusted	-4.28%	+70.18%	-2.05%	+34.31%
			Adjusted	+3.01%	+83.14%	+0.62%	+37.97%
		$A = 99$	Unadjusted	-9.58%	+60.76%	-0.01%	+37.11%
			Adjusted	-4.43%	+69.92%	+0.78%	+38.19%
	$N = 64$	$A = 1$	Unadjusted	-9.46%	+60.97%	-5.12%	+30.10%
			Adjusted	-0.60%	+76.73%	+0.38%	+37.64%
		$A = 99$	Unadjusted	-9.34%	+61.19%	+1.06%	+38.57%
			Adjusted	-5.13%	+68.67%	+1.53%	+39.22%
	$N = 96$	$A = 1$	Unadjusted	+0.39%	+78.49%	-5.63%	+29.40%
			Adjusted	+0.60%	+78.86%	-0.01%	+37.10%
		$A = 99$	Unadjusted	-5.80%	+67.49%	-1.04%	+35.69%
			Adjusted	-6.66%	+65.95%	+0.84%	+38.27%
PCA	$N = 16$	$A = 1$	Unadjusted	-3.95%	+70.77%	-3.01%	+32.99%
QPD+PCA				-11.71%	+56.97%	-7.52%	+26.81%
QPD				-4.48%	+69.83%	-4.69%	+30.68%
QPD		$A = 99$		-6.97%	+65.40%	-5.71%	+29.29%

Table 3. Performance of the best agents over the greedy policy in each of the 11 selected days. Note each cell represents a percentage difference (%), so the mean of the differences in daily income may not equal the difference in average income.

Income of the best agent		$S = 50, H = 25$	$S = 0, H = 0$
Average income		+3.01%	+1.53%
Income in each day	Driest	-8.59%	+2.19%
		+2.76%	+191.01%
		+0.08%	-1.27%
		+0.00%	+0.91%
		+0.00%	+1.81%
	Middle	+7.30%	+0.33%
		+1.33%	+0.37%
		+26.48%	+0.54%
		+5.29%	+1.49%
		+4.80%	+1.48%
	Rainiest	-0.81%	-0.05%

not only outperform the greedy policy on average, but also consistently earn a higher income on a daily basis.

10 Conclusions

In addressing the objectives outlined in Sect. 3, we arrive at the following conclusions:

1. The *intraday* multireservoir management problem can be modeled well within a Reinforcement Learning (RL) framework. This entails rewarding the agent based on generated income and providing it with past state information and future price forecasts, as detailed in Sect. 5.
2. Our results, as detailed in Sect. 9, indicate that the optimal approach for modeling the problem in a RL framework involves normalizing the state without additional transformations, adding the forecasted prices and inflows for the next half a day in the state, adjusting the reward, and using multi-step actions in the absence of operational costs.
3. The Soft Actor-Critic algorithm is a viable alternative to Q-learning for the hydropower optimization problem, capable of outperforming our benchmark method (greedy policy) with proper modeling choices.

In summary, our study sheds light on the effectiveness of RL techniques in solving the multireservoir optimization problem. By delineating the optimal modeling choices and exploring alternative algorithms, we have not only advanced our understanding of RL applications in this domain but also provided practical insights for real-world implementation. With the correct modeling choices, the RL algorithm outperforms the previous management method used by the hydropower system (greedy policy), rendering our method feasible for practical application.

Acknowledgments. This work was supported both by the project Project IA4TES (Advanced Intelligent Technologies for Sustainable Energy Transition) with file number TSI-100408-2021 from the 2021 AI R&D Missions Program, within the framework of the Spain Digital Agenda 2025 and the National Artificial Intelligence Strategy, funded by the Recovery, Transformation, and Resilience Plan and co-financed with European funds from the Recovery and Resilience Facility (RRF), Next Generation EU and the Spanish Agencia Estatal de Investigacin for the support provided by the Ministerio de Ciencia e Innovacin of Spain (Grant Ref. PID2022-137748OB-C31 funded by MCIN/AEI/10.13039/501100011033) and "ERDF A way of making Europe".

References

1. Pearson, K.: On lines and planes of closest fit to systems of points in space. Philos. Mag. **2**(11), 559–572 (1901). https://doi.org/10.1080/14786440109462720
2. Lee, J.-H., Labadie, J.W.: Stochastic optimization of multireservoir systems via reinforcement learning. Water Resour. Res. **43**(11), W11408 (2007). https://doi.org/10.1029/2006WR005627
3. Castelletti, A., et al.: Tree-based reinforcement learning for optimal water reservoir operation. Water Resour. Res. **46**(9), W09507 (2010). https://doi.org/10.1029/2009WR008898
4. Dariane, A.B., Moradi, A.M.: Comparative analysis of evolving artificial neural network and reinforcement learning in stochastic optimization of multireservoir systems. Hydrol. Sci. J. **61**(6), 1141–1156 (2016). https://doi.org/10.1080/02626667.2014.986485
5. Zarghami, M.: Short term management of hydro-power system using reinforcement learning. École de technologie supérieure (2018). https://books.google.es/books?id=xHscvwEACAAJ
6. Haarnoja, T., et al.: Soft actor-critic: off-policy maximum entropy deep reinforcement learning with a stochastic actor. arXiv preprint arXiv:1801.01290 (2018)
7. Matheussen, B. V. et al.: Hydropower optimization using deep learning. In: International Conference on Industrial, Engineering and Other Applications of Applied Intelligent Systems (2019). https://api.semanticscholar.org/CorpusID:195755546
8. Xu, W., et al.: Deep RL for cascaded hydropower reservoirs considering inflow forecasts. Water Resour. Manage. **34**(9), 3003–3018 (2020). https://doi.org/10.1007/s11269-020-02600-w
9. Xu, W., et al.: Deep RL for optimal hydropower reservoir operation. J. Water Resour. Plann. Manage. **147**(8), 04021045 (2021). https://doi.org/10.1061/(ASCE)WR.1943-5452.0001409
10. Raffin, A. et al.: Stable-baselines3: reliable reinforcement learning implementations. J. Mach. Learn. Res. **22**(268), 1–8 (2021). http://jmlr.org/papers/v22/20-1364.html
11. Towers, M., et al.: Gymnasium. Zenodo (2023). https://doi.org/10.5281/zenodo.8127026, https://zenodo.org/record/8127025
12. OMIE: Operador del Mercado Ibérico de Energía. OMIGROUP. https://www.omie.es/es/file-access-list?parents%5B0%5D=/&parents%5B1%5D=Mercado%20Diario&parents%5B2%5D=1.%20Precios&dir=Precios%20horarios%20del%20mercado%20diario%20en%20Espa%C3%B1a&realdir=marginalpdbc. Accessed 6 Feb 2023

Online Service Deployment in Edge Computing: A Comparison of Batching Techniques

Rosario Messana(✉), Alberto Ceselli, and Christian Quadri

Dipartimento di Informatica, Università degli Studi di Milano, Via Giovanni Celoria 18, 20133 Milan, Italy
{rosario.messana,alberto.ceselli,christian.quadri}@unimi.it

Abstract. We tackle an optimization problem related to the deployment of delay-sensitive services on computing facilities at the edge of the network. We consider an online setting, in which a service provider must dynamically decide where to activate services, in absence of information about upcoming users and deployment costs. We model the application as an online facility location problem under the assumption that serving each user yields a reward which is proportional to the network distance between the user and the serving facility. We propose an online algorithm to tackle such problem. It exploits mathematical programming in an online learning framework. As main feature, it updates the deployment decision after a batch of users has appeared, fitting two operational policies: to refresh decisions after batches of fixed cardinality, or to refresh at constant time intervals, thereby handling batches of variable sizes. We experiment our algorithm with simulations using a stochastic generator, built on real internet connection records. We compare the two operational policies in terms of applicability, profit and computational time. Our results indicate that using batches of variable size allow for easier policy application, with no significant decrease in profits, and only a minor increase in performance variability.

Keywords: Online optimization · Facility location · Edge computing

1 Introduction

Edge computing technologies allow service providers to offer delay-sensitive services with a higher level of quality of service (QoS) than cloud computing. By exploiting computing facilities distributed all over a geographical area, for example, a metropolitan area, service can be dynamically and elastically deployed in different facilities to optimize the overall QoS offered to final users. The high spatio-temporal dynamic of the service sessions leads the service provider to continuously adapt the service instances deployment to improve QoS and lower operational costs of using the computing capacity of the edge facilities.

A. A. Juan et al. (Eds.): DSA ISC 2024, LNCS 14778, pp. 393–403, 2025.
https://doi.org/10.1007/978-3-031-78238-1_35

In principle, to guide the optimization process, service providers can rely on instantaneous and historical information about the current and past service sessions. In practice, it is often not the case: a service provider can only observe and collect data related to the application level, having very limited or even no access to data of the layers below (e.g., edge facilities utilization, network congestion, radio access technology, and users' locations). Moreover, edge computing facilities are normally organized in a multi-tenant and multi-provider architecture, in which multiple service providers share the facility resources and at the same time can use computing resources offered by different edge infrastructure providers, and therefore retrieving disaggregated data is not possible. Finally, users may connect to edge facilities through different access network providers, i.e. telco operators, each having its own data infrastructure.

That is, in this complex network architecture, data ownership is distributed among multiple players, making it too difficult for the service provider to reconstruct an overall knowledge of the underlying systems. Data sharing among different players raises not only technological issues, but also privacy and security concerns, ultimately discouraging providers and operators.

We are interested in optimization in such an industrial setting of data scarcity at decision time. In particular, we aim to manage dynamic service deployment when user locations are unknown in advance and deployment costs are subject to frequent changes, which is a very common setting in applications. The optimization task takes place over a sequence of *deployment rounds*. At the beginning of each round, the provider identifies the computing facilities where instances of a service are to be deployed in order to account for users that will require connection in the same round. We assume potential connections to have a relative revenue value which is given by an unknown function of the distance between the user location and the computing facility. For example, revenue can be characterized in terms of quality of service (e.g. service provision time). In order to maximize the total revenue, each user is connected to the facility that yields the highest value. Each service deployment on a facility has an associated cost the provider must pay. The provider can observe revenue and cost values, but only after the service deployment has been implemented. Deployments are subject to budget constraints. In our setting, we assume a single linear budget constraint where each facility has an associate known and fixed budget consumption value. This setting is summarized in Fig. 1.

We remark that, without the support of specific optimization algorithms, there are no simple policies to adopt. A possible one would be over-provisioning, deploying service instances in many (or all) the available edge facilities. However, the budget constraints could make this option infeasible. Another option would be to perform a random deployment in arbitrary facilities, providing a possibly unacceptable quality of service to users. Instead, assuming to design suitable allocation algorithms, service providers have one main optimization lever: the deployment frequency strategy, that is the definition of the deployment rounds. In this paper we compare two policies: to defer a new deployment round until a fixed number of users appears, having therefore slots of different length, but

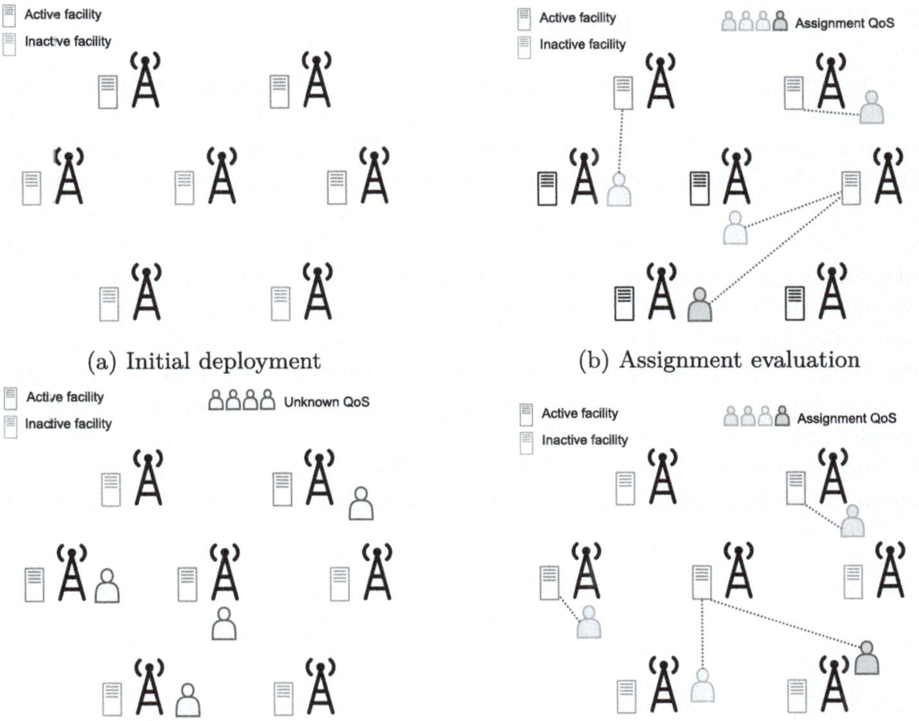

(a) Initial deployment (b) Assignment evaluation

(c) Re-deployment (d) New assignment evaluation

Fig. 1. Overall setting: the provider signs up contracts for the availability of computing power on a set of facilities. Where to actually deploy services is chosen dynamically (a); after deployment users appear, and connect to the closest service instance (b); at given intervals, deployment can be changed (c) and new users appear (d) in an iterative fashion.

corresponding to batches of users having a fixed size, or to schedule the deployment at fixed time intervals, each corresponding to a batch of users of variable length. Each strategy yields to a slightly different modeling of the optimization problem, and therefore to different algorithmic needs.

We show that the first policy maps into a Batch Online Facility Location problem (BOFL) [12]. The second policy, instead, has never been evaluated in the literature: we therefore adapt the techniques of [12] to the case of fixed length slots, containing batches of variable size. We also carry out a computational comparison of these two options. Our simulations are based on a data generator which builds on internet connection records of a mobile telecommunication network in the city of Milan. We investigate on how the algorithms behave with respect to the two batching techniques in terms of profit, running time and number of deployments per round.

2 Problem Statement and Background

The well known uncapacitated facility location problem [5] (UFLP) has been studied in a variety of online versions, see e.g. [6,7,9,15,16] and [2]. The variant we are interested in this paper is called Batch Online Facility Location (BOFL). In BOFL the decision maker is given a set $J = \{1, \ldots, n\}$ representing candidate facilities. At each round $t \in \mathbb{N}_+$, the decision maker chooses which facilities to open. The decision is represented by a vector $y \in \{0,1\}^n$ where 1 indicates an open facility and 0 a closed one. Subsequently, a set of clients requires service. We distinguish between BOFL with fixed batch size and with variable batch size, depending on the fact that the number of clients per iteration is constant or variable inside a certain range. Let us call m^t the size of the batch at iteration t. For each client $i \in \{1, \ldots, m^t\}$ and for each facility $j \in J$, a reward value r^t_{ij} is revealed. Moreover, for each facility $j \in J$, the cost c^t_j of opening that facility in the current iteration is revealed. Each client is assigned to the facility that gives maximum reward. The goal of the decision maker is to maximize the cumulative profit up to iteration $T \in \mathbb{N}_+$, given by:

$$P(T) := \sum_{t=1}^{T} \left[\sum_{i=1}^{m^t} max^n_{j=1} r^t_{ij} y^t_j - \sum_{j=1}^{n} c^t_j y^t_j \right]. \tag{1}$$

Each facility $j \in J$ has an associated consumption value $z_j \in [0,1]$, also called energy value. The sum of the consumption values of the open facilities cannot exceed 1. We call $z \cdot y \leq 1$ budget or energy constraint.

BOFL is a generalization of the single-client online facility location problem introduced in [16], where the authors propose an ad-hoc algorithm that provides theoretical guarantees in the worst case scenario. The problem has been generalized to the batch case with fixed size in [12], where the algorithm from [16] is extended to such a setting and empirically compared with both existing and novel methods for BOFL based on the iterative solution of mathematical programming models.

3 Follow The Leader with Clustering

In this section, we describe an algorithm for BOFL: Follow The Leader with Clustering (FLC). It is an adaptation of the fundamental online learning algorithm known as Follow The Leader [10], based on iterative solution of mathematical programming models, which has been adapted in different contexts to ensure good theoretical guarantees. In the combinatorial optimization setting, in particular, its variants in the worst case scenario are know as Follow The Regularized Leader [13,14,18]. In the stochastic setting, algorithms with similar characteristics have been proposed, see e.g. [3,8,11] and [4].

Let $R^t = \{r^\tau_i \mid i \in I^\tau,\ 1 \leq \tau \leq t-1\}$ be the set of all the reward vectors collected up to time step $t-1$, where I^τ identifies the clients that require service at iteration τ. It is reasonable to cluster such reward vectors so that different clusters identify different categories of clients. In particular, since reward values are proportional to the distance between clients and facilities, clustering reward vectors gives a spatial clustering of all past clients.

Let us call ℓ the number of clusters at each iteration and define $H := \{1, \ldots, \ell\}$. The clusters $C_1^t, \ldots, C_\ell^t$, which we assume to be non empty, are obtained by applying the K-means algorithm. Since the clustering operation is expensive, we compute the clusters ex novo less often as the time step increases, taking into account that any new reward vector in later iterations will have less impact on the resulting estimation quality, in general. In particular, we define a constant $\Delta \in \mathbb{N}_+$ and we execute the clustering only when $t = 1$ (trivial case) or when

$$t \equiv 0 \ (mod\ (\lfloor t/\Delta \rfloor + 1)).$$

In every other iteration, in order to maintain $\bigcup_{h \in H} C_h^t = R^t$, the reward vectors are assigned to clusters according to their predicted label.

We compute the surrogate rewards $\bar{r}_{hj}^t := \frac{1}{t-1} \sum_{r \in C_h^t} r_j$ for every $h \in H$ and $j \in J$. We also compute the average opening costs $\bar{c}_j^t := \frac{1}{t-1} \sum_{\tau < t} c_j^\tau$. Finally, we solve the following binary linear programming model, where each binary variable x_{hj} equals 1 if and only if cluster h is assigned to facility j.

$$\begin{aligned}
\max\ & \sum_{h=1}^{\ell} \sum_{j=1}^{n} \bar{r}_{hj}^t x_{hj} - \sum_{j=1}^{n} \bar{c}_j^t y_j \\
s.t.\ & \sum_{j=1}^{n} x_{hj} = 1 && \forall h \in H \\
& x_{hj} \le y_j && \forall h \in H\ \forall j \in J \\
& \sum_{j=1}^{n} z_j y_j \le 1 \\
& x_{hj} \in \{0, 1\} && \forall h \in H\ \forall j \in J \\
& y_j \in \{0, 1\} && \forall j \in J
\end{aligned} \tag{FLC}$$

A common feature of the Follow-The-Leader-like algorithms previously presented in literature is that the set of learning parameters in the objective function is fixed. As a consequence, such algorithms are more suited to address online facility location problems in which each client-related parameter, e.g. a reward value, refers to the same client throughout the entire optimization process. However, in BOFL the decision maker is not able to identify single clients over different iterations. As far as we know, FLC is the first algorithm specifically designed for BOFL.

4 Data Generation

We designed a rewards and costs generator based on real data. The data belongs to the Telecom Italia Big Data Challange Dataset [1]. According to its structure, the area of the city of Milan (Italy) was represented as a squared grid composed of cells with edge length of 235 meters approximately. We extracted from the dataset the data regarding the number of Call Detail Records (CDRs) for internet connections registered in the city from November 1st 2013 to January 1st 2014 by the telecommunication company Telecom Italia. Each day is represented as a sequence of 144 time intervals of 10 min length. Each record refers to a different cell inside the grid, indicating the origin of the connection request, and to a specific time interval within the day. The data was averaged by cell and time interval, then normalized to obtain a probability distribution over the grid for each time interval. The grid area was modeled using the set $[0, 1]^2$ and facility locations were placed in 20 sites, in 4 regular rows of 5 facilities each, as shown in Fig. 2. The locations of potential clients are the centers of the cells. For every client, the cell is sampled from the probability distribution corresponding to the current time interval and for each facility the respective reward is given by $\sqrt{2}$ minus the distance between the client location and the facility. The time interval τ is determined according to $\tau = (t \bmod 144)+1$, where t is the iteration counter.

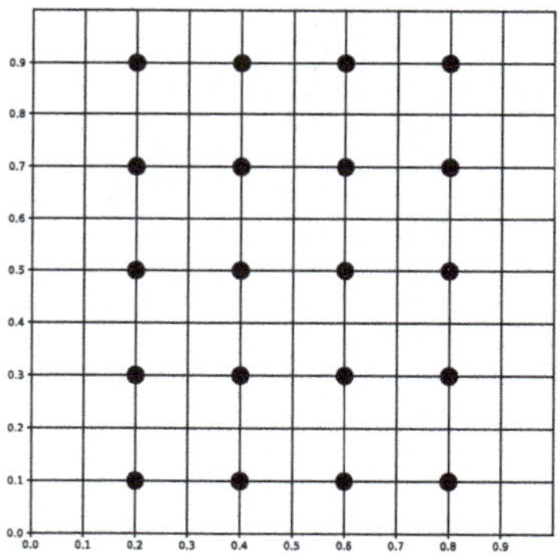

Fig. 2. Square grid over $[0, 1]^2$ representing the Milan area. Blue dots indicate the locations of the 20 facilities. In each time interval, every cell is characterized by a probability that a new client appears in the corresponding sub-region. (Color figure online)

A knapsack generator, a Python re-implementation of GEN2 [17], was used to generate 100 knapsack instances with 20 items each. For each instance (v, w, c), the value vector v was divided by the maximum possible value and used as the average opening cost vector of facilities. The opening cost vectors returned by the data generator are noisy versions of this average vector, where the noise is sampled from a uniform distribution over the set $[-0.3, 0.3]$ and the resulting values are clipped between 0 and 1. The weight vector w was divided by the capacity c and used to impose the budget constraint $w \cdot y \leq 1$ on the decisions.

5 Results

The code for the experiments was written and executed in Python 3.11 on a machine with Ubuntu 20.04.6 LTS, 32 GB DDR4 RAM and processor Intel(R) Xeon(R) W-1250P CPU @ 4.10 GHz. The binary linear programming models were solved using Gurobi 10.0.3. We executed FLC with both fixed and variable batch size for all the 100 instances described in Sect. 4, using a number of iterations equal to 200. The fixed batch size was set to 50. In the variable batch size case, the cardinality of the batch is drawn from the uniform distribution on $\{0, \ldots, 100\}$. FLC was applied with a number of clusters $\ell = n$ and with $\Delta = 50$. We considered profit per iteration and running time per iteration and we averaged the values over the 100 instances. We also applied a moving average smoothing to the resulting data, with a window of width of 10 iterations. We do not report explicit results concerning the number of deployments per round, because the results of the two batching policies are very similar. A selection of our results in terms of solutions quality and running times is shown in Figs. 3 and 4, respectively. The main outcome is the following: the behaviour of FLC is slightly more stable in terms of profit when the batch size is fixed (less than 1.5% in our experiments), but on average the performance obtained applying the two batching modalities is not significantly different. In order to explain the additional variability of the variable size batch policy, we evaluated correlations between batch size and reward. Figure 5 summarizes our analysis: it shows how, in the variable batch size setting, the number of clients in the batch and the profit of the iteration are strongly correlated. This feature is a consequence of the fact that the decision made by FLC are completely independent of any estimate about the number of clients per iteration. The choice to make clustering operations less frequent over time pays in terms of computing time: even though the algorithm takes a fraction of a second per iteration to run in both batching settings, its running time decreases on average.

In order to amplify the possible negative effect of opening too many facilities for small batches, we repeated the experiments by multiplying the costs for a factor 100. Our results are reported in Figs. 6, 7 and 8. The trend we observed is the same. Figure 8 still shows strong correlation; we argue that points below the main diagonal represent iterations in which a wrong number of facilites has been opened w.r.t those optimal for the batch size. That effect, however, has a minor impact on the overall revenue obtained by the variable batch policy. We finally mention that the same analysis was run also for 1000 iterations, still observing very mild differences in average revenue between the fixed and the variable batch policy, confirming the latter to be competitive.

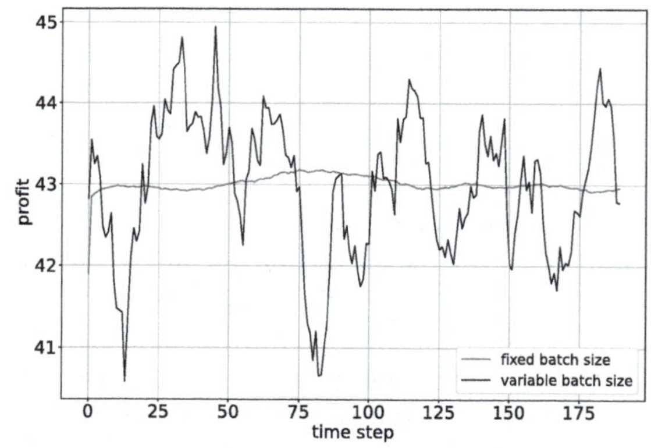

Fig. 3. Average Profit (y axis) per iteration (x axis).

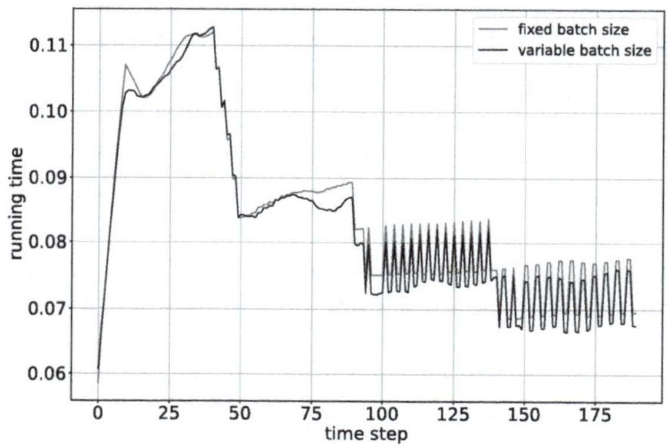

Fig. 4. Average Running time in seconds (y axis) per iteration (x axis).

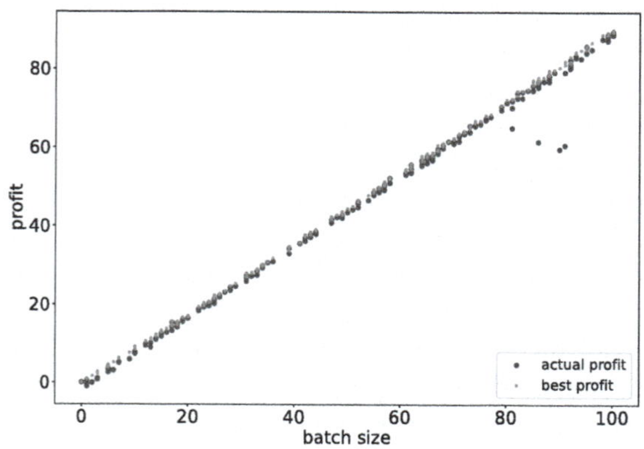

Fig. 5. Number of clients in the batch (x axis) and profit in the corresponding iteration (y axis), in the variable batch size setting for a representative instance. Batch size and iteration profit are strongly correlated.

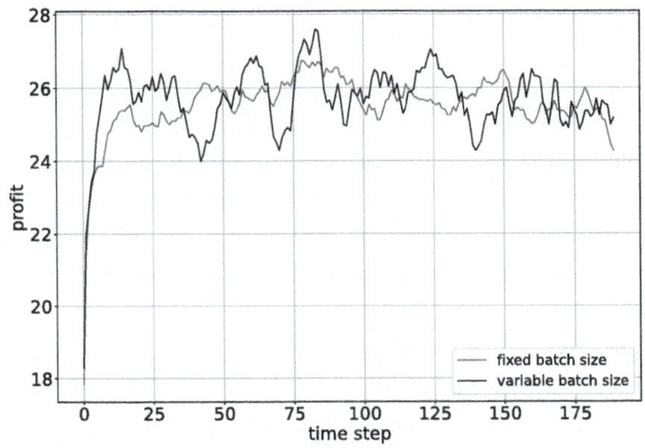

Fig. 6. Average Profit (y axis) per iteration (x axis) with 100x opening costs.

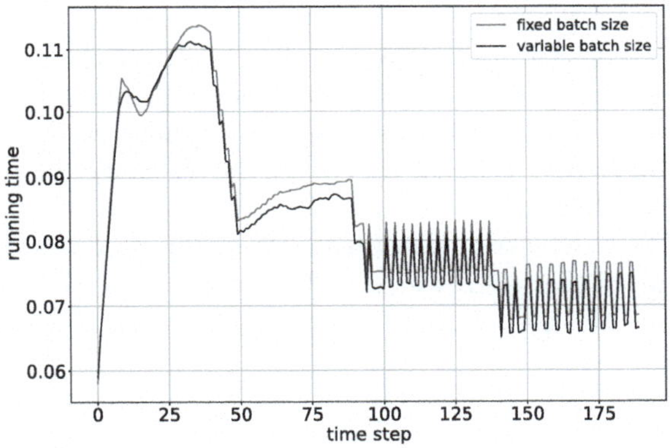

Fig. 7. Average Running time in seconds (y axis) per iteration (x axis) with 100x opening costs.

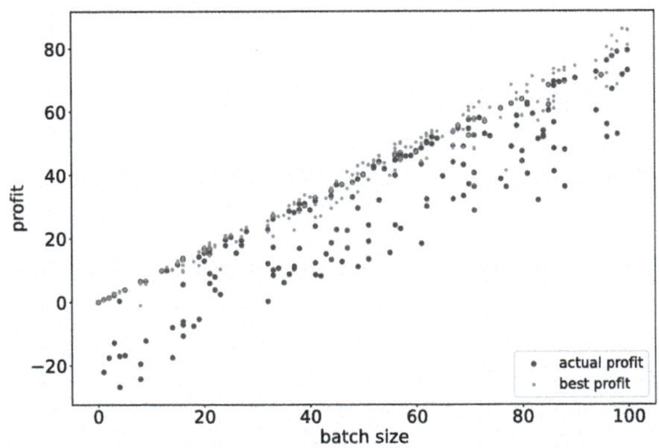

Fig. 8. Number of clients in the batch (x axis) and profit in the corresponding iteration (y axis), in the variable batch size setting with 100x opening costs for a representative instance. Batch size and iteration profit are still strongly correlated.

6 Conclusions

In this paper we have considered a relevant problem in network optimization, and more specifically in the optimal deployment of services in edge computing facilities. We have considered an online setting which proceeds in rounds. At each round the service provider needs to decide where to deploy services; batches of users appear after the deployment decision, and the service provider gets revenues for connections and costs for deployments which become known only after the decision time.

We cast this problem in an online learning framework, developing an algorithm which exploits the solution of a mathematical programming problem at each round, in the fashion of Follow The Leader methods. Our algorithm allows to handle batches of both fixed size and variable size. This makes it well suited for the common scenario in which deployment operations occur over time with a fixed frequency, and therefore the number of users in a batch appearing in a single slot is unknown.

Our simulations on data taken from Internet traces of the telecommunication network of Milan show that the additional flexibility of handling batches of variable size yields no substantial decrease in revenues, with a minor increase in revenue variability over rounds, which is however well explained by a strong correlation to the batch size.

Our Follow The Leader with Clustering algorithm is designed with an adversarial setting in mind. Future research will cover a deeper experimental analysis on scenarios where the distribution of user locations significantly change over time.

Acknowledgments. This work was partially supported by project SERICS (PE00000014) under the MUR NRRP funded by the EU-NextGenerationEU, and by the Italian MUR in the framework of the PRIN 2022 project "CAVIA: enabling the Cloud-to-Autonomous-Vehicles continuum for future Industrial Applications" grant 2022JAFATE.

References

1. Barlacchi, G., et al.: A multi-source dataset of urban life in the city of Milan and the province of Trentino. Sci. Data **2** (2015)
2. Chakraborty, A., Vaze, R.: Online facility location with timed-requests and congestion. ArXiv abs/2211.11961 (2022)
3. Chen, W., Wang, Y., Yuan, Y.: Combinatorial multi-armed bandit: general framework, results and applications. In: International Conference on Machine Learning (2013)
4. Combes, R., Talebi, M.S., Proutière, A., Lelarge, M.: Combinatorial bandits revisited. In: Conference on Neural Information Processing Systems (2015)
5. Drezner, Z., Hamacher, H.W.: Facility Location Applications and Theory. Springer, Heidelberg (2002)
6. Feldkord, B., auf der Heide, F.M.: Online facility location with mobile facilities. In: Proceedings of the 30th on Symposium on Parallelism in Algorithms and Architectures (2018)

7. Fotakis, D., Kavouras, L., Zakynthinou, L.: Online facility location in evolving metrics. Algorithms **14**, 73 (2021)
8. Gai, Y., Krishnamachari, B., Jain, R.: Combinatorial network optimization with unknown variables: multi-armed bandits with linear rewards and individual observations. IEEE/ACM Trans. Networking **20**, 1466–1478 (2012)
9. Jiang, S.H.C., Liu, E., Lyu, Y., Tang, Z.G., Zhang, Y.: Online facility location with predictions. ArXiv abs/2110.08840 (2021)
10. Kalai, A.T., Vempala, S.S.: Efficient algorithms for online decision problems. J. Comput. Syst. Sci. **71**, 291–307 (2005)
11. Kveton, B., Wen, Z., Ashkan, A., Szepesvari, C.: Tight regret bounds for stochastic combinatorial semi-bandits. In: International Conference on Artificial Intelligence and Statistics (2015)
12. Messana, R., Ceselli, A.: Comparing ad-hoc and MIP-based algorithms for the online facility location problem. In: Cologne-Twente Workshop on Graphs and Combinatorial Optimization, pp. 123–134. Springer (2023)
13. Neu, G., Bartók, G.: Importance weighting without importance weights: an efficient algorithm for combinatorial semi-bandits. J. Mach. Learn. Res. **17**, 1–21 (2016)
14. Neu, G., Valko, M.: Online combinatorial optimization with stochastic decision sets and adversarial losses. In: NIPS (2014)
15. Pasteris, S., He, T., Vitale, F., Wang, S., Herbster, M.: Online learning of facility locations. In: International Conference on Algorithmic Learning Theory (2021)
16. Pasteris, S., Vitale, F., Chan, K.S., Wang, S., Herbster, M.: MaxHedge: maximising a maximum online. In: International Conference on Artificial Intelligence and Statistics (2019)
17. Pisinger, D.: David Pisinger's optimization codes (1994). http://hjemmesider.diku.dk/~pisinger/generator.c
18. Yang, F., Chen, W., Zhang, J., Sun, X.: Follow the perturbed approximate leader for solving semi-bandit combinatorial optimization. Front. Comput. Sci. **15**, 155404 (2021)

Last Mile Drone Delivery: Complexity and Research Challenges

Fatos Xhafa[1(✉)], Cristian Domínguez[1], and Angel A. Juan[2]

[1] Universitat Politècnica de Catalunya, 08034 Barcelona, Spain
fatos@cs.upc.edu

[2] CIGIP, Universitat Politècnica de València, 03801 Alcoy, Spain
ajuanp@upv.es

Abstract. Drones or unmanned aerial vehicles technologies are experiencing a fast development pace propelled by a number of applications in diverse areas such as environmental monitoring, surveillance, military, smart cities, etc. In particular, the use of drones in smart city logistics is becoming increasingly relevant to the services of transporting small goods in urban neighborhoods. Hereby, a real life use case is the last mile delivery, which consists in completing the final phase of product delivery process from a truck to the customer's house. Prior to last mile delivery, products are shipped in a truck from a distribution center to a location in a city, which becomes the starting point of the last mile delivery. A key issue in last mile delivery is the drone route planning in order to guarantee the correct and efficient delivery service. Although the drone route planning in last mile delivery shares similar requirements with route planning in other delivery systems, there are complexities and challenges due to new requirements specific to drones such as limited flying distance, product weights, energy consumption, fleet costs, etc. In this paper, we analyze the complexity and research challenges of drone route planning in last mile drone delivery. The limitations of exact optimization methods versus heuristic methods are also highlighted through Gurobi solver results. The paper contributes to understanding the complex multi-objective nature of the drone routing in last mile delivery.

Keywords: Last mile delivery · Drone routing problem · Heuristic solutions · Mixed integer linear programming · Multi-objective optimization

1 Introduction

A drone is an unmanned aerial vehicle (UAV) capable of vertical takeoff and landing, as well as of following a previously defined route. The duration of a drone's battery is one of its most restrictive constraints. Therefore, small drones are typically considered for delivering only one lightweight parcel per trip. To effectively and efficiently employ drones for delivery, drone routing computation is indispensable. Drone routing from origin to destination must be computed

A. A. Juan et al. (Eds.): DSA ISC 2024, LNCS 14778, pp. 404–417, 2025.
https://doi.org/10.1007/978-3-031-78238-1_36

in advance, taking into account various restrictions and aiming to achieve one or more optimization objectives. Routing problems are a crucial component in many fields of commerce, logistics, and transportation at large. Nevertheless, each field has its own particularities, which are then translated into restrictions of these routing problems, such as the Traveling Salesman Problem (TSP), the Capacitated Vehicle Routing Problem (CVRP), the Vehicle Routing Problem with Time Windows (VRPTW), the Pickup and Delivery Problem (PDP), the Vehicle Routing Problem with Profits (VRPP), the Vehicle Routing Problem with Multiple Trips (VRPMT), etc. Indeed, drone routing shares some common features with other types of routing; however, there are constraints due to diverse limitations of drones.

The more restrictions are considered, the more complex the optimization model becomes due to inter-dependencies and/or contradictory objectives to optimize. For this reason, the optimal route is usually found by balancing its constraints (Pareto-like optimization). In some cases, it may be desirable to prioritize optimization for certain objectives, even if it results in worse performance for others (hierarchical optimization model). It should also be noted that finding optimal routes, beyond feasible routes, not only positively impacts the drone delivery system but also contributes to satisfying customer expectations for fast, comfortable, and safe delivery, which have become more demanding. Additionally, drone route delivery can improve traditional vehicle delivery by making the last mile delivery more agile in busy cities.

In this article, we analyze the Drone Routing Problem (DRP) for last mile delivery (Fig. 1), as well as its associated complexities, and research challenges. Depending on the requirements and restrictions, such as capacity restrictions, a single or multiple depots, delivery time window restrictions, and drone payload restrictions, different versions of the problem can be formulated. Among these, we distinguish between the case where drones are transported in a truck and used for the last mile delivery versus the model where drones are used for direct delivery. Being a version of the vehicle routing problem, DRP inherits the computational complexities of vehicle routing problems. The complexity of DRP can be even more challenging to address if weather conditions and the diversity of possible atmospheric phenomena impact drone power consumption, changes in battery performance, and therefore drone delivery. We also briefly discuss some technological solutions for drone delivery in retail and logistics companies. While traditional VRPs have been investigated for decades, now accounting for mature solutions both scientifically and technologically, last mile delivery using drones represents the latest form of vehicle-based delivery. Research and development in this area progress hand in hand, requiring the maturation of both research and technology for the successful implementation and deployment of last mile delivery using drones.

The rest of the article is organized as follows: In Sect. 2 we overview drone routing models. Section 3 presents drone routing models based on various requirements that take into account the technical limitations of drones and optimization objectives. Resolution methods are also discussed and some computational

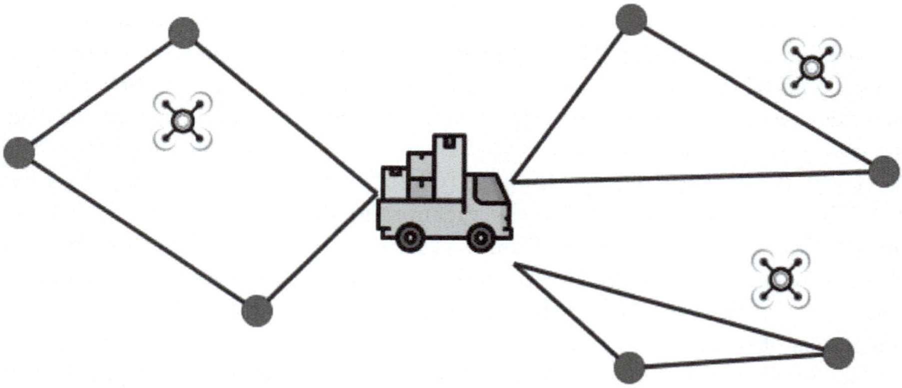

Fig. 1. Simple visualization of a DRP.

results using Gurobi[1] as optimizer solver are presented. Some research challenges are discusses in Sect. 4. We end the article in Sect. 5 with conclusions and an outlook on future work to fully address the optimization DRP.

2 Drone Routing Models

UAVs are a new type of vehicle based on novel technologies that can be fully or partially autonomous, but are more often controlled remotely by a human. Due to the novelty of drone technology, the DRP has recently attracted the attention of researchers and developers in the field [8]. Although it is not a new field of study – in fact, the DRP can be seen as a new version of the classical VRP – it comes with an array of new requirements and restrictions, making it worthy of investigation as a standalone problem. Additionally, the DPR is necessary in last mile drone delivery, with promising implications for future delivery systems in retail, logistics, postal, and parcel companies [10].

Among the research works available in the literature, we can find articles related to the DRP in combination with the traditional VRP. In such research, drones are assumed to work together with trucks to reduce delivery time. Drones use trucks as departure platforms from which they begin their routes, carrying the parcel, and returning to the truck once the delivery is completed. The position of the trucks can be assumed to be fixed in advance, or it can be changed during the delivery process. Therefore, the return point could be different from the starting point.

2.1 Combined DRP and VRP Models

Several articles in the literature have studied collaborative delivery using vehicles and drones [21,25,27,28]. Although they all used a combined model of vehicles

[1] https://www.gurobi.com/.

and drones for delivery, different features such as multi-depots, single versus multiple drones in the trucks, drone energy, delivery deadlines, time windows, etc., were considered. We briefly describe the main contributions of some of these research works. In [27], the authors formulated and solved the vehicle-and-drone cooperative delivery routing problem of blood products using exact methods. The advantage of vehicle-and-drone cooperative delivery over vehicle-only delivery is also discussed. In [25], the authors dealt with the mathematical formulation of the Multi-Depot VRP with Drones (MDVRP-D), where trucks are equipped with a UAV, dispatched from trucks while en route to make the last mile delivery. A meta-heuristic algorithm based on the Ant Colony Optimization (ACO) method is used to solve the problem. Similarly, in [28] the authors studied the VRP with drones, referred to as the two-echelon vehicle routing problem with drones, where multiple vehicles and drones work collaboratively to deliver to customers. Drones can perform multiple back-and-forth trips when their paired vehicle stops at a customer node, forming a two-echelon network. In the model, the number of drones taken by each vehicle is treated as a decision variable instead of a given parameter, providing more flexibility for planning vehicle and drone routes.

The NSGA-II algorithm is applied to the VRP with drones considering makespan and carbon emission as objectives in [14]. In [21], the authors presented models and a comparative study of three different approaches to a routing problem, namely a VRP using trucks, another one for DRP, and a third one – a hybrid model between the former two models – where trucks would carry drones and use them to deliver parcels. The comparative study showed that the combined solution presents the best overall results. The three delivery processes were studied in terms of transportation cost, CO_2 emissions, and congestion. The VRP resulted in being the less efficient delivery system, while the DRP was not able to fulfill all the delivery requests. The hybrid method presented lower costs, emissions, and congestion values than the VRP, and even if those values could be higher than those presented by the DRP, this delivery system meant a better approach in terms of the delivery requests. Coordinating platooning has also been studied for the case of VRP [17], and it could be a potential field of study in DRP as well, since not only could it contribute to energy saving but would also be useful to determine under which conditions drones could act in a swarm to carry heavier parcels. All in all, there are a variety of formulations of the combined VRP and DRP [1,22,26].

2.2 DRP Exclusively with Drones

Alternatively, in addition to the combined VRP and DRP, there are research works addressing delivery exclusively with drones. The main assumption here is that drones would depart from a depot with the parcel to deliver, transport it to the customer's address, and then return to the departure depot or to a different one. In routing problems, capacity constraint is an important aspect that impacts problem modeling. This is also the case in drone routing, whereby capacity is a feature to be considered. Drones can carry one or more parcels at once, and accordingly, they can visit one or more customers in the same trip. Nevertheless, unlike vehicle routing, weight is a highly restrictive constraint for drones, so it is often preferable to limit their deliveries to unique parcels. The requirement of unique vs. multiple parcel drones and the number of trips each drone is allowed to make lead to simple DRPs versus multi-trip drone routing problems (MTDRP), where drones can perform various trips, starting and returning from and to a depot [4,18,19].

Uncertainty is another important feature of the DRP, requiring more robust optimization approaches to obtain more accurate results. Uncertainties include drones' battery duration, which can change with the temperature of the air, leading to a variation in the available flying time [13]. In such models, the aim is to achieve a minimum threshold of functioning to ensure the return of the drones to the depot or, at least, to increase their probabilities, because other uncertainties can occur, which render any optimization framework not fully feasible. In this context, some studies even consider the possibility of having strategically distributed recharging stations to allow drones to reach further ranges [6,12].

2.3 Technological Solutions for Last Mile Drone Delivery

In recent years, delivery companies have conducted tests of drone delivery services. However, despite significant research efforts, software developments, and technological advancements, the implementation and deployment of drone delivery services remain limited. Next, we briefly describe some drone delivery services offered by world-leading companies in the field. Amazon introduced a last mile drone delivery service called *Amazon Prime Air*[2]. This service aims to ensure deliveries within a 30-minute timeframe. The service utilizes three different types of drones, depending on the delivery address. Most deliveries are tested in areas where customers' addresses have ample space, such as a courtyard, where customers can place a reference with a QR code or distinctive ground signal. This enables drones to detect and land the parcel accurately. Wing, a North American company dedicated to parcel delivery using drones and a subsidiary of Alphabet Inc., collaborates with the transportation company FedEx[3]. One notable feature of this service is that drones do not land on the ground to deliver parcels. Instead, they use an extendable piece of cord to lower the parcel closer to the ground or to the person awaiting its receipt. Drones utilize Wing maps of the delivery surroundings to avoid obstacles and accidents, allowing other flying objects to operate in the vicinity. In China, JD.com has accelerated deliveries using a drone service, particularly beneficial for reaching remote rural areas where e-commerce faces challenges. Drones automatically follow a fixed route, and once the parcel is delivered, they return to the departure point using the same route. In Spain, transportation companies such as MRW have collaborated with drone company HEMAV to provide delivery services.

---

[2] https://www.aboutamazon.com/news/tag/prime-air.

[3] https://newsroom.fedex.com/newsroom/global/elroyair.

Another aspect of drone delivery service capabilities has led some companies to explore more humanitarian applications. In Africa, the company Zipline[4] initiated delivering medical supplies in Rwanda using UAVs, which can quickly reach isolated areas and drop parcels along the way. The mechanism is simple: the drone departs from the pharmacy using a launching platform, enabling quick takeoff, and is then guided via GPS to the destination location, where it simply drops the parcel equipped with a parachute. To expedite the process, the GPS module is integrated into the drone's battery, ensuring it is always connected and ready to operate once equipped on the vehicle. Additionally, the wings are modular and easily installed when a delivery alert is received, allowing medical products to be swiftly loaded into the drone and sent to the customer. To facilitate a rapid launch phase, the UAV modules feature QR codes that, when scanned with a mobile application, prompt controllers to activate different parts of the wings and tail to ensure proper functioning. DHL also utilized its UAV, named Parcelcopter[5], to deliver medical supplies and other products to remote areas with difficult access. The Parcelcopter has four different versions, with the latest one tested in Mwanza and the Ukerewe Island, Tanzania, specifically for delivering medical supplies, boasting the advantage of vertical takeoff capability.

3 Requirements and Optimization Models

As shown in various research works in the literature, different versions of the DRP, including the combined VRP and DRP, can be formulated based on limitations of the drones, such as speed, energy, single versus multiple parcels, single versus multiple trips, payload, etc. Below, a list of the most common requirements on the DRP, arising from last mile delivery systems, is introduced.

3.1 Model Requirements

Assuming that programmable drones are capable of automatically following previously determined routes, some typical model requirements are:

- *Truck requirement:* Last mile routing, from a truck to the customers' address within a defined range of distance.
- *Drone requirement:* Last mile routing, from a single depot to the customers' address within a defined range of distance, considering the exclusive use of drones for the deliveries.
- *Depot requirement:* Last mile routing, from multiple depots to the customers' address within different ranges of distance, considering the exclusive use of drones for the deliveries.
- *Trip requirement:* Single-trip design, considering drones carry parcels from a warehouse to a single customer's address and return without cargo.

[4] https://www.flyzipline.com/.

[5] https://lot.dhl.com/saving-lives-one-drone-delivery-at-a-time/.

- *Multi-trip requirement:* Multi-trip design, considering drones carry parcels from a warehouse to different customer's addresses and return without cargo upon delivery completion (multiple customers on a single trip).
- *Single parcel requirement:* Unique-parcel carrier drones (unique payload).
- *Multiple parcel requirement:* Multiple-parcel carrier drones loading at a single depot (multiple payload).
- *Parcel and depot requirements:* Multiple-parcel carrier drones loading at different depots (multiple payload).
- *Charging stations requirement:* DRP with extended flight range by using charging stations.
- *Fleet of drones requirement:* Last mile delivery using an independent fleet of drones, i.e., the transport of parcels from a warehouse to relatively nearby locations using drones without coordination among them.
- *Swarm of drones requirement:* Last mile delivery using a swarm of drones, i.e., the transport of parcels from a warehouse to relatively nearby locations using drones with coordination among them.
- *Independence of VRP from DRP requirement:* Combined VRP and DRP, where the VRP optimization is independent of the DRP optimization.
- *Dependence of VRP from DRP requirement:* Combined VRP and DRP, where the VRP optimization is dependent on the DRP, aiming to optimize DRP.
- *Uncertainty requirements:* DRP subject to uncertainties such as weather conditions.
- *Topological environmental restrictions requirements:* DRP subject to environmental restrictions on flying areas of delivery.
- *Safety requirements:* DRP subject to safety restrictions on flying delivery areas.

3.2 Optimization Models

Both in the case of the DRP as in the case of the combined VRP and DRP, the first observation about optimization models is that they can be single- or multi-objective optimization problems. The possible objectives include time to delivery, cost of delivery, capacity (payload), and energy, among others. There is a twofold source of complexity for the various potential optimization problems. On the one hand, optimization models can be formulated to consider one single objective or two or more objectives simultaneously. Certainly, the more objectives considered, the more complex the resulting optimization model becomes. On the other hand, achieving the objectives under one or more requirements (listed in Sect. 3.1) adds further complexity due to their impact on the optimization landscape. Regarding the multi-objective optimization model, it is worth mentioning that three main resolution approaches are:

- *Reducing to a single objective:* In some cases, it could be possible to combine two or more objectives, say $f_1, f_2, \ldots, f_k$, into a single objective $f = \lambda_1 f_1 + \lambda_2 f_2 + \ldots + \lambda_k f_k$, with $\lambda_1 + \lambda_2 + \ldots + \lambda_k = 1$. This strategy reduces the number of objectives to a single one. Achieving this reduction through summing up different objectives requires objectives that can be expressed in the same unit (for instance, cost and time can be expressed in monetary units) and scaling to consistent range values. Additionally, this requires finding adequate values for the weights $\lambda_1, \lambda_2, \cdots, \lambda_k$. In some cases, this model can be used to prioritize certain objectives in advance by assigning larger values to corresponding λ weights.

- *Hierarchical optimization of multiple objectives:* This model can be seen as a combination of single optimization and priority optimization. It optimizes one objective at a time by establishing the optimization procedure in advance. Assuming there are n objectives, they are initially sorted according to priority criteria, from the most important objective to the least important one: $f_1 \succ f_2 \succ \ldots \succ f_n$. Then, at phase k, the f_k objective is optimized subject to not worsening the values of the objectives $f_1, f_2, \ldots, f_{k-1}$ optimized so far. This optimization model is particularly interesting in a business context where optimization objectives can be sorted according to their importance for the business.
- *Pareto optimization of multi-objectives:* This is a well-established approach to solving multi-objective optimization models in which objectives could be contradicting (improving one objective could worsen other objectives). A Pareto frontier is therefore a set of non-dominated solutions.

3.3 A Model with Two/Three Objectives

A first objective focuses on minimizing the total delivery time of the drones, allowing the delivery company to provide its fastest service possible to customers' orders. Delivery time takes precedence over costs in this scenario, as the company can allocate a budget for the fleet of drones. A second objective involves optimizing costs, considering that the delivery company operates within a limited budget. A third objective aims to minimize environmental impacts such as noise and pollution. A single sum-up function, like the one proposed in Eq. (1), can be calculated as the sum of three terms. The first term corresponds to the cost of the fleet of M drones, each having a drone cost denoted by dC. The second term accounts for cost reduction through drone reuse, considering that drones may need to change their batteries for each reuse, thus incurring a battery cost denoted by bC. The third term represents the power consumption of the drone. Authors in [7] noted that energy consumption is not the only factor influencing costs; the weight of the batteries also plays a role. While energy consumption leads to an increase in the required budget, battery weight helps reduce costs. This cost reduction can be significant depending on the circumstances, hence eC represents energy cost and P denotes the total power consumed according to Eq. (1).

$$cost = M \cdot dC - (dC - bC) \cdot n_{reusals} + eC \cdot P \tag{1}$$

The third term would necessitate finding a linear expression for power consumption, which would undoubtedly increase the problem's complexity (for a linear approximation to the power consumption of a certain type of drone, refer to [7]). Summing up only the first two terms would yield a function suitable for any mixed integer linear programming (MILP) solver.

The cost function considered is therefore of only two terms, as shown in Eq. 2.

$$cost = M \cdot dC - (dC - bC) \cdot n_{reusals} \tag{2}$$

Applying the hierarchical optimization approach entails setting the cost as the parameter of the model to minimize initially. As finding the optimal solution to the MILP can become complex, the program is halted when a reasonably good feasible solution is reached, indicating the use of the minimum number of possible drones. It is noteworthy that changing batteries is more cost-effective than expanding the fleet, and ensuring the maximum delivery time remains within limits. Subsequently, in a second step, the model is solved again with the obtained cost set as an upper bound, and delivery time becomes the new objective to optimize. This results in a solution with either the same or reduced cost compared to the previously obtained solution, while decreasing total delivery time, with variables such as the number of drones in the fleet possibly changing. The MILP model for the DRP includes a list of constraints, which are divided into five different categories: route validity constraints, reusability constraints, payload constraints, time constraints and cost constraints. The reader is referred to [7] for additional details.

3.4 Exact Versus Heuristics Solution Methods

DRP problem formulations inherit complexities from routing problems in general, and particularly from the VRP, making them NP-hard optimization problems – computational time to optimally solve moderate to large/very large instances of the model exceeds practical time limits. Some resolution methods explored in the literature for optimality include branch-and-price-and-cut algorithms [26]. Advancements in software and hardware have led to reduced computation times for computing optimal solutions. Various software developed for optimization purposes can be beneficial for the DRP. For instance, the IBM CPLEX Optimization Studio utilizes mathematical and restrictive programming to provide optimal solutions to complex optimization models. Other commercial optimization solvers include FICO Xpress and Knitro. Additionally, competitive commercial solvers like Gurobi Optimizer offer free academic licenses for research purposes. Gurobi supports several programming languages, including C/C++, C#, Java, Python, MATLAB, and R. Notice that optimal solutions, even for limited instances of the problem, are valuable for establishing the performance of heuristic solutions through straightforward comparison of results.

Due to the intractability of optimally solving the DRP, heuristic solutions are a common approach. Some examples are: NSGA-II [14], Ant Colony Optimization [25], Adaptive Large Neighborhood Search [23], Artificial Bee colony [15], Greedy Randomized Adaptive Search Procedure [11], Adaptive Multi-start Simulated Annealing [1], Tabu Search [24], Randomized Variable Neighborhood

Descent Heuristic [9], Learnheuristics integrating metaheuristics and machine learning [2], etc.

A sample of graphical representations of exact solutions (obtained with the Gurobi solver) for the model of Eq. 2) is shown in Figs. 2 and 3, corresponding to concrete scenarios (number of customers, service delivery area, budget limit). In the figures, green lines connect the different locations of the customers' addresses (in blue) with the depot (in red). The locations are marked with a number in order to identify them, starting from 0 for the depot to N – the number of customers – for the rest. A black arrow covering half of the green line between locations indicates the direction of the drone, i.e., the position it departs from to where it goes. Since all the drones start and finish their routes at the depot, when they are visiting a unique location, the green line connecting it with the depot is fully covered by two opposite arrows. The axes represent the extension in meters covered by the drone service, aiding in visualizing the distribution of the customers' addresses.

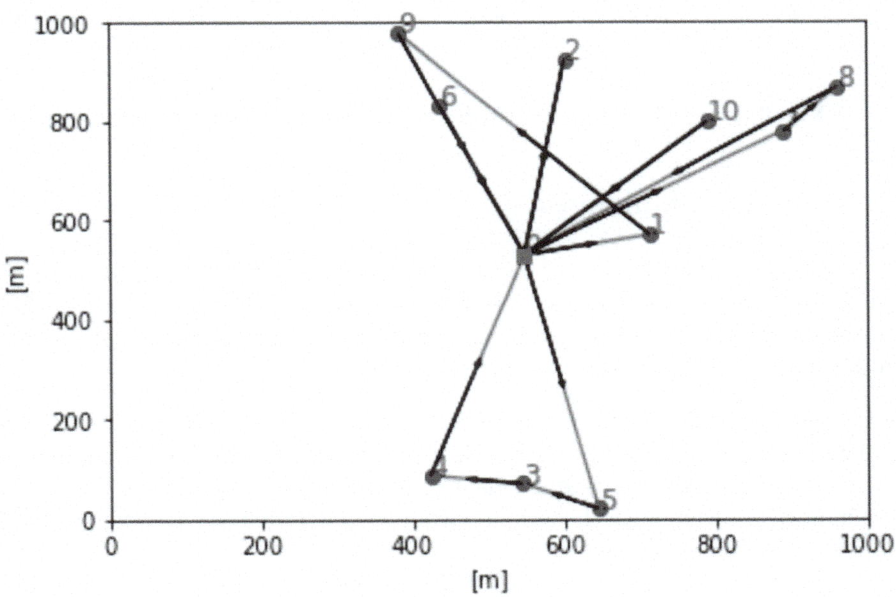

Fig. 2. Example of a route distribution for a simpler scenario.

We have not implemented any heuristic methods as external algorithms to the Gurobi's model solver, but it should be noted that Gurobi utilizes a catalog of sixteen heuristic techniques in its solving process. These heuristics are fast and, in general, they are able to capture the structure of a wide variety of problems. Rounding and relaxation induced neighborhood search are two examples of such heuristics used by Gurobi.

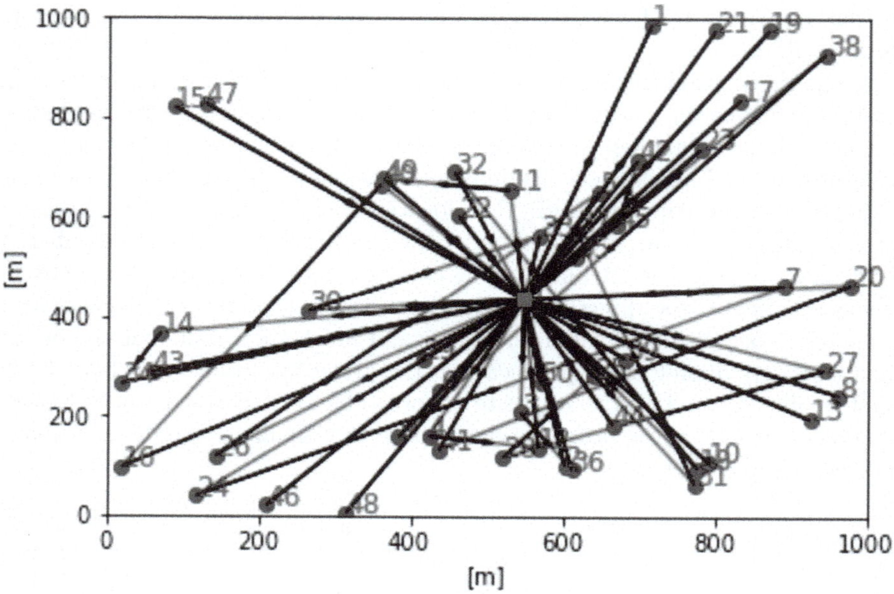

Fig. 3. Example of a route distribution for a more complex scenario.

4 Research Challenges

As discussed, solving DRP variants is computationally hard even for a single objective and a few constraints. Research challenges in this domain arise when addressing full versions of the problem. Additionally, there are problem versions not yet addressed due to their larger complexity:

- *Dependence of the VRP from the DRP optimization:* In current formulations of the combined VRP and DRP, the optimization of the VRP and that of the DRP are addressed independently. It would be interesting to solve the combined problem with the aim of optimizing DRP delivery time or the total joint delivery time of both.
- *Topological restrictions:* Current DRP formulations do not consider topological restrictions in cities regarding flying areas such as weather conditions, temporary safety restrictions, etc. Still, cities are complex areas and not homogeneous regarding their topological restrictions (barriers to flying).
- *Uncertainties and dynamics of environment:* Current DRP formulations do not account for uncertainties and dynamics of the delivery area, which are common in many real-life applications. It is challenging to include constraints from uncertainties and dynamics of the environment, which could lead to non-linear constraints (e.g., the drone energy consumption is a nonlinear function of the drone load that varies along the route) or drone flight time is not deterministic as assumed in most current approaches. In such uncertain sce-

narios, robust optimization or simheuristic approaches could be chosen for optimizing drone routing problems [5].
- *Safety regulations:* Safety regulations are imposed by city administrations and include various limitations on flying areas to ensure minimal impact of drone delivery on city areas. These regulations aim to ensure safety in living areas, exclude areas like children's playgrounds and schools, minimize acoustic pollution, avoid city services due to impacting critical infrastructures, etc.
- *Agile Optimization:* Limitations of current approaches for the DRP and its combination with the VRP arise from the fact that they require, especially for large instances, considerable resolution time. Such approaches compute routes in advance of the delivery and are not able to cope with uncertainties and dynamics of the delivery areas. Agile optimization has been proposed in the heuristics domain as an approach to address the highly complex nature of these problems in flexible and short resolution times, i.e., to be able to generate reasonably good solutions in less than a second and also re-optimize the system every few seconds [16]. This is achieved by combining parallel computing techniques with biased-randomized heuristics [3]. Agile optimization approaches have also been proposed for the optimization of vehicular networks in smart cities [20].

5 Conclusions and Future Works

The use of drones in logistics in smart cities is becoming increasingly relevant for transporting small goods in city neighborhoods. Last mile delivery, which involves completing the final phase of the product delivery process from a truck to the customer's house, has emerged as a key optimization problem in this context. A critical issue in last mile delivery is drone route planning to ensure correct and efficient delivery services. In this paper, we have analyzed the computational complexities of the last mile delivery problem, primarily due to new requirements specific to drones such as limited flying distance, product weights, energy consumption, fleet costs, etc. We have also highlighted the limitations of exact optimization methods compared to heuristic methods.

Furthermore, we have discussed several research challenges that need to be addressed in future works. These challenges include incorporating topological restrictions, uncertainties, and dynamics of the environment, as well as adhering to safety regulations. Additionally, we emphasized the need for new algorithmic solutions based on agile optimization to overcome the limitations of current approaches and to enable more efficient and adaptable drone route planning in last mile delivery scenarios. The results of the agile optimization methods will be compared to exact solutions obtained with Gurobi solver. Finally, we also plan to implement stand alone heuristic solutions (such as local search SA/Tabu Search and evolutionary algorithms such as GAs) and compare their performance to exact solutions obtained with Gurobi solver.

Acknowledgments. This work has been partially supported by the Spanish Ministry of Science and Innovation (PID2022-138860NB-I00 and RED2022-134703-T).

References

1. Amine Masmoudi, M., Mancini, S., Baldacci, R., Kuo, Y.H.: Vehicle routing problems with drones equipped with multi-package payload compartments. Transp. Res. Part E: Logist. Transp. Rev. **164**, 102757 (2022)
2. Bayliss, C., Juan, A.A., Currie, C.S., Panadero, J.: A learnheuristic approach for the team orienteering problem with aerial drone motion constraints. Appl. Soft Comput. **92**, 106280 (2020)
3. do C. Martins, L., Hirsch, P., Juan, A.A.: Agile optimization of a two-echelon vehicle routing problem with pickup and delivery. Int. Trans. Oper. Res. **28**(1), 201–221 (2021)
4. Cheng, C., Adulyasak, Y., Rousseau, L.M.: Drone routing with energy function: formulation and exact algorithm. Transp. Res. Part B (2020)
5. Chica, M., Juan, A.A., Bayliss, C., Cordón, O., Kelton, W.D.: Why simheuristics? Benefits, limitations, and best practices when combining metaheuristics with simulation. SORT **44**(2), 311–334 (2020)
6. Coelho, B.N., et al.: A multi-objective green UAV routing problem. Comput. Oper. Res. (2017)
7. Dorling, K., Heinrichs, J., Messier, G., Magierowski, S.: Vehicle routing problems for drone delivery. Natural Sciences & Eng. Research Council of Canada (2016)
8. Eskandaripour, H., Boldsaikhan, E.: Last-mile drone delivery: past, present, and future. Drones **7**(2), 77 (2023)
9. de Freitas, J.C., Penna, P.H.V.: A randomized variable neighborhood descent heuristic to solve the flying sidekick traveling salesman problem. Electron. Notes Discrete Math. (2018)
10. Garg, V., Niranjan, S., Prybutok, V., Pohlen, T., Gligor, D.: Drones in last-mile delivery: a systematic review on efficiency, accessibility, and sustainability. Transp. Res. Part D: Transp. Environ. **123**, 103831 (2023)
11. Gómez-Lagos, J., Rojas-Espinoza, B., Candia-Véjar, A.: On a pickup to delivery drone routing problem: models and algorithms. Comput. Industr. Eng. **172**, 108632 (2022)
12. Hong, I., Kuby, M., Murray, A.T.: A range-restricted recharging station coverage model for drone delivery service planning. Transp. Res. Part C (2018)
13. Kim, S.J., Lim, G.J., Cho, J.: Drone flight scheduling under uncertainty on battery duration and air temperature. Comput. Industr. Eng. (2018)
14. Kuo, R., Edbert, E., Zulvia, F.E., Lu, S.H.: Applying NSGA-II to vehicle routing problem with drones considering makespan and carbon emission. Expert Syst. Appl. **221**, 119777 (2023)
15. Lei, D., Cui, Z., Li, M.: A dynamical artificial bee colony for vehicle routing problem with drones. Eng. Appl. Artif. Intell. **107**, 104510 (2022)
16. Li, Y., Peyman, M., Panadero, J., Juan, A.A., Xhafa, F.: IoT analytics and agile optimization for solving dynamic team orienteering problems with mandatory visits. Mathematics **10**(6) (2022)D
17. Luo, F., Larson, J., Munson, T.: Coordinated platooning with multiple speeds. Transp. Res. Part C (2018)
18. Luo, Z., Gu, R., Poon, M., Liu, Z., Lim, A.: A last-mile drone-assisted one-to-one pickup and delivery problem with multi-visit drone trips. Comput. Oper. Res. **148**, 106015 (2022)
19. Meng, S., Guo, X., Li, D., Liu, G.: The multi-visit drone routing problem for pickup and delivery services. Transp. Res. Part E: Logist. Transp. Rev. **169**, 102990 (2023)

20. Peyman, M., Fluechter, T., Panadero, J., Serrat, C., Xhafa, F., Juan, A.A.: Optimization of vehicular networks in smart cities: from agile optimization to learnheuristics and simheuristics. Sensors **23**(1), 499 (2023)
21. Plugiese, L.D.P., Guerriero, F., Macrina, G.: Using drones for parcels delivery process. In: International Conference on Industry 4.0 and Smart Manufacturing (2020)
22. Raj, R., Murray, C.: The multiple flying sidekicks traveling salesman problem with variable drone speeds. Transp. Res. Part C (2020)
23. Sacramento, D., Pisinger, D., Ropke, S.: An adaptive large neighborhood search metaheuristic for the vehicle routing problem with drones. Transp. Res. Part C: Emerg. Technol. **102**, 289–315 (2019)
24. Schermer, D., Moeini, M., Wendt, O.: A hybrid VNS/tabu search algorithm for solving the vehicle routing problem with drones and EN route operations. Comput. Oper. Res. **109**, 134–158 (2019)
25. Stodola, P., Kutěj, L.: Multi-depot vehicle routing problem with drones: mathematical formulation, solution algorithm and experiments. Expert Syst. Appl. **241**, 122483 (2024)
26. Xia, Y., Zeng, W., Zhang, C., Yang, H.: A branch-and-price-and-cut algorithm for the vehicle routing problem with load-dependent drones. Transp. Res. Part B: Methodol. **171**, 80–110 (2023)
27. Yin, Y., Qing, L., Wang, D., Cheng, T., Ignatius, J.: Exact solution method for vehicle-and-drone cooperative delivery routing of blood products. Comput. Oper. Res. **164**, 106559 (2024)
28. Zhou, H., Qin, H., Cheng, C., Rousseau, L.M.: An exact algorithm for the two-echelon vehicle routing problem with drones. Transp. Res. Part B: Methodol. **168**, 124–150 (2023)

Author Index

A. A. Juan et al. (Eds.): DSA ISC 2024, LNCS 14778, pp. 419–421, 2025.
https://doi.org/10.1007/978-3-031-78238-1

The manufacturer's authorised representative in the EU is Springer Nature Customer Service Centre GmbH, Europaplatz 3, 69115 Heidelberg, Germany. If you have any concerns regarding our products, please contact ProductSafety@springernature.com

Printed and bound by CPI Group (UK) Ltd, Croydon, CR0 4YY
30/09/2026
02225560-0014